The Ethics of Insurgency

As insurgencies rage, a burning question remains: How should insurgents fight technologically superior state armies? Commentators rarely ask this question because the catchphrase "we fight by the rules, but they don't" is nearly axiomatic. But truly, are all forms of guerrilla warfare equally reprehensible? Can we think cogently about *just* guerrilla warfare? May guerrilla tactics such as laying improvised explosive devices (IED), assassinating informers, using human shields, seizing prisoners of war, conducting cyber strikes against civilians, manipulating the media, looting resources, or using nonviolence to provoke violence prove acceptable under the changing norms of contemporary warfare? The short answer is "yes," but modern guerrilla warfare requires a great deal of qualification, explanation, and argumentation before it joins the repertoire of acceptable military behavior. Not all insurgents fight justly, but guerrilla tactics and strategies are also not always the heinous practices that state powers often portray them to be.

Michael L. Gross is a professor in, and the head of, the School of Political Science at the University of Haifa, Israel. His articles have appeared in *Political Studies, Social Forces,* the *New England Journal of Medicine, Political Research Quarterly, Journal of Applied Philosophy,* the *American Journal of Bioethics,* the *Journal of Military Ethics,* the *Journal of Medical Ethics,* and *Political Psychology.* His books include *Ethics and Activism* (Cambridge University Press, 1997), *Bioethics and Armed Conflict* (2006), *Moral Dilemmas of Modern War: Torture, Assassination, and Blackmail in an Age of Asymmetric Conflict* (Cambridge University Press, 2010), and an edited volume, *Military Medical Ethics for the 21st Century* (2013). He serves on regional and national bioethics committees in Israel and has led workshops and lectured on battlefield ethics, medicine, and national security for the U.S. Army Medical Department at Walter Reed Medical Center, the U.S. Naval Academy, the International Committee of Military Medicine, the Dutch Ministry of Defense, and the Medical Corps and National Security College of the Israel Defense Forces.

The Ethics of Insurgency

A Critical Guide to Just Guerrilla Warfare

MICHAEL L. GROSS

University of Haifa

CAMBRIDGE
UNIVERSITY PRESS

32 Avenue of the Americas, New York, NY 10013-2473, USA

Cambridge University Press is part of the University of Cambridge.

It furthers the University's mission by disseminating knowledge in the pursuit of education, learning, and research at the highest international levels of excellence.

www.cambridge.org
Information on this title: www.cambridge.org/9781107684645

© Michael L. Gross 2015

First published 2015

Printed in the United States of America

A catalog record for this publication is available from the British Library.

Library of Congress Cataloging in Publication data
Gross, Michael L., 1954–
The ethics of insurgency : a critical guide to just guerrilla warfare /
Michael L. Gross, University of Haifa.
    pages   cm.
Includes bibliographical references and index.
ISBN 978-1-107-01907-2 (hardback) – ISBN 978-1-107-68464-5 (pbk.)
1. Guerrilla warfare – Moral and ethical aspects.   2. Irregular warfare – Moral and ethical aspects.   3. Insurgency – Moral and ethical aspects.   4. Just war doctrine.
5. Military ethics.   I. Title.
U240.G67   2015
172'.42–dc23   2014020943

ISBN 978-1-107-01907-2 Hardback
ISBN 978-1-107-68464-5 Paperback

From Ada to Ayala

Contents

List of Tables	*page* ix	
Preface	xi	
List of Abbreviations	xv	

1   Just Guerrilla Warfare: Concepts and Cases   1
   Guerrillas and Insurgents   3
   The Ethics of Insurgency: A Brief Overview   6
   Cases and Examples: A Few Words   14

PART I   THE RIGHT TO FIGHT

2   The Right to Fight: Just Cause and Legitimate Authority
    in Guerrilla Warfare   21
   The Right to Fight: Just Cause   23
   Just Cause, Self-Defense, and the Right to Fight   29
   Legitimate Authority and Guerrilla Warfare   37
   The Right to a Fighting Chance   45
   Going to War   49

3   The Right to Fight: Who Fights and How?   50
   Conscription: Building a Guerrilla Organization   51
   Fighting Well: The Right to Shed Uniforms   61
   Fighting Well: Noncombatant Immunity and Civilian Liability   64
   The Incentive to Obey the Law   72

PART II   HARD WAR

4   Large-Scale Conventional Guerrilla Warfare: Improvised
    Explosive Devices, Rockets, and Missiles   81
   Guns, Bombs, and IEDs   82
   Improvised Explosive Devices (IEDs)   86

Rockets and Missiles   92
Conventional Warfare and a Fighting Chance   100

5   Small-Scale Conventional Guerrilla Warfare: Targeted
    Killing and Taking Prisoners   102
    Assassination and Targeted Killing   103
    Prisoners of War   114
    Targeted Killing, Prisoners of War, and International Norms of Conduct   124

6   Human Shields   127
    Human Shields: Theory and Practice   130
    Harming Human Shields: Who Is Responsible?   141
    Human Shields and Free Riding   148

PART III SOFT WAR

7   Terrorism and Cyberterrorism   153
    Enemy and Compatriot Terrorism   155
    Permissible Attacks on Liable Civilians   162
    Sublethal, Nonlethal, and Cyber Weapons   167
    Rethinking Attacks on Civilians in Modern Warfare   180

8   Economic Warfare and the Economy of War   184
    Economic Warfare: Sanctions, Siege, and Blockade   185
    The Economy of Guerrilla War   196
    The Economic Dimensions of Just Guerrilla Warfare   210

9   Public Diplomacy, Propaganda, and Media Warfare   213
    Public Diplomacy, Propaganda, and Public Works   214
    Guerrilla Public Diplomacy   219
    Public Diplomacy, Ethics, and Jus in Bello   225
    The Status of Journalists and Media Facilities   234
    The Allure of Public Diplomacy   237

10 Civil Disobedience and Nonviolent Resistance   240
    Nonviolent Resistance   241
    Nonviolent Resistance in Guerrilla Warfare   245
    Hunger Striking and the Problem of Consent   261
    Last Resort, Nonviolence, and Guerrilla War   263

PART IV CONCLUDING REMARKS

11 Just War and Liberal Guerrilla Theorizing   271
    The Practice of Just Guerrilla Warfare   273
    Enforcement and Compliance   278
    The Prospects for Just Guerrilla Warfare   279

References   283
Index   315

Tables

6.1. Human Shielding, Intent, and Immunity   *page* 132
6.2. The Immunity of Involuntary and Voluntary Shields
       Indirectly and Directly Participating in the Hostilities   134
6.3. Deterrence and Human Shields: The Moral Distinctions   140
7.1. The Liability of Civilian Actors in Asymmetric War   156
7.2. The Direct and Collateral Effects of Sublethal, Nonlethal,
       and Cyber Weapons   168
11.1. Just Guerrilla Warfare: Practices and Provisos   274

Preface

Writing about war, I often mistype the word "casualties," leaving me to wonder what is casual or causal about the harm befalling combatants and noncombatants. Similarly, as a student of armed conflict, I often wonder what is civil about civilians or civil war. Casual suggests the chance or accidental nature of wartime injuries and deaths. Causal, on the other hand, directs our attention away from chance and toward a discernible sequence of events that result in injury or death. Civil connotes a measure of respect for normative behavior and, therefore, responsibility on the part of all participants, including soldiers, civilians, and bystanders, for the goings on in wartime.

Responsibility and liability do not change much whether one considers war from the perspective of states or insurgents. In many ways, therefore, *The Ethics of Insurgency* is a sequel to *Moral Dilemmas of Modern War*. Both books question the moral and legal limits imposed on state and non-state actors in modern warfare. In *Moral Dilemmas* I asked how states may fight successfully against guerrillas who employ terrorism and fight from within civilian populations. My answer, I thought, was rather modest. I did not advocate dogmatic adherence to existing law, nor did I advocate jettisoning the law in its entirety. Rather, I hoped that the ethical principles that protect the basic rights of combatants and noncombatants could guide me as I threaded my way through the demands of ethics and the exigencies of modern battle. The result was to lend qualified support to targeted killing and various nonlethal weapons and to lower the bar on harming civilians who provided significant support to their side's war-fighting efforts.

The response was spirited. Some reviewers condemned any attempt that they thought might weaken the law and erode the already meager protections that noncombatants enjoy. Many others, however, were happy for any effort to give state armies some additional maneuvering room to battle insurgents. This played well to a certain "realist" and maybe hawkish community. But it also came with many caveats about just war that the hawks ignored. While the history of international humanitarian law (IHL) and the law of armed conflict (LOAC) is sufficiently dynamic to make room for change, however belated, attempts to fiddle with the existing rules of war must always be taken with care and only in the context of just war: wars of self-defense, self-determination, or humanitarian intervention. This caveat is important because the slippery slope is always present. During a workshop with military and law enforcement officers, I once discussed the constraints that the rules of engagement pose for NATO. It was not long before officers from less enlightened domains – Nigeria, China, and Zimbabwe – jumped up and complained about the restrictions that the law of war imposes. When I tried to point out that it was a long and inadmissible jump from fighting Al Qaeda to suppressing internal dissent, they admonished me for my hypocrisy: "We are fighting terror too," they staunchly declared.

Addressing the rules of war that states must follow is only half the project because the very same concerns bedevil guerrilla warfare. Guerrillas and insurgents, too, want to know how they can fight against superior state armies, and I try to provide an answer guided by the moral principles that protect the rights of combatants and noncombatants. The result is to think about *just* guerrilla war and here, too, I am inclined to offer qualified support for human shields, rockets and missiles, hostage taking, cyber-warfare, media manipulation, and efforts to disable civilians who take an active role in armed conflict. Now, the same hawkish community that liked the first project is unlikely to be happy. This brings me back to NATO officers who complain loudly about how unfair things are: "We," they declare, "have to obey the law of war while guerrillas and terrorists flout it openly." But broaching the same subject to, say, a group of Palestinian Israeli lawyers only brings derision. For them, the law of war is also discriminatory and obstructionist, but in quite the opposite way that states perceive. LOAC, they say, only condemns guerrilla tactics while leaving plenty of room for strong state armies to do whatever they want.

Now it might be that both projects are pointless. By making concessions to states and insurgents, it may be that the rule of law will garner no respect and eventually fall by the wayside. But that argument is a little like preaching abstinence to teenagers when the right answer is to go out and buy them a bigger bed. Buying a bigger bed for belligerents means reexamining the ground between what the law forbids and what moral principles permit, thereby allowing aggrieved parties the space they need to pursue just cause with greater chances of success.

In this endeavor, I am grateful to many colleagues – Yitzhak Benbaji, Daphna Canetti, Cecile Fabre, George Lucas, Ben Mor, Cian O'Driscoll, and Paul Schulte – who took the time to read and offer critical comments on many parts of this manuscript. I am especially indebted to Tamar Mieisels who set things aside not only to read the entire manuscript but also to confront me vigorously with objections on the many matters on which we disagreed. The book is certainly better for it. Students from my graduate seminars, particularly Ameer Fakhourey, Nora Kopping, and David Reis, were extremely helpful as they struggled with some of the unorthodox arguments in this book and offered incisive suggestions. My thanks to the Israel Science Foundation for providing funds for part of this research and to the University of Haifa for the opportunity to take leave and spend a semester in Beijing. China, as one might imagine, is not the easiest place to study war and ethics. Many Internet sites are blocked, the people are reticent, and ethnic tensions boil beneath the surface. Tibet, for instance, is an especially sad place, and the casual visitor is struck by how deeply the people miss their Dalai Lama. It will be enormously interesting to see what happens when he is gone and Tibetans have to confront the Chinese alone. There must be better options than self-immolation.

Back in the Middle East there *are* other options: missiles, human shields, public diplomacy, and cyber-warfare, just to name a few. In July 2014, just as this book landed on my desk for final editing, war once again erupted in Gaza. The summer also found me teaching a graduate seminar on Thucydides and, as jets buzzed overhead, I spent my days toggling between the local news, my manuscript, and the *Peloponnesian War*. To say this was surreal is an understatement. While *The Ethics of Insurgency* can only offer a modest assessment of how guerrillas might fight, Thucydides furnishes trenchant and enduring lessons for states. One stands out. Speaking to the Athenians after a disastrous plague decimates their city,

Pericles is frighteningly candid as he encourages his compatriots to persevere. "To recede," he says, "is no longer possible. For what you hold is, to speak somewhat plainly, a tyranny; to take it perhaps was wrong, but to let it go unsafe." Throughout their very long war the Athenians wrestled with justice, expediency, and no small measure of aggrandizement. As of this writing, I don't know how the current conflict will end, but the fate of Athens is well known and ignored at significant peril.

Abbreviations

API	Additional Protocol I, 1977
APII	Additional Protocol II, 1977
EPLF	Eritrean People's Liberation Front
FALANTIL	Forcas Armadas de Libertacao Nacional de Timor-Leste (Armed Forces for the National Liberation of East Timor)
FRETILIN	Frente Revolucionária de Timor-Leste Independente (Revolutionary Front for an Independent East Timor)
GAM	Gerakan Aceh Merdeka (Free Aceh Movement)
HRW	Human Rights Watch
ICRC	International Committee of the Red Cross
IHL	International Humanitarian Law
IO	Information Operations
ISAF	International Security Assistant Force (Afghanistan)
KLA	Kosovo Liberation Army
LDK	Lidhja Demokratike e Kosovës (Democratic League of Kosovo)
LOAC	Law of Armed Conflict
LTTE	Liberation Tigers of Tamil Eelam (Tamil Tigers)
NGO	Nongovernmental Organization
PHR	Physicians for Human Rights
PKK	Partiya Karkerên Kurdistan (Kurdistan Workers' Party)
PLO	Palestine Liberation Organization
POW	Prisoner of War

SPLA	Sudan Peoples' Liberation Army
UN	United Nations
UNGA	United Nations General Assembly
UNSC	United Nations Security Council
WHO	World Health Organization

Selected Contemporary Guerrilla Wars

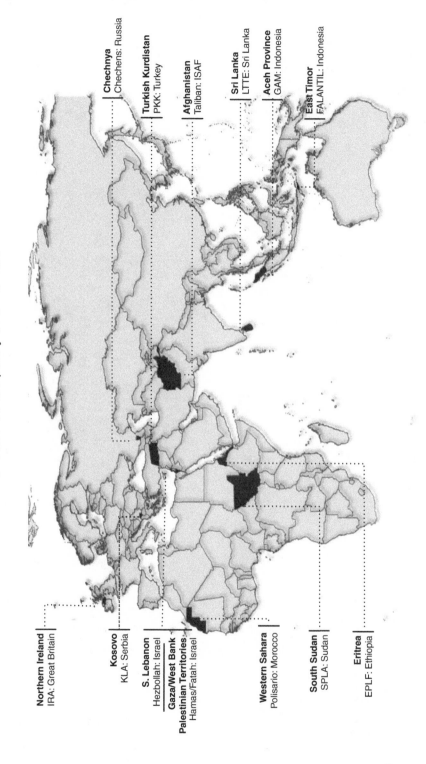

Northern Ireland
IRA: Great Britain

Kosovo
KLA: Serbia

S. Lebanon
Hezbollah: Israel

Gaza/West Bank
Palestinian Territories
Hamas/Fatah: Israel

Western Sahara
Polisario: Morocco

South Sudan
SPLA: Sudan

Eritrea
EPLF: Ethiopia

Chechnya
Chechens: Russia

Turkish Kurdistan
PKK: Turkey

Afghanistan
Taliban: ISAF

Sri Lanka
LTTE: Sri Lanka

Aceh Province
GAM: Indonesia

East Timor
FALANTIL: Indonesia

1

Just Guerrilla Warfare

Concepts and Cases

Writing in 1976, Walter Laqueur confidently predicted that guerrilla warfare was nearing its end. Post–World War II wars of decolonization had wracked the international system but would wane in the years following the ratification of the 1977 Protocols to the Geneva Conventions. By 1998, however, Laqueur reversed course and noted a resurgence of small wars in Afghanistan, Sri Lanka, Turkey, Chechnya, and the Middle East (Laqueur 1998: ix–xiii, 404–409). This trend had only intensified in the years following the breakup of the Soviet Union. The 1993 Oslo peace accords between Israel and the Palestine Liberation Organization (PLO) disintegrated, and decades of terror, civil unrest, and open warfare in Gaza and the West Bank ensued. American and Coalition forces waged war in Afghanistan against a Taliban enemy that claimed to fight foreign intervention and a corrupt central government. The 1994 Chechen war turned out to be only the first, while the second (1999–2009) proved a far more bloody and vicious affair that still left Chechnya's demand for independence unaddressed. In the Western Sahara, Polisario guerrillas and Moroccan forces have locked horns since Spain departed Africa in the mid-1960s. This conflict continues to simmer unresolved. In short, guerrilla organizations are still very active. And while some reports suggest a steady decline in intrastate violence, there is no doubt that new wars brew as citizens rise up against autocratic regimes in North Africa and the Middle East (Human Security Report Project 2013). On the other hand, some conflicts, thought intractable when Laqueur wrote, resolved after prolonged guerrilla war. Thanks to international military intervention, East Timor finally rid itself of Indonesia in 2002, while NATO made it possible for Kosovo to achieve de facto independence from Serbia in 2008. In 2011, following fighting that caused some of the worst casualties since World War II, South Sudan gained independence from its northern neighbor.

Many of these conflicts are national insurgencies – wars of liberation or secession waged by an armed group against a sovereign state. And, in fact, this study is confined largely to national insurgencies predominant in the post–Cold War period and includes conflicts in Afghanistan, Chechnya, Eritrea, Indonesia, Kosovo, Lebanon, the Palestinian territories, Sri Lanka, Sudan, and Turkey. Insurgencies did not end with European decolonization or with the collapse of the Soviet Union – quite the contrary. Since the breakup of the old Soviet bloc, guerrilla warfare has moved out of the shadows to increasingly occupy state forces and the international community. Modern media has put these conflicts on the front burner and in full view while growing humanitarian concern among Western nations has brought the United States and its allies to commit men and materiel as never before.

Today, there is often a tendency to tar all guerrilla movements with the brush of global terrorism, especially because many of the remaining national insurgencies pit sovereign states against Islamic movements. This is unfortunate and skews our understanding of guerrilla warfare and insurgency. Many guerrilla organizations indeed resort to terrorism, but most are neither terror organizations nor a party to global terrorism. Commentators are, nevertheless, so preoccupied with the global war on terror that there is a misguided tendency to see many national guerrilla organizations as nothing but a prop for Al Qaeda. As such, we overlook important questions of justice that surround many struggles for national self-determination.

There is no doubt that as insurgencies rage, one of the burning questions remains: How should a state army battle an adversary that uses human shields and wages war from among the civilian population? This is an important question, one that I and others have addressed in recent years (Gross 2010a). As crucial and interesting as this subject is, it also raises another, equally compelling question: How should guerrilla armies fight a sophisticated and technologically superior state army? This question is rarely asked because it is widely assumed that human shields, attacks on civilians, and kidnapping soldiers violate international and humanitarian law in the most flagrant way. The catchphrase "we fight by the rules but they don't" is nearly axiomatic.

But is it true? Are all forms of guerrilla warfare equally reprehensible? Can we think cogently about *just* guerrilla warfare? Guerrilla armies do not go to war to defend the borders of a sovereign state, but aim instead for independence, autonomy, or regime change. Can these causes, broadly defined as self-determination, rather than traditional notions of territorial self-defense provide just cause for war? Can such guerrilla tactics as using human shields to protect military installations, recruiting civilians to provide vital military services, capturing and exchanging soldiers in lopsided prisoner swaps, laying improvised explosive devices (IEDs) in populated areas, conducting cyber strikes against civilians, waging economic warfare, or pursuing a campaign of deceitful public diplomacy prove acceptable under the changing norms of contemporary warfare? The short answer is "yes," but guerrilla warfare requires a great deal of qualification, explanation, definition, and argumentation before it joins the repertoire of acceptable military behavior. There *is* room to speak of just guerrilla warfare. It is not always what guerrillas practice, but neither are its tactics and strategies the heinous practices that state powers often portray them to be.

GUERRILLAS AND INSURGENTS

Throughout this book, the words "guerrilla" and "insurgent" are used interchangeably. Unlike armed militants who mount sporadic challenges to the state, insurgents and guerrillas sustain hostilities for relatively long periods of time and take control of significant territory (Wilson 1988:23). "Guerrilla," moreover, depicts an actor, not unlike a rival state, that is independent of and alienated from the state they are fighting. Originally, notes Laqueur (1998:xvi), the term guerrilla "describes military operations carried out by irregulars against the rear of an enemy army or by local inhabitants against an occupying force." Largely confined to conventional war, guerrillas comprised those soldiers who were not part of the regular army (hence "irregular") and who fought rearguard actions that often complemented the set piece confrontations between regular armies. Post–World War II guerrilla warfare, however, is a far more sprawling affair. Guerrillas do not usually fight alongside regular forces or confront occupying, but transient, enemy forces. Instead, guerrillas represent local peoples struggling against entrenched colonialism or a repressive regime. Twenty-first-century guerrilla warfare embraces all of these causes: independence, secession, and regime change. Its aims are, therefore, revolutionary; insurgents do not seek to restore or maintain the status quo ante bellum, but to change it dramatically.

Writing in 1975, Sam Sarkesian (1975:7) defined revolutionary guerrilla warfare as "the forcible attempt by a politically organized group to gain control or change the structure and/or policies of the government, using unconventional warfare integrated with political and social mobilization." This definition usefully demonstrates how guerrilla warfare is defined by its organization, tactics, and aims. A guerrilla movement is organizationally complex and includes both a military cadre of trained fighters and a political wing that provides an array of supporting services. The military wing is usually the smaller of the two and cannot wage war effectively without the support of a vast political organization that can raise money; buy arms; secure allies; provide health, welfare, and education to insurgents, their families, and the surrounding population; wage a successful public relations campaign; and pursue diplomatic and legal campaigns at home and abroad. Guerrilla tactics, therefore, are multifaceted and include military operations, economic warfare, and public diplomacy. Each tactic aims at a different audience. Locally, insurgents are constantly trying to inculcate their ideology, recruit support, and generally "win the hearts and minds" of the people they claim to represent. Nationally, they are struggling to gain independence or sweep aside an entrenched regime. Internationally, they cultivate support among the world community. Today, more than ever, international backing is crucial for success and necessary for insurgents to overcome the inherent asymmetry of guerrilla warfare.

Compared to interstate war, guerrilla wars are materially and legally asymmetric. Material asymmetry reflects the disparity of arms between the opposing sides and is common in any war. Nations, after all, go to war when they feel they have the upper hand. But in guerrilla war the material asymmetry is glaring, indeed monopolistic, as the weaker side often lacks sophisticated weaponry, tanks, a navy, an air force, or an air defense system. Legal asymmetry points to the disparate status of the parties to the conflict. On one hand, sovereign nation-states are the building blocks of the international order and the only legitimate purveyor of armed force. They confront, on the other hand, an array of non-state actors that include guerrillas representing national groups (e.g., Palestinians, Chechens, or Kosovars), organizations wielding some governmental authority (e.g., Hamas or Hezbollah), or the remnants of a defeated government (e.g., Taliban) fighting occupation and their own state government.

These definitions of "guerrilla" and "asymmetric war" are justice neutral. They say nothing about "just" guerrilla warfare. Following the traditional dichotomy in just war theory, it will be useful to define just guerrilla warfare in terms of its just ends (*jus ad bellum*) and its just means (*jus in bello*). This is not just a philosophical exercise. Consider how the 1981 OAU Charter on Human and Peoples' Rights sets the stage for the conduct of just guerrilla war. Article 20 of the Charter proclaims:

1. All peoples shall have the right to existence. They shall have the unquestionable and inalienable right to self-determination ...
2. Colonized or oppressed peoples shall have the right to free themselves from the bonds of domination by resorting to any means recognized by the international community.
3. All peoples shall have the right to the assistance of the States' parties ... in their liberation struggle against foreign domination, be it political, economic or cultural.

This proclamation is laddered in an interesting way. Paragraph 1 stipulates a universal right of self-determination for all *peoples.* Paragraph 2 describes a narrower right: all peoples may possess the right of self-determination but only an oppressed people throwing off the bonds of domination may resort to "any means" to secure their rights. This paragraph introduces the moral asymmetry of just guerrilla warfare and the power of just cause. The sides to an asymmetric conflict are not morally equal. In wars of humanitarian intervention and the war on terror, moral asymmetry favors the stronger side, reinforcing its material and legal advantage. In wars of national liberation, the moral advantage shifts to the weaker side, thereby offsetting its material and legal disadvantage by allowing qualified recourse to armed force and, as I argue, to practices sometimes unlawful under international law. As a result, the obvious candidates for just guerrilla warfare are national liberation and secessionist movements together with recent urban revolts and popular uprisings against repressive, autocratic regimes. Candidates do *not* include pan-national Islamic terrorist groups fighting to undermine the international system or the many civil wars motivated by greed, looting, and predation. Paragraph 3 of Article 20 turns from victims to bystanders: if victims have a right to wage war, then bystanders may have a duty to aid or, at the very least, to refrain from hindering or obstructing their efforts. These three points are, of course, contentious and shrouded in ambiguity. Conveniently enough, they also echo the structure of this book.

THE ETHICS OF INSURGENCY: A BRIEF OVERVIEW

Part I of this book addresses questions of *jus ad bellum* and *jus in bello*. Chapter 2 analyzes the legitimate ends of just guerrilla war. In the minds of many, guerrillas are little more than criminals or terrorists. This view requires far greater nuance. While traditional just war theory broadly accepts a state's authority to wage war, ethics and law deny the same forbearance to non-state actors. Guerrillas and insurgents must prove their worth. They must establish the justice of their cause on the same basis of national self-defense that gives states the right to fight and must prove themselves the legitimate representatives of their people's national aspirations. What conditions underlie a guerrilla movement's right to fight? May guerrilla leaders who gain less than the complete consent of their people go to war? What role should the international community play as a people strive to realize its right to self-determination? These questions occupy Chapter 2. Some short answers include: (a) a people often, but not always, enjoy the right to wage an armed struggle to gain self-determination, whether independence or autonomy, and/or to secure their compatriots' right to a dignified life; (b) insurgents gain legitimacy when they best serve their people's interests and enjoy some degree of popular consent; and (c) the international community is duty bound to refrain from enacting or enforcing laws that unduly restrict guerrillas' right to fight and from supplying arms to repressive states. When insurgents cannot prevail, the international community may be bound to actively intervene.

"Who Fights and How" is the subject of Chapter 3 and addresses the means of just guerrilla war. The means of war are not only its tactics (as many theorists assume) but speak to the resources necessary to wage war. Before asking how to fight, one must first ask with what. How may guerrillas raise troops and enlist material support from the civilian population? Like states, guerrillas require coercive means to conscript men and materiel. Lacking the coercive institutions of state, however, guerrillas may resort to a wide range of measures short of violent coercion and physical intimidation. Permissible measures include social pressure, ostracism, fines, or imprisonment. Compared to state institutions, guerrilla institutions are weak, leaving insurgents to go to war with limited resources and with erratic support from their compatriots. Once committed to an armed struggle, the second question arises: How may militants fight and what rules govern their use of armed force? This is the concern of *jus in bello*.

International law is not entirely clear about the rules of war insurgents should follow. By law, insurgents enjoy various rights under the 1977 Protocols I and II to the Geneva Conventions. Protocol II protects the fundamental human and legal rights of participants in a non-international, that is, internal armed conflict, but does not grant insurgents the rights of ordinary combatants. Insurgents fighting a civil war, for example, may be held criminally liable and punished for fighting. Protocol I, on the other hand, offers guerrillas fighting "colonial domination, alien occupation and racist regimes" the same rights and status as state soldiers. Nevertheless, these codified laws of war restrict the practice of war and can make it very difficult for guerrillas to conduct an armed struggle effectively. As insurgents chafe under the laws of war, the challenge is to carve out a space that, while often unlawful, is nonetheless morally permissible. This moral floor, as it were, rests on humanitarian principles that protect combatants from inhuman treatment and noncombatants from direct harm, bodily injury, and loss of life.

Combining the premise of the OAU statement together with elaborations yet to come, a provisional definition of just guerrilla warfare denotes *sustained, unconventional military and political operations that utilize armed violence, non-kinetic force, and soft power to realize a people's right to national self-determination and/or protect their fundamental right to a dignified life by means that do not violate the rights of civilians and enemy combatants.* This is a tall and complicated order. Many will assume that no guerrilla organization can meet its requirements and, under many common interpretations of the law and ethics of war, they may be right. Certainly, the law of armed conflict prohibits direct attacks on civilians, hostage taking, human shields, and the exploitation of prisoners of war in the most categorical way imaginable. How is it possible that anything resembling such tactics does not violate the rights of civilians and enemy combatants? Can one cogently argue that some guerrillas not only enjoy the right to fight by such unconventional means, but that the international community has the concomitant duty to either refrain from interference or to actively assist insurgents? The answer can only lie in qualifying each tactic by the humanitarian principles that must constrain it.

Taking such principles as a yardstick to measure moral compliance, Parts II and III investigate a wide array of tactics that guerrilla and insurgents employ. Part II addresses hard, kinetic warfare: improvised explosive devices, rockets, targeted killing, and human shields. Part III examines the little explored field of soft, non-kinetic warfare: cyberterrorism, economic warfare, public diplomacy, and nonviolent resistance. Soft war lies largely outside the domain of international law and requires a moral going-over that just war theory has rarely provided.

Part II: Hard War

Chapter 4: Large-Scale Conventional Guerrilla Warfare: Improvised Explosive Devices, Rockets, and Missiles

"Conventional" guerrilla warfare refers to the tactics guerrillas employ to confront state armies. Regardless of the emphasis placed on terrorism and the growing use of soft, non-kinetic force, military action is the dominant business plan. While guerrillas typically avoid set piece confrontations (Eritrea's war of independence being a major exception), a host of deadly tactics remain. Improvised explosive devices (IEDs) are high explosives set to disable troops, tanks, and convoys. They are easy to build, difficult to detect, and among the most devastating weapons insurgents can employ. State armies decry their use because IEDs cause many civilian casualties when left unattended at the roadside. Similarly, guerrillas employ relatively unsophisticated missiles. These weapons, too, are difficult to control and may bring excessive civilian casualties when guerrillas target military sites. In the worst cases, guerrillas stand charged of deliberately using inaccurate and undiscriminating weapons to terrorize civilian populations. Whether utilizing IEDs or rockets, insurgents often respond with the same refrain: they have no other weapons to fight state armies effectively.

These charges and countercharges require a careful sorting out. Some civilian casualties may, indeed, comply with the conditions for permissible collateral harm. But this requires evidence that guerrillas seek military targets. Despite their avowed intentions, many insurgents abandon roadside bombs unsupervised, ready to detonate when the first car drives by.

Supervision and controlled detonation, regardless of the additional risk this poses to guerrillas, seem to be a necessary condition for permissibly employing IEDs. Discrimination, too, regulates the deployment of any missile system so that combatants hit the military targets they aim at. This requires, first, that guerrillas launch rockets with the intent of destroying a military target and not simply wreak collateral harm and, second, that they deploy missiles reasonably capable of hitting their targets. The latter point is particularly contentious given the vast disparity between the missile capabilities of insurgents and states. Nevertheless, guerrillas may equitably demand consideration for less accurate armaments when necessary to wage a just war. Related concerns guide targeted killings and the taking of prisoners.

Chapter 5: Small-Scale Conventional Guerrilla Warfare: Targeted Killing and Taking Prisoners

IEDs and missiles are big hammers employed to disable large military targets. Targeted killings and taking prisoners are more like surgical pliers, often utilized to pick out specific individuals. For states, targeted killing refers to a detailed process of identifying, locating, and killing ununiformed insurgents. For insurgents, however, the process is multi-faceted. On one hand, insurgents seek out enemy military officers and high-ranking political leaders. On the other, they also target compatriot informers and collaborators. In each case, permissible targeting turns on liability. In the first instance, international law and just war theory are slowly making room for disabling civilian political leaders on the assumption that these figures – the head of a guerrilla organization's political wing, for example – contribute significantly to war-making operations. These arguments resonate on both sides, and offer guerrillas the same latitude. In the second instance, guerrillas often develop rudimentary institutions to try and punish informers and collaborators. Unfortunately, their proceedings sometimes lack discipline and smack of summary execution. Just guerrilla warfare challenges these adverse outcomes by demanding due process and proportionate punishment while proscribing any attempt to intimidate or terrorize the local civilian population.

Prisoner taking is targeted killing without the killing. While states armies take prisoners all the time, two issues dominate when insurgents seize enemy soldiers. First, states and guerrillas must reevaluate the conditions necessary to prevent ill treatment. While there is a firm baseline to provide medical care and to prevent torture, mutilation, and execution, other demands of international law might be revisited and modified. As evidenced by Guantanamo Bay and elsewhere, states no longer hurry to declare an end to hostilities and repatriate their prisoners. As states modify the norms surrounding prisoners of war, there is also room to think about how guerrillas may legitimately leverage the benefits of the few prisoners they hold. Guerrillas often deny access to prisoners and bargain information for the return of their own in lopsided prisoner exchanges that states denounce as grossly unfair. These practices speak to the second peculiarity of taking prisoners. Taking prisoners to bargain for the release of compatriots raises the specter of kidnapping and hostage taking, distinctly odious practices that, nonetheless, may find a place in just guerrilla war. Using human shields raises similar hackles.

Chapter 6: Human Shields
When state armies confront guerrillas, human shields are not far from their minds and lead to vocal complaints about violations of international law. For guerrilla armies, on the other hand, enlisting civilians to shield military operations is very effective and demands a closer look. The literature on human shields is relatively sparse. Distinguishing between voluntary shields (civilians who agree to shield) and involuntary shields (those coerced to shield), commentators raise – but do not resolve – a range of questions about intentionality, liability, and the obligation to protect civilians from harm. While many understand that those who intentionally shield military operations are culpable and may suffer death or injury as enemy forces attack a shielded site, the reality is more complex.

   Civilians are a constant feature of the landscape and difficult to avoid. Some work for an organization's political wing and provide significant war-sustaining aid. Their presence shields many vital facilities from direct attack thereby leaving states to search for alternative means to disable these sites. At the same time, guerrillas often mount their attacks from within civilian population centers while enlisting or, perhaps, conscripting other civilians to shield command centers and weapon depots. These civilians also provide cover, but their intentions are difficult to discern. Some are conscripted, some volunteer, and some are entirely ignorant of ever acting as shields. All this makes it difficult for states to target human shields directly and leaves guerrillas with a very useful tactic to deter attacks. While the use of human shields is not inherently unethical, it is subject to certain constraints. These require guerrillas to secure consent, to employ human shields only when they provide a military advantage unobtainable by other means, and to remain vigilant about exposing their own citizens to unnecessary harm.

Part III: Soft War

Part III pushes past hard, kinetic warfare into soft warfare and, an area largely neglected by just war theory. While Chapter 7 opens with terrorism, it quickly repudiates kinetic terrorism and turns to nonlethal and cyber-warfare as a means to disable civilians to gain military advantages. In this regard cyber-warfare is little different from economic warfare, the subject of Chapter 8. Both target civilians directly. In contrast, Chapter 9 directs our attention to the power of public diplomacy, and the last chapter describes how some guerrilla organizations renounce violent resistance altogether.

Chapter 7: Terrorism and Cyberterrorism
Chapter 7 addresses one of the core issues of just guerrilla warfare – the use of terror – but offers no sustained discussion of kinetic terrorism, a reprehensible practice that leaves no room for deliberation. Instead, the chapter focuses on a different variant: direct but *nonlethal* attacks on civilian populations. Guerrillas, like states, require some means to disable civilian, war-sustaining targets that provide crucial support for war. Just as state armies target the political infrastructures of such organizations as Hamas, Hezbollah, or the Taliban, insurgents, too, may try to disable the vast civilian bureaucracy that runs the state machinery of war. The permissible means that guerrillas have at their disposal, however, are limited. While states can utilize electromagnetic weapons or precision guided missiles to destroy or disable financial, telecommunications, electric, and media facilities, guerrillas can only resort to small-scale kinetic attacks or, increasingly, to cyber strikes.

Cyber-warfare describes the use of computer technology to disable critical infrastructures. Aimed at disabling military and civilian computer systems, cyber-warfare may nonetheless cause considerable hardship to civilians. Currently, sophisticated hackers can disrupt financial systems, steal proprietary data, crash social networks, block internet access, and organize widespread identity theft thereby sowing panic and havoc in a society increasingly dependent on computerized networks. Guerrillas also may use cyber-warfare to disseminate propaganda, disable military facilities or gather intelligence. Future scenarios depict hackers disrupting or destroying vital medical, transportation, and water control facilities and causing widespread death, injury, and devastation. Whether we call this warfare or terrorism depends upon the targets guerrillas choose and the harm they cause. Cyber-warfare strives for a military advantage; cyber-terror sows panic and fear. But because cyber-attacks rarely (and, to date, never) cause physical injury, many declare "no harm, no foul" and reject any attempt to regulate cyber war by the law of war. This conclusion is premature. Cyber-warfare varies tremendously. Depending on which systems are disrupted, damage may be localized or widespread, transient or permanent, costly or inexpensive, and directed, with varying degrees of intensity, at military or war-sustaining civilian infrastructures. In the absence of physical harm, psychological harm may be, nonetheless, widespread and debilitating. Considerations of mental suffering suggest that cyber-warfare, like ordinary warfare, is subject to constraints of necessity, discrimination, and proportionately.

Chapter 8: Economic Warfare and the Economy of War
The economic dimensions of war are exceedingly important. Economic warfare – sanctions, boycotts, and blockades – holds a special place in just war theory because it often represents the penultimate resort, the last step that states must weigh before going to war. As a result, perhaps, international law is curiously lenient about the death and destruction economic warfare may intentionally bring upon the civilian population. Philosophers, on the other hand, condemn economic warfare for targeting civilians directly. In response to the hundreds of thousands of children who died following sanctions against Iraq, many observers turned their attention to "smart sanctions" to embargo arms or seize financial assets. Guerrillas, however, cannot impose blockades or seize a state's financial assets. Instead, they strike economic targets and attack oil pipelines (e.g., Southern Sudan), tourist sites (e.g., Basque militants), business districts (e.g., the IRA), airline equipment (e.g., Tamil Tigers), or organize international boycotts (e.g., Palestinians).

Like many other tactics of guerrilla war, economic warfare may be permissible insofar as it is effective and does not cause direct or excessive harm to the civilian population, a caveat, sadly enough, that states do not usually respect.

How states and guerrillas raise the money they need to fight presents challenging moral questions about the economy of war. States impose taxes, sell bonds, and print money. Guerrillas, too, impose local taxes and tolls and accept foreign aid. At the same time, however, many insurgents veer toward unlawful and predatory practices: counterfeiting, extortion, drug trafficking, and smuggling. Such a war economy raises two questions. First, what coercive means may guerrillas employ to ensure the payment of taxes and tolls given the absence of any formal institutions of state to compel compliance? The same question arises when we ask about conscription and leads to similar recognition of moderate, but not violent coercion or physical intimidation: fines, incarceration, expropriation of property, and ostracism. Second, what unlawful practices may guerrillas employ to raise funds? The answer turns on the imperative to protect the rights of those who may be harmed. Here, each practice – drug production, drug trafficking, kidnapping, smuggling, looting, and counterfeiting – deserves separate consideration. Not all merit sweeping condemnation.

Chapter 9: Public Diplomacy, Propaganda, and Media Warfare
State armies often wring their hands in frustration when they win military battles but lose the media war. Media warfare, characterized by an aggressive campaign of public diplomacy, is emerging as an increasingly effective means for guerrilla organizations to capitalize on their underdog status and take advantage of deadly mistakes by state armies. Digitization, communication networks, and growing internet access are the pivotal developments that allow insurgents to quickly disseminate opinion-shaping information. Images of civilian casualties and wrecked infrastructures in East Timor, Lebanon, Gaza, Iraq, and Afghanistan enlist support for guerrilla organizations while disrupting state army military operations. Media warfare can be extremely effective, affecting current military operations, future military planning, government policy, world public opinion, and the deliberations of international organizations.

However effective it is, public diplomacy elicits harsh criticism when guerrillas turn to "gray" propaganda: staged media events, doctored photos, half-truths, lies, and bald-faced manipulation of the media. With the laudable exception of banning states from inciting genocide and war crimes, neither the ethics of war nor humanitarian law address media warfare or public diplomacy. Nevertheless, "information" ethics imposes constraints of truth telling, media accuracy, fair access, and unfettered debate that have significant bearing on the ethics of media warfare and public diplomacy. Not all of these constraints are compelling, however, and thereby leave guerrillas to pursue a wide range of otherwise suspect practices.

Chapter 10: Civil Disobedience and Nonviolent Resistance
A recurring question today is, where is nonviolent civil disobedience in the Middle East and elsewhere? Chapter 10 evaluates the role of civil disobedience and nonviolent resistance in guerrilla warfare by examining general strikes, prison strikes, hunger strikes, demonstrations, protests, and boycotts on a local, regional, and international scale. In many cases, nonviolent resistance is extraordinarily effective. But even as practiced by such activists as Mohandas Gandhi and Martin Luther King, nonviolent resistance may harm protesters substantially when organizers pursue "backfire." Long recognized as one of the most effective tactics of nonviolent resistance, backfire describes how protesters successfully provoke a brutal and disproportionate response from their adversary that will solidify support at home and swing world opinion to the guerrilla's side. Nonviolent resistance, it turns out, is not so nonviolent, and backfire tactics are deeply problematic. As with all other tactics of just guerrilla warfare, insurgents continue to walk a line between the unlawful and morally permissible even when they do nothing more seemingly benign than undertake nonviolent resistance.

CASES AND EXAMPLES: A FEW WORDS

Although many historical cases animate these chapters, this book is neither an empirical nor exhaustive study of contemporary guerrilla warfare. Rather, it uses contemporary cases to illustrate the moral principles behind just guerrilla warfare. In contrast to social scientists who have explored the origins and causes of contemporary civil war in remarkable detail, just war theorists have made but isolated references to guerrilla war. Algeria (circa 1960), Al Qaeda, and the Palestinians predominate. In between, however, are a vast number of compelling historical cases from which to draw lessons about just guerrilla war.

As guerrillas strive for self-determination, independence, or autonomy some enjoy significant legal support from the international community and some do not. In the former category are such struggles as East Timor-Indonesia (2002) (date of resolution in parentheses); The Western Sahara-Morocco (ongoing); Kosovo-Serbia (2008); Southern Sudan-Sudan (2011); and Palestine-Israel (ongoing) (See Map 1). In each case, insurgents could appeal to international resolutions that affirmed their right to national self-determination. In the second category, legal grounds for independence or autonomy are tenuous and subject to the vagaries of war: Eritrea-Ethiopia (1993); Chechnya-Russia (2009); Tamil-Sri Lanka (2009); Northern Ireland-Great Britain (1998); Aceh-Indonesia (2005); and Turkish Kurdistan-Turkey (ongoing).

These cases provide a wealth of information for studying just guerrilla warfare. Citing historical, religious, linguistic, or cultural differences, secessionist movements press for statehood against a sovereign state of which they form an integral part. Aiming for independence, secessionist movements, nevertheless, sometimes settle for something less. Southern Sudan declared its independence in 2011 and basked in international recognition. Following the breakup of Yugoslavia and Serbia's unilateral abrogation of Kosovar autonomy in 1989, however, Albanian Kosovars initially conducted a nonviolent but unsuccessful campaign for national independence. Turning to an armed struggle in the late 1990s and aided by the international community, Kosovo finally broke free from Serbia and declared independence in 2008 but still finds its bid blocked by many nations. Following two fierce separatist wars with Russia in 1994 and 1999, Chechnya eventually settled for a precarious autonomy secured by a strong-arm government in Grozny and an authoritarian regime in Moscow. For three decades, the Aceh fought Indonesia for independence until settling for political autonomy in 2005. Like the Chechen's, the Aceh settlement did not differ significantly from what they were offered early in the fighting, before the civilian population was decimated by death, disease, and forced migration. Fighting to evict the British from Northern Ireland since the late 1960s, the Provisional

Irish Republican Army employed varying degrees of terrorism, military confrontation, economic warfare, and public diplomacy until agreeing to a power sharing agreement in 1998. Like the Irish Catholics, the Aceh, and the Chechens, Kurdish Turks under the leadership of Abdullah Öcalan and the PKK (including the Kurdish Workers Party and its military wing) conducted a long running, violent insurgency against Turkey. Since its founding in 1978, the PKK oscillated between demands for independence and regional autonomy until agreeing, at least for now, to push for autonomy and renounce armed force. The Tamil fared less well than many contemporary guerrilla movements as they fought to secede from Sri Lanka in 1983. After maintaining a sophisticated military and political organization for nearly thirty years, Sri Lankan forces decisively defeated the Liberation Tigers of Tamil Eelam (LTTE) in 2009. Political and economic reconstruction continues.

Many contemporary independence movements remain rooted in twentieth-century colonial warfare and prove less amenable to autonomy-based resolution. Eritrea, a former Italian colony, found itself federated with Ethiopia in 1952, wholly annexed in 1962, and abandoned to its own devices by the international community thereafter. The result was another thirty year war for independence that only ended when the Eritrean People's Liberation Front (EPLF) decisively defeated Ethiopian forces in a conventional war. East Timor was one of Portugal's last colonies. It was slated for independence, but was violently annexed by Indonesia in 1975 as the Portuguese departed. While the UN affirmed the national rights of the Timorese to self-determination and independence in 1975 (United Nations General Assembly 1975, 1976), it took another quarter century and aggressive foreign intervention before the people of East Timor finally gained independence in 2002. The festering situations in the Western Sahara and Palestine are also relics of colonial warfare. Both the Spanish and British were prepared to accept the independence of their territories in the Western Sahara and Palestine in 1966 and 1947 respectively. The international community supported these moves and endorsed independence *before* widespread fighting erupted (United Nations General Assembly 1947, 1966, 1972, 1974). In contrast to secessionist movements that hope for recognition after claiming independence, independence movements often gain recognition prior to armed resistance. Nevertheless, events intervened to scuttle independence in each case. The Moroccans blocked every attempt to hold a referendum on independence in the former Spanish colony, while the Arabs rejected the partition plan of Palestine and went to war with Israel. The 1967 Arab-Israel war and subsequent occupation of the West Bank and Gaza have only exacerbated the plight of the Palestinians.

Other guerrilla wars do not fit the anti-colonial, liberation, mold as well as some of those described. The Taliban, for example, is not an international terrorist organization but espouses a struggle against foreign occupation, which resonates among many Afghans. True, its fight against Western forces bears aspects of terrorism and a civil war, but it also embraces aspects of an independence movement. It is, therefore, not without reason that the United States now refrains from calling the Taliban "unlawful combatants" and seeks negotiations to end the fighting. Finally, consider proxy guerrilla wars. Proxy guerrilla wars occur as state actors solicit the aid of non-state proxy agents to initiate conflict in their stead. In the 2006 Lebanon War, for example, Syria and Iran enlisted Hezbollah in their fight against the United States and Israel. In all of these cases there may be some semblance of *casus belli.* Nationalist guerrillas fight for independence, self-determination, and/or protection of human rights while proxy guerrillas and their patron states use armed force to deter, distract, or weaken their enemies.

Fledging states, insurgencies, and liberation movements are as varied as the circumstances surrounding their birth, and no single repertoire of fighting typifies these struggles. Among the examples cited, only Eritrean guerrillas could pursue anything like a conventional set piece war against its enemy. The Eritreans fielded large armies, attacked the Ethiopians with infantry and armored units, captured significant numbers of prisoners of war, and maintained a vast array of statelike institutions to provide social welfare, education, and medical care. Elsewhere, the means of war varied considerably. Some movements eschew terrorism entirely, while others embrace it wholeheartedly. Some take many prisoners and some take few. Some treat captives well while others abuse them horribly. Weaponry is growing increasingly sophisticated as IEDs give way to missiles and cyber strikes, and guerrillas turn to carefully orchestrated campaigns of public diplomacy and strategically effective campaigns of nonviolent resistance.

In short, conflicts of the last twenty-five years teach us much about the prosecution of just war. As these conflicts resolve or persist, the world community has shown a willingness to recognize the importance of human rights regimes and makes way for nations fighting for national liberation. Humanitarian intervention and military support are two avenues that the international community is ready to examine closely. Yet, many insurgents are also prepared to fight alone, and they require a firm measure of tolerance for some of the tactics they might usefully employ. The norms of contemporary warfare are slowly changing as state armies insist upon exceptional means for exceptional times. Cannot guerrilla armies demand the same forbearance as they pursue just cause? The beneficiaries of these changes will be the remaining peoples who continue to struggle for independence or autonomy, as well as the growing number of peoples who are hoping to rise up and democratize tyrannical and oppressive regimes.

PART I

THE RIGHT TO FIGHT

2

The Right to Fight

Just Cause and Legitimate Authority in Guerrilla Warfare

While it is vital to ask how state armies should wage just war against guerrillas, it is equally important to ask whether guerrillas and the people they claim to represent can wage just war against states. It is often thought that they cannot, that guerrilla armies systematically violate the principles of just war and leave state armies to wring their hands in frustration. If this dismal perspective dominates the thinking of state governments, it is decidedly narrow and, ultimately, unconvincing. Some, though not all, guerrilla organizations enjoy the right to press their claims by force of arms. To establish their right to fight, guerrillas must meet two challenges. First, they must show just cause and establish legitimate authority; second, they must fight justly. The first challenge is the subject of this chapter; the second challenge is explored in the following chapter.

Just cause reaches beyond traditional notions of territorial self-defense and affords some insurgents the right to use armed force as they strive for national self-determination. Similarly, the principle of legitimate authority – who, exactly, may authorize the use of armed force – broadens in the context of guerrilla warfare to include liberation movements and guerrilla organizations. Not everyone with a political grievance can pick up a gun and fight. Absent just cause and legitimate authority, guerrilla organizations indeed face charges of criminal behavior. Armed with just cause and legitimate authority, these same organizations enjoy the right to fight, the right to shed their uniforms, the right to combatant status, and, in some cases, the right to aid from the international community.

Unlike states, guerrilla organizations cannot ignore the demands of just cause and legitimate authority. For states, the *ad bellum* principles that define the grounds for starting a war have never been very popular.

States recognize their supreme moral status in the international order. As a result, they rarely contest another state's right to wage war when necessitated by national interest or seldom condemn a fellow state for aggression.[1] It is not surprising then that international humanitarian law and the law of armed conflict – strict legal regimes dedicated to mitigating the destruction of modern war – generally eschew anything having to do with just cause. For Jean Pictet (1985:13–14), the eminent jurist and midwife of post–World War II humanitarian law, the very idea of *just* war represents nothing less than a "malignant doctrine" that has "hampered humanitarian progress for centuries." By offering one side (or both!) the moral anchor of justice, just war theory relegates one's adversary, its soldiers and civilians alike, to criminal status. With right on one's side, conquering an unjust foe permits and even demands the unrestrained use of armed force, thereby stripping unjust soldiers of their rights and just soldiers of their duties.

These prospects warranted the wrath of legal scholars. They understood that any successful attempt to introduce restraint into armed conflict would require each side to bracket the justice of its cause and confer rights and privileges on enemy soldiers and civilians as if they were moral equals, untainted by criminality. And, in fact, this is what they did. Putting *ad bellum* principles aside, jurists and philosophers directed their attention to the just and legitimate conduct of war – *jus in bello*. But whereas *ad bellum* principles had been of scant concern when states fought one another, they stand front and center when states fight guerrillas. In asymmetric war, just cause and legitimate authority are everything. Without them, non-state actors – guerrillas, insurgents, rebels, and/or freedom fighters – are without standing, deprived of combatant rights and treated as criminals. With just cause and just authority, these same non-state actors regain their standing as combatants and their right to fight. Only when guerrillas enjoy the right to fight may they choose to go to war, recruit, and, if necessary, conscript soldiers and civilians, levy taxes, fight without uniforms, and employ a host of tactics that today meet with considerable disdain among state armies.

The right to fight can be formulated as follows:

---

[1] Prominent examples include condemnation of China and North Korea (1950–1951), the Soviet Union (1980), and Iraq (1990).

When fighting for a just cause and enjoying legitimate authority, insurgents have a right to pursue their claims by force of arms as a last resort and by means that are (1) effective and necessary and (2) do not violate the basic rights and protections due combatants and noncombatants.

If guerrillas meet these conditions, neither international law nor practice can abrogate the right to fight by making its exercise impossible. States, therefore, have a duty not to impede or interfere with the exercise of the right to fight and may have a positive duty to aid guerrillas fighting for a just cause.

   This chapter and the next elaborate each of these ideas. Just cause for war is anchored narrowly in the right to self-determination and the right to live a dignified life. Legitimate authority requires a peoples' consent or trust, which guerrillas gain through a range of formal and informal political and social institutions. While necessary, just cause and legitimate authority are insufficient to wage an armed struggle justly. Indeed, I suggest that guerrillas do *not* always enjoy the right to fight no matter how just their cause may be and no matter how firmly their legitimacy is entrenched. Additional conditions – last resort, necessity, effectiveness, and respect for combatant and noncombatant rights – are central to just war theory. These conditions, however, take on a different meaning in the context of guerrilla warfare. Last resort is paramount: unlike states, guerrillas and insurgents are expected to diligently pursue means short of war. While effectiveness demands a stringent cost-benefit analysis relative to guerrillas' goals, these goals often fall short of anything resembling military victory and often turn on nothing more than symbolic gains. Finally, combatant and noncombatant rights constrain the means of war but leave room for tactics the law does not always permit.

THE RIGHT TO FIGHT: JUST CAUSE

For a people, rather than a state, just cause is often – but not exclusively – framed in terms of the right to national self-determination. Since World War II, the international community has slowly acknowledged the right of certain ethno-national communities to expel colonial powers and establish their own nation. In places such as Algeria, Eritrea, East Timor, and Mandatory Palestine, peoples sought independence in the wake of colonial disintegration. Elsewhere – in the former Yugoslavia, Sudan,

Chechnya, Northern Ireland, and Sri Lanka, for example – ethno-national groups sought to secede from existing nations of which they had been an integral part. In many instances, the struggle for national self-determination was accompanied by grave human rights abuses and terrorism on both sides. Perusing the scene, observers can ask two questions. First, which peoples enjoy the right to self-determination, whether in terms of statehood or some lesser form of autonomy? Second, of those that do, which enjoy the right to pursue these goals by armed force? There is considerable discussion about the first question but far less about the second.

The Right to Self-Determination

Writing in 1919, Woodrow Wilson defined self-determination as the "right of every people to choose the sovereign under which they live" (Cass 1992:23). This is both an individual and collective right. Article 27 of the International Covenant on Civil and Political Rights (1976) emphasizes an individual right: "In those States in which ethnic, religious, or linguistic minorities exist, persons belonging to such minorities shall not be denied the right, in community with the other members of their group, to enjoy their own culture, to profess and practice their own religion, or to use their own language." Two points are important. First, Article 27 reflects an individual, not a group right. Second, it does not entail political, economic, or social autonomy, much less independence. The right to self-determination only requires that minority group *members* receive the means necessary to maintain their unique identity (Cassese 1995:61).

There are, nevertheless, collective aspects of self-determination. International instruments, for example, acknowledged the right of certain groups of people – those listed in "trust or non-self-governing territories" – to self-determination and, indeed, independence (Wilson 1988:56–87). These were colonies and territories of the United States and European states, and most received independence by 2002. Moreover, Additional Protocol I (1977:Article 1.4), recognizes the right to self-determination of peoples fighting against colonial domination, alien occupation, and racist regimes. Although the terms "colonial domination," "alien occupation," and "racist regime" were tailored solely to grant combatant status to guerrillas fighting against Portugal, Israel, and apartheid-era South Africa, respectively, the gross injustice associated with "alien occupation" continues to resonate as *casus belli* today among guerrillas fighting in Palestine, Chechnya, Afghanistan, the Western Sahara, Turkey, and elsewhere.

These citations – from positive, international law – do not tell us *why* certain groups have the right to self-determination. Here, too, individual and collective explanations are important. At the individual level, the right to self-determination reflects an exercise of freedom and autonomy necessary for a person to realize a dignified life. A dignified life delivers more than subsistence; it speaks to an autonomous life characterized by a level of self-worth, fulfillment, and, in Hugo Bedau's (1968:567) elegant words, "the opportunity for the release of productive energy." Such opportunities allow persons and their community to formulate, pursue, and realize a life plan that fosters an individual's capabilities as a human being. Human capabilities are physical (physiological functioning), intellectual (rationality, creative thinking, and imagination), and social (the capacity for interpersonal and community ties) (Barilan 2012:131–136; Fabre 2012:19). The conditions necessary to realize these capabilities include human rights that protect individuals from murder, rape, assault, servitude, torture, cruelty, and indignity; material or welfare rights to forestall crushing poverty, illness, and ignorance; cultural rights to preserve a people's community and its ethnic, linguistic, and historical heritage; and civil rights that safeguard personal autonomy through political representation, employment opportunities, property ownership, freedom of press and assembly, and so forth. Within this constellation, the right to self-determination joins these human, material, cultural, and civil rights to facilitate a dignified life.

These rights are both hierarchically ordered and intertwined. At the top of the hierarchy are human rights that satisfy those needs necessary for any semblance of a dignified human life. As such, these rights are universal, and the systematic violation of them provides just cause for war. At the other end of this rights hierarchy are civil rights. Civil rights generally demand widespread respect but might reasonably be set aside during extreme emergency. Censorship or limits on public assembly come to mind. In between human and civil rights are those material, economic, and cultural rights whose inviolability turns on the harm right holders suffer when these rights are abused. Material deprivation may be life threatening or inconvenient; infringements upon cultural identity may be emasculating and humiliating or just constricting. In the former instance, such deprivations and infringements make a dignified life impossible. In the latter, they may simply reflect legitimate budgetary constraints. The former may be cause for just war; the latter only cause for concerted political action. Self-determination, too, admits of similar degrees of injury when violated.

Self-determination provides a political framework to safeguard security, protect human rights, and nurture the essential ethnic, cultural, linguistic, and historical components of self-identity (Caney 1998; Nielson 1998). In this sense, self-determination is not national but individual. At minimum, individual self-determination requires a decent, stable society whose government, as Rawls (1999:64–68) suggests, respects human rights, offers opportunities for public consultation, and provides for law, order, and justice. When ethnic and national minorities populate a state, self-determination may also require relatively humble institutions like community centers, private schools, and religious organizations or provisions as complex as local and regional autonomy, federalism, or independence. Lack of self-determination in its many forms can endanger physical well-being, foster anomie, and cripple self-development (Gilbert 1998; Margalit and Raz 1990).

A collective right to national self-determination looks beyond the benefits that self-government affords individuals and highlights the intrinsic rights of groups as organic units whose value exceeds the sum of its members. While national and other communities are instrumentally valuable, they are also valuable in their own right as an expression of a collective ethos, norms, and ways of life without which humanity would be a fundamentally poorer place. For communitarians, these ideas speak to an ideal where rights-respecting cultures of separate peoples interlace to contribute to an ever-perfecting world order.

Thus, the "people" who enjoy the right to self-determination are variegated to the extreme. International law takes a narrow view, limiting the right to national self-determination to only three kinds of groups: colonial peoples, peoples subjected to foreign occupation who are entitled to political independence, and racial and religious groups entitled to "internal" self-determination – autonomy, adequate political representation, and/or minority protections (Cassese 1995:319; Cass 1992). However, as theorists from the "nationalist" school claim, there is no obvious reason to deny the right to self-determination to any people defined by ascriptive traits such as culture, history, ideology, language, ethnicity, and/or territory (Kymlicka 1998; Philpott 1998; Wellman 1995). Such a broad and encompassing view of peoplehood, however, is too permissive and, ultimately, unworkable. Nationalist interpretations of the right to self-determination yield far too many claimants for the international community to bear efficiently. Cognizant of these difficulties, a more judicious "just cause" or "remedial" theory of self-determination reserves self-determination for those peoples suffering from egregious human rights abuses at the hands of their state (Buchanan 1997; Moore 1998). Generally, these abuses must be severe and correspond to either genocide or crimes against humanity. Such offenses include "murder, ethnic cleansing, enslavement, forced deportation, torture and disappearance and widespread and deliberate rape and sexual abuse" (Rome Statute of the International Criminal Court 2002:Articles 6, 7). Remedial rights theorists further extend the right to self-determination to a people living under unjust and unlawful military occupation, regardless of whether or not they experience human rights abuse (Buchanan 1997:37).

No matter how one construes the right of self-determination, it remains at best a necessary condition of just guerrilla warfare, and two questions still remain. First, which peoples merit statehood, and which merit less independent forms of political organization? Second, which among these enjoy the right to fight? In answering these questions, national and remedial claims do not compete but complement one another.

Self-Determination: Autonomy or Statehood?

While some peoples require a state of their own, autonomy or self-rule *within* a viable sovereign state will satisfy the right to self-determination when a state can guarantee security and the material, political, and cultural conditions for a dignified life. As such, the right to self-determination, whether grounded in nationalist aspirations or remedial rights, does not entail the right to political independence. Generally, the broader the right to self-determination, the more restricted the right to political independence. Thus, the many national groups defined by ascriptive criteria must usually seek self-determination through various avenues of autonomy short of independence. In practice, only territorially defined national, racial, ethnic, or cultural groups that are targets of political repression garner support for statehood and, occasionally, international intervention. As such, national peoplehood *and* political repression constitute the necessary and sufficient conditions for demanding statehood. National ambitions or political persecution alone are insufficient to ground the right to political independence. Instead, the political aspirations of an unpersecuted people are usually steered toward various forms of autonomy and/or constitutional protections of minority rights. The Basque, Quebec French Canadians, or Iraqi Kurds are prominent examples. A persecuted non-national group, on the other hand, might turn to national or international legal remedies or demand refugee status should they decide to flee. Persecuted religious groups in some Middle Eastern nations – Egyptian Copts or Iranian Baha'i, for example – demand legal redress and/or political asylum but not secession.

The situation in the field, of course, does not make for a normative argument. Self-determination nurtures individual and collective well-being by offering institutions to sustain ethnic identity and assure a dignified life. There are, therefore, good moral reasons to accord any group – whether national, cultural, or otherwise self-defined – a right to self-determination. Practical limitations, however, prevent the international system from granting all of these rights holders an independent state. Many would not be viable, and many, although perhaps viable, may significantly impinge upon the welfare of the states from which they secede, thereby leaving everyone worse off. When a national group suffers some degree of persecution or discrimination short of genocide or severe human rights deprivation that would deny them the possibility of a dignified life, self-determination turns on those provisions that are minimally disruptive to the international order. These include local provisions – autonomy, federation, and/or legal safeguards to protect minority rights – that do not infringe upon state sovereignty (Cassese 1995:350). Independence is warranted only in extreme cases: when existing political arrangements deny the possibility of a dignified life and when alternative political arrangements or protections are not feasible.

To avoid undermining existing sovereign nations, a demand for independence must legitimately repudiate another's claim to the same territory. This is easily accomplished when an aggrieved group is colonized or when an aggrieved people has a strong legal case for independence. East Timor should have gained independence as soon as the Portuguese relinquished their claims in 1974, and the UN could have done better with

Italy's former colony, Eritrea, than federating it with Ethiopia in 1950. Instead, both peoples would suffer "secondary colonialism" for decades (Weldemichael 2013:35). Similar circumstances left Palestine and the Western Sahara occupied by their larger neighbors (Daadaoui 2008; Morris 2011; Solà-Martín 2007). Establishing a case for independence is considerably more difficult when the legal and historical grounds for sovereignty are weak, as was the case in Northern Ireland, Kosovo, and Chechnya. In the absence of feasible alternatives that preserve a people's cultural, political, and human rights, a territorially bound group that can substantiate its legal and historical right to the territory it claims may demand political independence. Whether a people merit statehood or autonomy and whether this right is remedial or national, the right to self-determination does not entail recourse to force of arms. Considering just cause for guerrilla warfare, this is the crucial question: Which groups enjoy the right to fight?

JUST CAUSE, SELF-DEFENSE, AND THE RIGHT TO FIGHT

The Legal Terrain

Were a persecuted or occupied people a sovereign nation state, their right to employ armed force when facing invasion, occupation, or severe economic sanctions would be uncontested. For nations, Article 51 of the UN charter recognizes "the inherent right of individual or collective self-defense if an armed attack occurs ... " Under these circumstances, the right of self-defense permits the use of armed force when necessary, proportionate, and exercised in accordance with the *in bello* provisions of international law. The working definition of an "armed attack" speaks to both the means and the consequences of an attack. The means are usually kinetic – bombs, tanks, missiles, and guns – while the consequences must include "territorial intrusions, human casualties or considerable destruction of property" (Dinstein 2005:193). While jurists debate the legality of preemptive or preventive wars, there is no doubt that once an attack occurs, nations may legitimately resort to armed force regardless of the magnitude of harm an attack causes. A border fracas permits an armed response no less than a full-scale invasion does. Although last resort is an oft-cited condition of just and lawful war, nations frequently turn to military force as their first and only course of action. And unless defensive measures are wildly disproportionate or indiscriminately target noncombatants, defending nations suffer no reproach.

For guerrillas, the situation is very different. While legal analyses of a people's right to attain independence by armed struggle are inconclusive, strands of this argument are instructive. Some insurgents shoehorn the rights of liberation movements into Article 51. Chechens, for example, assert the right to turn back Russia's colonization of their nation in 1722 (Kasymov 2011), while the Algerian FLN claimed they were fighting delayed wars of self-defense against long-dead invading forces that first arrived in 1830. Regardless of the merits of this argument, it can only apply to guerrilla movements that represent once sovereign states – clearly a minority of cases. For the majority of cases, an appeal to Article 51 might construe colonialism as a form of ongoing aggression that justifies an armed response in self-defense. Still, this does not cover cases of benign colonialism where occupying powers guarantee welfare, security, and some semblance of political participation. Legal instruments are sufficiently elastic to permit disparate interpretations. The 1970 UN Declaration on Principles of International Law, for example, emphasizes the duty of every state "to refrain from any forcible action which deprives peoples ... of their right to self-determination and freedom and independence," while allowing such peoples to take "actions against, and resistance to, such forcible action in pursuit of the exercise of their right to self-determination ... " Well-placed ambiguities surround the meaning of "forcible action" and "resistance." Does the former refer to all forms of colonialism or just violent deprivations of a people's right to self-determination? Does resistance mean "armed resistance" or something less?

The 1977 Protocols did not clarify the issue. These amendments to the Geneva Convention extend combatant rights to peoples fighting against colonial, alien, and racist regimes. While the Protocols seem to tacitly endorse the rights of peoples to fight against these regimes, they neither explicitly grant nor prohibit the right to pursue self-determination by armed force. This has led Antonio Cassese to conclude their right to use armed force is only a *license,* "intended to encapsulate the idea that wars for self-determination are not ignored by international law ... Rather, legal rules take these wars into account, without however upgrading them to the status of manifestations of *jus ad bellum*" (Cassese 1995:153; also Wilson 1988:130–135). On this view, a nation that forcibly denies a people's right of national self-determination is no longer immune from attack. An offended people, therefore, have a license or permission from the international community to respond with armed force. The international community, in turn, has the duty to refrain from taking "any forcible measures to repress the action undertaken by libera-tion movements to realize self-determination," while third party states are "legally authorized to grant assistance to liberation movements [and] strictly forbidden from granting any military or economic assistance to the oppressive State ... " (Cassese 1995:153).

Even construed as a license, the right to fight only extends as far as the legal instruments permit – that is, to those national liberation move-ments fighting for independence against colonial, alien, and racist gov-ernments (Daboné 2011). Many more people who enjoy the right to self-determination lack a firm legal right to fight when their grievances go unmet. What of peoples facing imminent genocide or ethnic cleans-ing? Do they enjoy the right to fight? What about those who do not face imminent danger but confront a recalcitrant regime that denies a people autonomy or minority rights? Do these peoples enjoy the right to fight? While some legal instruments recognize many groups' right to self-deter-mination, only a few groups enjoy the legal right to fight. Can we con-struct a more expansive argument to anchor the right to fight?

The Right to Fight: The Moral Terrain

International instruments address a people's right to fight when they face a colonial or racist regime. Such people need not necessarily face armed aggression to undertake an armed struggle; it is sufficient that occupation or colonization obstruct their legal rights to a state of their own. Moral arguments, however, turn on self-defense and ask when a people facing aggression may turn to force of arms. Here, the magnitude of the threat they face matters.

Consider the extreme case: genocide and ethnic cleansing. Recognizing the responsibility to protect, the international community will override state sovereignty and protect an endangered people through armed mili-tary intervention "should peaceful means be inadequate and national authorities are manifestly failing to protect their populations from geno-cide, war crimes, ethnic cleansing and crimes against humanity" (United Nations General Assembly 2005). In these instances, aggressor nations are liable to attack because they have grossly infringed upon the rights

of others. State sovereignty offers no protection against military intervention when necessary to rectify such wrongs. Surely if the world community can intervene on their behalf, then the same people may also
fight to protect themselves. Yet if an intervening military force is akin to
a police force, then a persecuted people are akin to endangered citizens
who do not usually have the right to take the law into their own hands. To
exercise armed force against an aggressor, a person must face an imminent and deadly threat, avoid using disproportionate force, and, indeed,
refrain from using any force whatsoever if other means of saving oneself
are readily available.

Nations are not bound by some of these conditions. While nations
may resort to force in the face of an armed attack, the threat these attacks
pose need not always be imminent or deadly. And while nations must
avoid disproportionate force, they may, for reasons of state, choose war
as a first option. One reason for this latitude is the "self-help" dimen-
sion of warfare. States reside in an anarchic system, leaving no one to
call upon for help; endangered citizens, however, can usually call the
police. Peoples facing brutal aggression (genocide, war crimes, ethnic
cleansing, and crimes against humanity) fall somewhere in between a
state facing aggression and an individual facing a deadly threat. They are
not players in an anarchic international system, but neither do they live
in rights-respecting, decent societies.

More so than nations going to war, a people fighting aggression draws
its right to self-defense directly from the aggregation of each person's
individual right to live free of wanton bodily harm and indignity. Lacking
the status of states, a people's grounds for fighting must be tighter than a
state's both to avoid precipitous fighting that will destabilize surrounding
states already weakened by internecine warfare and to allow the international community to engage on its behalf. A people, in other words,
may only go to war in extreme conditions. And, with the exception of
those few peoples with pre-existing legal claims, most peoples cannot
respond to any infringement of sovereignty. Because the idea of territorial integrity remains the domain of states laying claim to internationally recognized borders, a non-state people must settle for something
less than armed attacks against their sovereign borders. This "something
less" ultimately determines their right to go to war. In some cases, peoples may be defending themselves against states practicing genocide or
crimes against humanity. In other cases, the abuses may be systemic and
indicative of long-term discrimination that denies equal representation

or equal access to resources. Thus, occupation, for example, is typically a grave affront to the dignity of native peoples. Lodged in a particularly blatant form of paternalism and exploitation, occupation perpetuates inequality and radically impedes a people's ability to obtain the rudiments of minimal self-government necessary for individual and collective well-being. Under these circumstances and deprived of any other feasible means of redress, an occupied people may resort to armed force. In other instances, however, a people may not suffer so egregiously but simply desire a greater level of control and autonomy than membership in the state where they currently reside affords.

Although a people's right to self-determination may suffer in each case, only aggression, colonization, or occupation that keeps a people below the threshold of a dignified life can anchor the right to fight. Like an endangered civilian, an endangered people must face an imminent and deadly threat that denies it the conditions necessary for a dignified life, and avail itself of international aid and other strategies before resorting to armed force. Unlike an endangered civilian, however, a people engage in collective action and must eventually build a leadership cadre that enjoys the authority to mobilize resources and wage war. I return to legitimate authority in the following section.

Exercising the Right to Fight

Understanding a people's right to resist severe aggression in terms of a person's right to self-defense raises a number of questions that do not arise in the context of interstate warfare.

1. When is a Threat Sufficiently Imminent and Severe to Trigger the Right to Fight?

While genocide and related threats are sufficient to trigger the right to fight, one may ask when and how this threshold is determined. In 2008, The International Criminal Court indicted the president of Sudan for crimes against humanity. Did guerrilla organizations purporting to represent the Darfurian people thus gain the right to wage an armed struggle only in 2008 or at some earlier date coincident with extensive rights violations? In the very least, evidence of an imminent threat is necessary for a people to gain the right to fight. And, while this right might include preemptive warfare, one should be careful about expanding the notion of "imminent threat" to include preventive guerrilla warfare. While an immediate, overwhelming, and otherwise unavoidable danger justifies preemptive war, preventive war only addresses a distant and still indistinct threat. Extending the right of preventive warfare to nascent guerrilla movements who may have scores to settle or only desire a military crusade to mobilize support could be severely destabilizing. Insurgents, then, have two choices: wait for firm evidence of aggression, such as that documented by international governmental organizations or human rights groups; or collaborate with the international community who will signal their support by military intervention or by material aid for guerrillas. Just cause for guerrilla war is often tied to a proactive international community.

2. What Other Rights Violations Justify a People's Right to Fight?

Genocide and crimes against humanity are sufficiently grave that they not only deny the conditions necessary for a dignified life; they make no life possible at all. But a dignified life means more than physical survival. It is a life that offers an opportunity for individuals to flourish and requires a range of civil rights, material entitlements, and cultural protection. Violations of which other rights, then, constitute sufficiently severe aggression to justify war?

First, lack of attention to rights that allow people to die triggers the right to self-defense, just as killing them does. This includes severe and perpetual poverty, lack of basic medical care and sanitation facilities, and the lack of military and police protection necessary to provide the minimal security people need to lead a dignified life. Long before the Germans began to murder the Jews, they enacted systematic exclusionary policies that gradually reduced Jews to penury and denied them any opportunity for a dignified life. Any people abandoned in this way not only owe no obligations to its government (for it has failed to provide even the most the basic provisions) but also gain the right to secure these goods by armed force, if necessary.

Nevertheless, things get muddy when revolutionary groups claim just cause to fight against a corrupt, rapacious, or reactionary regime. While systematic torture, abduction, and execution may certainly trigger just cause, other abuses are more difficult to evaluate. Rather than subjecting just cause to the whims of political ideology, it might be better evaluated using the indicators of failed states. These include measurements of health care, political discrimination, violence and public security, poverty and employment, welfare and education, governance and corruption, human rights, and rule of law (The Fund for Peace 2010). A very poor score on some or all of these indices may offer guerrillas just cause for armed struggle in the absence of genocide or crimes against humanity. However, there is no easy formula, and these indicators must be used with care. Many, for example, pointed to human rights abuses in Libya as justification for the West's support of an armed insurrection in 2011. Indeed, Libya scored very low on the human rights measure in 2010, nestled between Egypt and Ivory Coast. Libya, however, scored relatively high on economic measures and its overall score put Libya ahead of Mexico, China, and Israel/West Bank.

Second, consider that discrimination and inequality admit of many degrees, not all of which are severe. Limitations on freedom of expression or the right to work may not stand in the way of a dignified life. While a people may enjoy the right to self-determination, depriving them of an autonomous educational system, cultural autonomy, or home rule may impair their ability to maintain their cultural heritage but not rise to the level of severity to justify armed violence. Under such circumstances, the costs of an armed struggle overwhelm the benefits of redress leaving an aggrieved group recourse to relentless judicial, legislative, and electoral activism; nonviolent resistance; and public diplomacy. And, if alternative means of struggle are unavailable, a minority will be left to suffer until conditions change for the better or deteriorate to the point of severe deprivation.

3. Is Aggression Conditional?

Sometimes a threat is conditional stemming only from a person's reluctance to surrender. Consider a robber who wants a person's watch but threatens no harm until the victim resists. David Rodin (2004:132–138) terms this a "conditional threat" and questions the victim's right to defend himself with force. Similarly, human rights abuses are not always severe until a people press its claims for independence. Had Chechnya accepted the comprehensive autonomy that Russia offered the other autonomous regions instead of turning to armed violence, the massive human rights abuses that accompanied the First and Second Chechen Wars might have been avoided. Similarly, the Aceh rebuffed Indonesian offers for autonomy and went to war (Schulze 2004). In neither case could the aggrieved minority offer compelling grounds for self-determination that autonomy could not satisfy. As a result, just cause for war was absent. Once war erupted, however, atrocities were rampant. As the Russians carried out indiscriminate shelling of civilians, torture, and assassination (Human Rights Watch [HRW] 1995), the Chechens gained a remedial right to fight and the spiraling escalation makes it soon impossible to separate national from remedial claims. The Aceh, too, suffered gravely at the hands of the Indonesian army (Aspinall 2009:123–129; Schulze 2003), and it did not take long for the Chechens or the Aceh to reframe their demands for statehood. Understanding that autonomy might be taken to satisfy their national claims, insurgents boldly argued that only independence could satisfy their remedial grievances. Demanding "freedom from fear, violent acts and barbarism," Aceh insurgents contended that "Indonesian injustice meant the state had forfeited its social contract with the Acehnese" (Aspinall 2009:14–15, 144–145). The argument is not persuasive. While the crimes committed against the Chechens and the Aceh offered cause to fight, they did not inflate their right to self-determination to include statehood. Finding their adversaries unwilling to yield, both the Chechens and the Aceh eventually settled for an autonomy arrangement little different than was first offered before they went to war.

In East Timor and Kosovo, on the other hand, national claims to independence were interwoven with remedial demands from the beginning. Just as the Indonesians rejected East Timor's legitimate claim to statehood, the Serbians abrogated Kosovo's long-established and constitutionally guaranteed autonomy. These affronts establish the nationalist claim. Accompanying each state's repudiation of separatists' political claims was a fierce campaign of intimidation and terror. In East Timor, Weldemichael (2013:128) describes "bodily and psychological harm; depopulation, including forcible transfer of Timorese children; and allegations of preventing births." In Kosovo, the slide toward repression was gradual, beginning with mass layoffs of Albanians by the government in Belgrade, severe restrictions on Albanian media and educational institutions, and suppression of dissent in 1990 (Besnick 2004) before culminating with a campaign of ethnic cleansing, indiscriminate civilian casualties, starvation, torture, and assassination (Clark 2000; HRW 2001;

Perritt 2008:54). Under these conditions, national and remedial claims reinforce one another allowing aggrieved groups like these to demand statehood and to turn to armed force.

In summary, it is important to distinguish the right to self-determination from the right to fight. National and remedial claims establish the right to self-determination whether in terms of statehood or autonomy; rights deprivations and aggression anchor the right to fight. National claims turn on a variety of ascriptive ethnic, cultural, linguistic, and related traits. Remedial grievances reflect pervasive human rights abuses that undermine any chance of a dignified life or unjustly thwart solid legal claims to autonomy or independence. Citing all or some of these claims, aggrieved groups must make their case before the international community. The most suitable political arrangement to guarantee self-determination for a particular people is a function of several conditions, including territorial control; historical and legal claims to sovereignty; prevailing respect for human rights, and the institutional needs of an aggrieved group; and the impact of secession on surrounding states. In many cases, self-determination will require some degree of autonomy, if not statehood. However, the right to self-determination alone is insufficient to establish the moral right to fight. A people may only fight for the rights they are denied, whether autonomy or statehood. Peoples deserving of autonomy have no right to battle for independence. No people, whether entitled to statehood or autonomy, may turn to armed force in the absence of aggression and the conditions necessary to maintain a dignified life. Nor may they fight if less bloody alternatives are feasible or if the cost of an armed struggle is disproportionate.

Despite its prominence, just cause alone is not enough to go to war. War is a collective enterprise that requires mass mobilization, conscription, arms acquisition, taxation, administration, and, ultimately, orders to kill. Only a legitimate authority can do this. States come by this naturally. Guerrillas and insurgents must work for it.

LEGITIMATE AUTHORITY AND GUERRILLA WARFARE

Just cause only supplies a necessary condition for fighting. As important for insurgents is legitimate authority. Although the idea of legitimate authority is central to classical just war theory, it is generally of little concern in conventional war. By law and custom, internationally recognized

sovereign states and members of the United Nations enjoy the authority necessary to wage war. While this excludes unrecognized states such
as Taliban Afghanistan, it does not appreciably affect a criminal regime
in Sudan or terror-sponsoring states in Syria or Iran. While these states
may find their sovereignty curtailed and their regimes subject to military
intervention or sanctions, these states do not lose their standing in the
world community. Should one or another go to war in defense of legitimate political interests and in accordance with international law, their
right to do so would not be questioned.

Non-state actors do not enjoy such forbearance. Just as guerrillas must
energetically prove they have just cause on their side, they must also prove
that their nascent military and political organizations, of which there are
sometimes more than one, have the authority to wage an armed struggle.
Authority is essential to mobilize support, raise funds, conscript soldiers,
and gain international support. The right to authorize wartime killing is
a heady one and, ordinarily, reserved only for states facing aggression.
International law grants a state the positive right to arm its citizens and
send them off to kill another nation's defenders. Anyone killing without the state's permission is a criminal. For guerrillas, however, no state
exists yet, but the cause for war is substantial, and an independent state
or sweeping autonomy is the goal.

The Face of Legitimate Authority

What does legitimate political authority look like? Cassese (1995:147),
for example, is explicit: "racial groups are entitled to claim self-determination and vindicate their rights only if there is a representative
organization capable of acting on behalf of the entire group." National
liberation movements must meet similar criteria (Abi-Saab 1979:407–
415; Schindler 1979:133–144). Cassese demands two conditions: representation and capability. Others, less sanguine about the possibility of
representation, only demand capability. Fabre (2012:142), for example, grants legitimate authority to those "best placed to put a stop to
the wrongdoings which provide agents with a just cause for war." The
test then is not one of representation but of efficacy. From Fabre's cosmopolitan perspective, any person or persons whose basic rights are
threatened may turn to force assuming the actions meet the other condition of just war: just cause, necessity, effectiveness, proportionality,

and so forth. And while Fabre uses this argument to extend legitimate authority to guerrilla organizations (and indeed, any organization or individual capable of fighting unjust aggression effectively), her claim is problematic for two reasons. First, there is no reason to assume that an effective organization enjoys legitimate authority. A warlord criminal gang, which lives from drug trafficking, pillaging, and extortion and recruits its fighters with the promise of booty, might effectively rout a repressive regime in a failed state and maintain no small measure of stability. But warlords lack the legitimacy to wage war. Their authority comes from a venal mix of patronage, brutal threats, and material rewards and is neither embedded in people's respect for widely shared values nor derives from popular consent (Henriksen and Vinci 2007; Marten 2006/2007). Warlords are not revolutionaries, visionaries, or state builders but thrive in the absence of a strong state. They seek no just cause, only personal aggrandizement. Second, decentralization is often dangerous and destabilizing in guerrilla organizations (Zehr 2013). Granting a license to any group that thinks it is effective is a recipe for chaos, confusion, and, at worse, fratricide among those vying for "best placed." A single force seems to be a much more effective alternative, and, if efficacy is Fabre's principle requirement, one might easily stipulate that a persecuted people must, at one point, coalesce around a single military and political organization. Otherwise, no one is best placed to end the wrongdoing.

But must this organization be representative in any particular way? The world community, after all, does not demand representative government from its member nations. The link between legitimacy and representation deserves closer scrutiny among non-state actors. For state actors, legitimacy comes with sovereignty, beneficence, and effectiveness, not necessarily with democratic representation. Such a government need not be liberal, but only decent insofar as it assures the possibility of a dignified life by maintaining an efficient bureaucracy, rule of law, social welfare, and respect for shared social norms. An illegitimate government is criminal, terrorizes and persecutes all or some of its people, wages war without cause, and offers none of the benefits that legitimate governments strive to provide. Only a few governments are illegitimate in this sense. Most governments, then, even those that deny democratic representation, enjoy legitimate authority and recognition from their people and the international community. What then of non-states?

Legitimate Authority among Insurgents

As guerrillas and insurgents strive to establish legitimate authority, they often build upon charismatic leadership and traditional sources of authority before evolving to what Weber calls "legal authority." As these labels suggest, charismatic authority leans upon particular leaders or strong personalities, while traditional authority is grounded in respect for widely shared norms, often associated with religious and patrimonial bodies and their leaders (Blau 1963). In contrast, legal or rational authority stems from formal bureaucracies that often draw their right to govern from some mechanism of popular consent (Spencer 1970).

The evolution of authority varies greatly among guerrilla movements. South American guerrillas began by trying to impose a revolutionary ideology. Divorced from the community, most of these groups failed to secure authority and eventually collapsed (Asprey 1994:711–715; Moreno 1975). More successful groups build upon pre-existing patterns of authority that imbue the tight social networks of clans and tribes. Here traditional authority is anchored in long revered customs, law, and kinship and is critical for the genesis of guerrilla authority. Insurgents managing to bring local elders, chiefs, priests, and patriarchs over to their cause can then build upon their authority to mobilize support. Guerrilla organizations as culturally disparate as the Kosovo Liberation Army, the Free Aceh Movement (Indonesia), and the Taliban (Pakistan) relied heavily on existing social networks and family ties to recruit soldiers and civilian supporters (Kraja 2011; Humphreys and Weinstein 2008; Marten 2012; Qazi 2011; Schulze 2004). As recruitment increased, legitimacy and authority strengthened. Slowly, and with the ability to show moderate success in the field, guerrilla organizations gain momentum.

To maintain legitimacy, guerrilla organization must evolve in the direction of rational authority. They must build efficient and transparent bureaucracies to provide essential social services and human security and open up their organization to some degree of representation to gain international support. For these reasons, warlords, for example, can never be effective agents of just cause. Representation is not a Weberian criterion of legitimate authority, and, in fact, guerrillas operating in traditional and patrimonial environments do not require democratic representation to *establish* legitimate authority. But the growing need for efficient bureaucracies will require transparency. Transparency, in turn, requires dissent, dissent requires a voice, and voices require some form of representation

(Blau 1963). Ultimately, these demands coalesce around the nascent institutions of a decent society. Representation might be democratic – that is one person, one vote – but may also include group or corporate representation whereby local leaders speak for their community.

Satisfying these conditions – effectiveness, domestic recognition, representation, and international approval – is not easy. Some organizations will provide services and enjoy domestic recognition but not international approval. First elected to office in Gaza in 2005, Hamas was blackballed as a terror organization and could not gain international legitimacy much less national self-determination. Nevertheless, its authority was far reaching and stood upon traditional foundations anchored in religion and clan and on a rational-legal footing that reflected their leaders' positions within a formal bureaucracy that provided institutions of governance, education, welfare, judicial services, charitable works, and law enforcement (Gunning 2009; Levitt 2006). The Hamas bureaucracy also highlights the critical importance of formal and informal consent schemes. These schemes fortify legitimate authority and include elections, public consultation, and a practical procedure for advice and dissent among Hamas members. The process is not entirely democratic: Hamas elections depend upon carefully supervised elections that limit open competition, generate intense pressure to conform, and curb opposition (Gunning 2009: 95–142).

Formal channels of representation, however, are not always possible. Instead, activists rely on sensitivity. Commenting on the establishment of a new umbrella organization of resistance parties, one East Timor activist writes:

Given the state of war we existed in, it was impossible to have any sort of public consultation about the reorganization. But I know, through my personal contacts, that the vast majority of the population accepted and agreed with the changes. There were increasing numbers of young people who had never been part of any political party, such as FRETILIN [Revolutionary Front for an Independent East Timor], and who grew up during the war. If people didn't want to be part of FRETILIN, they should be free to make that choice. We had to put the national interest over the interest of a political party. We have to be open to everyone's ideas, and the CNRM (National Council of Maubere Resistance) provided the necessary space. That's why people accepted the CNRM as the legitimate organization incorporating all of the groups or organizations or political parties in the struggle for the liberation of East Timor. (Pinto and Jardine 1997:123)

Lack of formal representation or avenues of sufficient consultation is a common problem. While the CNRM was the blueprint for East Timor's democratic institution after independence, many guerrilla organizations govern paternalistically at the early stages because obtaining consent is not always feasible. For some theorists, lack of consent need not undermine their authority. "Just as a patient's consent is not *always* necessary for doctors permissibly to operate on him," writes Fabre (2012:155), "the consent of victims of rights-violations is not *always* necessary for insurgents permissibly to wage war on their behalf."

The analogy is not quite accurate. To avoid running roughshod over patient autonomy, bioethicists often supplement "best interest" with the assumption that a patient would have consented to a medical intervention if asked. Similarly, guerrillas must also assume that the people whose best interests they purport to represent, or their community leaders, would consent to an armed struggle if asked. Like the unconscious patient, an endangered people may be unable to voice consent and require paternalistic intervention. But this can only be a temporary expedient. Unlike those of lifesaving surgery, the benefits of armed struggle are not always obvious. Surgeons are happiest when obtaining active, informed consent, and insurgents must do likewise. To maintain their authority, guerrilla organizations require avenues of representation and consultation. Initially, these may be informal and include demonstrations, rallies, and irregular episodes of public discourse (Finlay 2010; Gunning 2009:61,109,156–157). Eventually, however, the process or representation must evolve to embrace formal institutional structures. To do otherwise, insurgent movements risk their legitimacy.

Movements that fail to heed popular demands saw their authority deteriorate and were forced to take remedial action. After waning support amid accusations of unresponsive leadership and human rights abuses, the Sudan People's Liberation Army (SPLA) convened its first national convention in 1994. The outcome was to democratize the movement; separate military from the civilian authorities; create a legislative, executive, and judicial branch of government; and strengthen the local courts. Thereinafter, notes Metelits (2004:71), "the SPLA experienced little difficulty marshaling support for its cause ... " (also Johnson 1998). The Eritrean People's Liberation Front (EPLF), understood this from the beginning and built one of the most open and successful liberation movements by instituting village democratization, local and regional elections, minimum quotas for women's representation, land reform, and universal suffrage (Connell 2001; Woldemikael 1991).

Domestic legitimacy entwines with international legitimacy. Unlike Hamas, many guerrilla organizations gain solid international support. Eleven national liberation movements recognized by the League of Arab States and the Organization of African Unity (OAU), including the Palestine Liberation Organization (PLO) and the African National Congress, joined the deliberations leading to Protocol I and II (Suter 1984:137). In recent years, the international community has recognized guerrilla movements or their governments in exile in Kosovo, Libya, and East Timor. States and international organizations recognize guerrilla organizations for many reasons. Out of affinity for their brethren and concern about encroaching colonial powers, African and Arab states recognized local colonial struggles in the 1970s. Humanitarian concern loomed large as the Western nations undertook military intervention to alleviate human rights abuses in the Balkans, Sudan, East Timor, and Libya. Recognition turned on a mix of humanitarian concern, geopolitical interests, and fears about regional stability and helped legitimize insurgents' authority at home and abroad.

Unlike states, then, guerrilla organizations must constantly struggle to gain legitimate authority. Guerrillas lack the coercive institutions of a state to conscript soldiers and tax their citizens. Without legitimate authority, guerrillas will find it impossible to do either. The sources of authority are as varied as the circumstances surrounding national liberation. They are also fluid. Movements start with the barest of authority to wage war on behalf of their people. Authority begins with the exercise of power but only acquires legitimacy if married to an ideology of just cause that gains the support of local religious, tribal, or community leaders. Rarely, do liberation movements come to exercise legitimate authority without attaching themselves to pre-existing loci of power. In time, they may turn in several directions. Some movements may push toward democratization and popular participation or other forms of consultation and, thereby, gain a measure of authority commensurate with a decent state. Other movements, however, may descend into warlordism and champion violence, nepotism, patronage, and a predatory war economy and, thereby, lose their popular support and the authority that goes along with it (Marten 2011; Mukhopadhyay 2009).

Effectiveness, Necessity and Proportionality

Imbued with just cause and legitimate authority, the right to fight remains provisional. Armed struggle must meet other conditions of just war, including effectiveness, necessity, proportionality, and last resort. Effectiveness is one of the first rules of any war and demands a reasonable chance that war will achieve its aims at costs that do not exceed its benefits. Translating costs and benefits in wartime is notoriously difficult because there is no common currency. "Lives" are an obvious choice, but few wars save more lives than they cost. The difference is made up in honor, freedom, territory, deterrence, and a host of other intangibles. Guerrilla warfare complicates this calculation further because its goals do not always translate into military victory, an aim most guerrilla organizations are ill-equipped to achieve. Instead, guerrillas and insurgents often hope to harass their enemy and weaken their resolve to fight, raise morale at home, and enlist the support of the international community. Very often, acts of guerrilla warfare are symbolic, of little weight in their own right, but which resonate loudly among dispirited compatriots and sympathetic bystanders. Guerrillas adapt their tactics accordingly, and effectiveness, therefore, must often be measured against these modest goals rather than victory on the battlefield. Assessing effectiveness at the tactical level is a central theme of the following chapters.

Cognizant of a goal and with a vague idea of what means might realize it, guerrillas must then ask themselves if an armed struggle is necessary and proportionate. Necessity demands that that no less costly means are feasible to achieve their chosen objective. Again, costly might be measured in terms of compatriot lives, enemy lives, money, or international support and good will. Proportionality is commonly reserved for the tactics a guerrilla army chooses and will warrant a full discussion in the following chapter. However, proportionality also applies to the strategies insurgents adopt. Beyond satisfying the demand of cost effectiveness, armed struggle is also constrained by *excessive* cost. War may prove necessary and effective but still prove disproportionate relative to its gains. One would not necessarily permit a guerrilla movement to sacrifice its multitudes to pursue a territorial enclave. Nevertheless, just cause allows guerrillas considerable room to maneuver. Fighting for self-determination and a dignified life, guerrillas are striving for goods that may allow them to cause many causalities before their struggle is disproportionate.

Necessity and proportionality come together in the principle of last resort, a particularly taxing requirement for guerrilla organizations. The demand for last resort is necessary to protect the world order from constant disruptions by insurgents and guerrillas. Resorting to strategies other than war is expedient if guerrillas are to gain any sympathy and support from the international community. Before going to war, therefore, guerrilla organizations are expected to conduct long negotiations over referendum procedures, as in the Western Sahara, or over borders, as in the Palestinian territories. Even then, war is only permissible when the threat is imminent and life threatening. This reflects a more general principle I noted earlier: the right to fight is strongest when human rights are on the line. When civil, cultural, and political rights are at stake, the right to fight attenuates considerably, although the right to self-determination remains undiminished. In these cases, war may very well be disproportionate. As such, war is not a last resort; it is no resort at all.

To this point, just guerrilla warfare depends entirely on the actions of the aggrieved. Dispossessed and tortured people must prove to the international community that their cause is just, their authority legitimate, and their strategies effective and necessary before undertaking an armed struggle. If and when they do, what does the world community owe them?

THE RIGHT TO A FIGHTING CHANCE

It is often observed that a right without remedy is no right at all. The right to self-determination would suffer grievously if those enjoying the right did not have recourse to armed force under extreme conditions. Yet the right to fight is no right at all if other nations may obstruct the efforts of a people fighting for self-determination. For this reason, Cassese (1995:153) emphasizes that the license to fight prohibits "any forcible measures to repress the action undertaken by liberation movements to realize self-determination." Forcible measures include blockades or arms embargoes. But the right to fight is also no right at all if legal barriers prevent any reasonable chance of fighting successfully. Legal impediments are just as restrictive as material impediments. Both unduly restrict, if not abrogate, an aggrieved people's right to fight. For this reason, international law relaxed provisions requiring combatants to wear uniforms or insignia. This was a huge concession to guerrilla armies, but consistent with their right to fight. Commenting on this very provision of Protocol I, the French jurist, Charles Chaumont writes:

In order to remain objective and credible, humanitarian law must allow every party an equal chance in combat. If a norm of this body of law is incompatible with this principle and makes it impossible from the outset for one of the parties to have any prospect of victory, it is better not to draft such a norm at all. (Additional Protocol I, Article 44, Commentary, p. 519, n. 40)

In one sense, these words only reiterate the reasonable stipulation that the international community not interfere with another party's right to defend itself. But by stating things as he does, Chaumont's comments are particularly provocative and invoke what may be called "the right to a fighting chance." Working through this right, it is necessary to answer two questions. First, what prospect of victory do guerrillas and insurgents deserve? Second, what might they demand from the international community?

One can imagine two kinds of opportunities guerrillas should enjoy. On one hand, they might demand an equal or overwhelming prospect of victory. This might then require that guerrillas receive the men and materiel to wage war, a requirement that critics rightly point out is well beyond any legitimate demand during war (O'Driscoll 2012). On the other hand, some minimal prospect of victory is a more reasonable demand because it does not impose overwhelming costs on third party states. It only requires that a state or group going to war secure for itself the resources necessary for a reasonable chance of victory and that other states do not interfere as they exercise their right of self-defense. Consider, first, the legal impediments to waging just war and then the material obstacles.

The Right to a Fighting Chance: Legal Impediments

Legal noninterference goes to the heart of Chaumont's remarks. When guerrillas fulfill the conditions of just war, the international community cannot promulgate laws that make it impossible for an oppressed people fighting for a dignified life to wage war. Chaumont speaks directly to international humanitarian law and its effects on the wherewithal to undertake a just struggle. Commenting on human rights law, William

Abresch (2005:750) expresses a similar concern. "Human rights law," he writes, "must be realistic in the sense of not categorically forbidding killing in the context of armed conflict or otherwise making compliance with the law and victory in battle impossible to achieve at once."

One must be careful here to avoid the impression that guerrillas may simply disregard the rules of war when it suits them. To obviate this claim, it is important to remember that apart from just cause and legitimate authority, the right to fight requires effective and necessary strategies that do not violate the basic rights of combatants and noncombatants. War, and the tactics it employs, may be effective but unduly destructive nonetheless. Torturing prisoners or deliberately killing noncombatants may be an effective and even necessary means to win a war but impermissible because they egregiously violate the rights of combatants and noncombatants. Combatants have both rights and duties. Their rights protect them from unnecessary suffering and superfluous injury, while their duties require soldiers to protect the welfare of noncombatants. Noncombatants, on the other hand, have only rights and no duties as long as they do not participate in the fighting. Noncombatant rights, enshrined in the principle of noncombatant immunity, protect them from unnecessary, direct, and disproportionate harm. I expand on these rights in the following chapter. Suffice it to say, guerrillas may neither repudiate the rights of others, nor exercise their right to fight at the expense of combatant and noncombatant rights. This restraint will figure crucially in subsequent discussions.

Thus, Chaumont's stipulation is necessarily self-limiting. The international community may not enact laws that make it impossible for a people to fight for self-determination unless these laws are necessary to safeguard the rights of participants. This is an important caveat that demands lawmakers and philosophers navigate between the rights of aggrieved groups and the rights of combatants and noncombatants on both sides. Protecting the latter is critical, but as the following chapters demonstrate, the right to a fighting chance also leaves room for provisions to shed uniforms, employ human shields, target civilian infrastructures, hold prisoners incommunicado, and manipulate the media, all in defiance of the legal norms of wartime conduct. The same right to a fighting chance also demands that other nations refrain from imposing material impediments to success or, in some cases, assure guerrillas the material means to prevail.

The Right to a Fighting Chance: Material Impediments

Because the right to fight demands a chance of success when the cause is just, states have a duty not to impede or interfere with guerrillas as they exercise this right. And as the "responsibility to protect" attests, states may also have a positive duty to aid the most desperate and defenseless peoples. While nonintervention only facilitates the right to fight for self-determination, humanitarian intervention can go a long way to guaranteeing it. In general, the international community may only intervene militarily in those instances outlined by their responsibility to protect: genocide, war crimes, and crimes against humanity. In these cases, one can invoke a rule of rescue to obligate states to tender military aid at reasonable costs. In other cases where rights violations are less pressing, bystander duties extend to noninterference. Absent overwhelming danger to others, there are no grounds for third parties to risk life or limb. The timing, however, can be tricky. Repression is often gradual, suggesting that early or, as was the case with Libya in 2011, preemptive intervention may be warranted.

The duty of humanitarian aid rests on an oppressed group's claim that the international community must ensure fundamental human rights for all peoples when the cost to bystander nations is reasonable (Gross 2010a:205–229). Human rights and universal principles of justice are pivotal for justifying armed humanitarian intervention. "If we deny the moral duty and legal right to [intervene]," writes Fernando Tesón (2003:129), "we deny not only the centrality of justice in political affairs, but also the common humanity that binds us all" (also Brown 2003). Of course, humanitarian intervention takes many forms; resort to military force is the extreme form. Less violent means include economic sanctions, diplomatic isolation, protection and transport of refugees, and humanitarian aid. All of these measures fall short of putting boots on the ground. The right to fight and concomitant duty to tender military aid is strongest when guerrillas fight for fundamental human rights and a dignified life. As the right to fight weakens, all bystander states owe guerrillas is noninterference. Material noninterference demands that bystanders neither provide an oppressor nation with military support nor impose sanctions against an oppressed people fighting for self-determination. Legal noninterference requires that international law refrain from provisions that make victory impossible. Neither concession is trivial during asymmetric war.

GOING TO WAR

Taking stock of local support, material resources, men under arms, and international backing, guerrillas, like states, estimate their chances of victory. Victory is an elusive concept, particularly because guerrillas and insurgents do not usually aim for a decisive military triumph. Rather, they often fight sporadically while working vigorously to draw others to their side. To warrant international support, guerrillas must wage just war. And, as guerrillas fight for self-determination there is no reason to think that they cannot do so justly. Just guerrilla warfare, more so than just interstate warfare, demands close attention to just cause and legitimate authority, without which guerrillas cannot establish their right to fight. But just guerrilla warfare also demands a right to a fighting chance, some accommodation from the international community that avoids placing unwarranted obstacles in the path of insurgents, and aids those who might otherwise fail and face grave abuse. In return, insurgents must scrupulously respect the rights of combatants and noncombatants. Some noncombatants are their own – compatriots whose rights guerrillas must respect as they mobilize support and raise an army. Other noncombatants reside among the enemy and are those whose rights of immunity guerrillas must safeguard even as we admit that guerrillas, no less than states, must target some civilians directly to ever prevail. The right to a fighting chance forces some fundamental changes to the *in bello* principles of war. These are the subject of the next chapter.

3

The Right to Fight

Who Fights and How?

East Timor's struggle against Indonesia was asymmetric in every way possible. Materially, the sides were wildly mismatched. The FALANTIL guerrillas of East Timor fielded small arms while Indonesia bought sophisticated armaments from the West (Klare 1992/1993). Legally, Indonesia twice flouted international resolutions to block independence for East Timor: first in 1975, as Portugal decolonized, and second in 1999, as the UN shepherded East Timor toward statehood. Morally, Indonesia undertook indiscriminate bombing, torture, assassination, and a brutal scorched-earth policy that left the infrastructure devastated, thousands killed, and 60 percent of the population displaced by 1999 (Dunn 2001; Kilcullen 2009:206–208; UNICEF 1999). For many years, East Timorese leaders, whose widespread authority turned on charismatic and traditional leadership backed by the steady evolution of formal elections and ad hoc consultation (Niner 2007), had few options other than an armed struggle.

While it is always difficult to establish just cause and legitimate authority unambiguously, the justice of East Timor's struggle should be relatively uncontroversial. Just cause remedies human rights violations and enforces internationally recognized legal claims. Last resort and the prospect of reasonable success confirm the right to fight. Legitimate authority afforded competent and respected leadership. Having satisfied these conditions, how may a guerrilla movement, like that of East Timor, wage an armed struggle?

The principles that govern the conduct of fighting – discrimination, noncombatant immunity, proportionality, and freedom from superfluous injury and unnecessary suffering – are easy enough to articulate. Yet, as the subsequent discussion demonstrates, severe difficulties arise in practice. Distinction – or the obligation to discriminate clearly between combatants and noncombatants – is confounded by the inability of many guerrillas to fight successfully without shedding their uniforms. Noncombatant immunity – the obligation to refrain from unnecessary, disproportionate, or direct attacks on noncombatants – runs afoul of the crucial role that civilians play in contemporary armed conflict. Aware of this fact, state armies constantly search for ways to disable war-sustaining civilian infrastructures, a tactic that insurgent armies may also consider. The protection from serious injury and unnecessary suffering that combatants enjoy is often undercut. State armies routinely employ aggressive interrogation techniques and targeted killing, leaving guerrillas to reasonably consider assassinating some enemy soldiers and kidnapping others. The familiar principles of *jus in bello* change during just guerrilla war, leading some to suggest that guerrillas simply disregard the rules that do not suit them. While one must resist this conclusion, there is, nevertheless, a measure of latitude that guerrilla forces enjoy as they exercise their right to fight. This begins as they face the challenges of building an armed force and comes to the fore as they refine their tactics.

CONSCRIPTION: BUILDING A GUERRILLA ORGANIZATION

Unlike a state, insurgents do not obviously enjoy the right to employ coercive measures to conscript an army. A state gains a right to coerce its citizens through legal and political institutions that are often, but not necessarily, rooted in representation and backed up by force. In contrast, a guerrilla organization may initially mobilize compatriots by appealing to "obedience, allegiance ... respect or trust" (Friedman 1990:61) and only later by exercising its nascent legal and political apparatus. As such, insurgents must depend on voluntary compliance to a far greater degree than must a state.

Guerrilla military organizations are relatively small. Hezbollah forces were estimated at 5,000–15,000 (including reserves) during the Second Lebanon War (2006), and Hamas numbered about 10,000 when it fought Israel for several weeks at the end of 2009 (Arkin 2007:74; Harel and Issacharoff 2008; International Institute of Strategic Studies [IISS] 2006). David Kilcullen (2009:48–49) estimated Taliban forces at 8,000–10,000 regulars and 22,000–32,000 part-time guerrillas in 2008. The Kosovo Liberation Army weighed in at 18,000 soldiers in

1998 and dropped to about 5,000 a year later (Perritt 2008:56–57). The East Timor military organization, the FALINTIL, mustered nearly 20,000 soldiers in 1975, but this number fell drastically to 600–800 by 1997 (Nevins 2005:29; Pinto and Jardine 1997:246). Estimates for the Western Saharan Polisario range from 3,000 to 6,000 individuals under arms in 2000 (Cordesman 2003:24). Among the largest guerrilla militaries, the Eritrean People's Liberation Front (EPLF) fielded 45,000 armed fighters in 1989 (Pateman 1990:81), while the Sudanese People's Liberation Army (SPLA) grew from about 12,500 combatants in 1986 to more than 50,000 in 1991 as fighting with Sudan intensified (Federation of American Scientists [FAS] 2000). Chechen forces reached 15,000 by 1995, while a 30,000–40,000 volunteer Home Guard provided additional personnel (Hughes 2007:83).

Alongside each guerrilla military organization, a political wing provides an extensive range of social services, local law enforcement, and judicial institutions to safeguard the local population, maintain order, oversee commercial activities, and otherwise sustain the war effort. Guerrilla armies, no less than foreign invaders, require the means to win the hearts and minds of the local population and secure their support for armed struggle (Ly 2007; Grynkewich 2008). In Kosovo, the LDK (Democratic League of Kosovo) emerged as the political and diplomatic force behind the struggle for independence. Financed largely through donations of diaspora Albanians, the LDK operated an entire "parallel structure" that provided for education, medical care, sports, and commercial activities (Clark 2000:95–121; Judah 2000:70–72). Eritrean insurgents similarly provided health, education, and social services while undertaking extensive land reform (Cliffe 1989:142–143; Sabo and Kibirige 1989). Apart from its social service institutions, Hezbollah maintained sophisticated media facilities (Love 2010). In East Timor, two organizations operated alongside the guerrillas: a "Diplomatic Front" and a "Clandestine Front." The former publicized the East Timorese cause through diplomatic initiatives and lobbying, while the latter enlisted civilians to "relay messages, smuggle out reports and photographs to Indonesian and international human rights organizations, and launch a number of daring protests" (Stephan 2006:61; also Weldemichael 2013:195–217).

Often insurgent activities put civilians in the middle of the fighting by choice. During the Second Lebanon War (2006), local Lebanese farmers stored, serviced, and operated weapons on a part-time basis (Bar Joseph 2007). Others provided transport and logistical and intelligence support, donated money, hid guerrillas, or, as I describe in subsequent chapters, willingly acted as human shields. Civilians are not only *physically* close to and outwardly indistinguishable from guerrillas; many civilians perform war-related functions. The political and military guerrilla organizations augment and support one another – and, as I explain, also render civilians liable to attack. To staff their military and political wings, guerrilla armies must enlist volunteers and employ recruitment strategies that rely heavily on positive and negative incentives.

Mobilizing Compatriots for War: The Right to Conscript

Nation-states conscript civilians to guarantee security, protect national interests, and, sometimes, pursue humanitarian intervention. Failure to comply with conscription usually incurs heavy penalties. To establish a state's right to conscript, most liberal thinkers consider just cause and equity. Rawls (1971:380–381), for example, permits a state to conscript its citizens in the defense of liberty and only when conscription distributes risk evenly, avoids class bias, and secures the agreement of its citizens through a representative legislative process. Similar caveats enjoy wide support (Bedau 1971; Gewirth 1982:251–254; Monro 1982) and reinforce the importance of just cause and legitimate authority. For guerrillas, just cause and legitimate authority not only establish their right to fight; these principles lay the foundation of their right to conscript civilians.

Challenges abound. Like states, guerrilla organizations face a collective action problem when they conscript compatriots. Consider a citizen contemplating military service:

YOSSARIAN: I don't want to be in the war anymore.
MAJOR MAJOR: Would you like to see our country lose?
Y: We won't lose. We've got more men, more money and more material. There are ten million men in uniform who could replace me. Some people are getting killed and a lot more are making money and having fun. Let somebody else get killed.
MM: But suppose everybody on our side felt that way?
Y: Then I'd certainly be a damned fool to feel any other way. Wouldn't I?

(Heller 1995:129)

Yossarian's self-interested logic is impeccable. Security is a *public* good, one that anyone can enjoy whether or not they risk their lives to attain it. Like any collective good, security remains unattainable unless a critical mass of participants cooperates. At the same time, however, no single participant can appreciably affect the outcome. This is Yossarian's dilemma. If enough young people are prepared to fight and die, then no one needs Yossarian. If, on the other hand, too few people are prepared to risk their lives, then Yossarian's contribution is for naught because all will die in vain. In either case, and regardless of how his compatriots behave, Yossarian should never volunteer to fight. If many individuals think this way, collective action fails despite the fact that most individuals, including Yossarian, understand that they will benefit from the security that collective action can provide. Guerrillas face exactly the same problem.

To avoid calamity, a state must coerce individuals into risking their lives the state can fulfill its duty to protect the lives of its citizens. And, to supply the army, nations must tax and punish recalcitrant citizens if they do not contribute their share. Each of these problems – marshalling recruits for war and the money to pay them – also beset guerrilla organizations. I address the question of taxation in Chapter 8. Here, I want to ask: How may a legitimate guerrilla organization conscript men and women to fight on its behalf?

Recruitment Strategies: Ideological, Social, and Material Incentives

To recruit soldiers and civilian supporters, political movements utilize positive and negative ideological, social, and material incentives (Gross 1997:90–125; Knoke 1988). Material incentives provide tangible benefits: payment or loot for joining the movement; punishment for defecting. Social or solidarity incentives reflect the fraternal benefits of working with likeminded friends and family or the disincentive of social exclusion that comes with failing to participate. Ideological incentives are moral or "extra-rational" and motivate individuals by appealing to fairness, religious or civic duty, or altruism in lieu of a careful analysis of costs and benefits (Elster 1989; Hardin 1982). Responding to ideological incentives are political entrepreneurs, those precursors who build an organization from scratch and whose devotion to the cause overwhelms all other costs. For many others, however, ideological commitment is neither necessary nor sufficient. Certainly, social justice, self-determination, and human rights fire the minds of dissatisfied or underprivileged people. But unless participation can provide some more tangible material or solidarity benefit, many people will prefer to sit on the sidelines until things get going.

Once the original cadre takes root, entrepreneurs begin to recruit additional members. Describing the Kamajor militia fighting in Sierra Leone, P. Muana explains how "these fighters are conscripted with the approval and consent of the traditional authority figures, maintained and commanded by officers loyal to those chiefs. This ensures a high level of commitment on their part and an insurance against atrocities on the civilian population on whom they rely for sustenance, legitimacy, and support" (Muana 1997:88 cited in Humphreys and Weinstein 2008:441).

This description could just as easily portray recruitment in Kosovo, Chechnya, Afghanistan, or Indonesia. As guerrillas enlist fighters and civilian supporters, they diligently cultivate clan and kinship networks by appealing directly to local authority figures: parents, relatives, elders, and chiefs. Individuals linked by strong familial and social ties make for dedicated fighters. They trust one another, recognize deeply ingrained obligations to protect and provide for each other, and often respond willingly to appeals from their local authorities or families:

Khalil, who was built like a tough little boxer, introduced me to his fighting unit. He did not introduce them by rank but by blood, pointing to each of the armed men around him and saying: "My nephew, my cousin, my brother, my cousin, my nephew, my son, my cousin ... " Free Syrian Army units are often family affairs. (Friedman 2013)

Social networks and small communities also provide the context for fighting. National causes become local as young people take up arms to avenge clan members injured, killed, or left destitute by government policy and counterinsurgency warfare. Elsewhere, young militants hope to continue a fighting tradition begun by elders in earlier campaigns (Aspinall 2009:120, 161–166; Galeotti 2002; Kraja 2011; Qazi 2011).

Security concerns also weigh heavily. As a guerrilla presence grows, government policy may turn increasingly brutal and leave many people no choice but to seek safety among the insurgents (Eck 2009; Kalyvas and Kocher 2007). Routine material incentives are also effective. Often, guerrilla groups pay recruits to join or trade on future benefits that promise jobs or land reform. Employing material disincentives, guerrillas resort to brutal intimidation or punishment that leaves some to fear the consequences of defecting or failing to join. When all else fails, abduction is not unknown.

As a rule, solidarity incentives are more effective than material incentives. While "opportunistic joiners" readily support guerrillas for the material incentives they offer, many leave once resources run out. Opportunists are also the most difficult to discipline and control. Favoring easy gains and low costs, they often prefer to plunder soft civilian targets than to fight. Ideology and kinship networks, on the other hand, provide the basis for recruiting the most dedicated and disciplined followers and, in contrast to material incentives, require the least resources (Weinstein 2007:96–126, 139). So which of these varied methods may guerrillas and insurgents legitimately employ to fight a just war?

Recruitment Strategies: Moral Perspectives

Intuitively, some recruitment strategies – a fee for services, for example – are morally straightforward while others, such as abduction, are well beyond the pale. But conditions in the field are muddy. Payment for volunteers, for example, seems unproblematic until one considers that funds may come from pillage, extortion, and drug trafficking. It therefore seems reasonable to demand that recruits receive wages from noncriminal sources. And while abduction seems categorically wrong, Humphreys and Weinstein (2008:48) note "the ambiguity of the notion of abduction" as they describe how rebels in Sierra Leone paid reasonable wages to abductees and thereby employed "carrots and sticks ... simultaneously."

Elsewhere, education sometimes accompanies abduction. In Nepal, for example, Maoist guerrillas could successfully enlist civilian support by promising land reform and exploiting thick social networks to enlist supporters. As guerrillas gained control of rural territory, they broadened their base through indoctrination and political education. Convening mass gatherings, guerrillas would defend their tactics, publicize their successes, and trumpet the cause. This kind of dialogue was revolutionary. "By addressing the villagers, discussing their problems, and requesting their assistance," writes Kristine Eck (2010:44), "the Maoists encouraged the villagers to be active political agents, a radical departure from villagers' previous experiences of marginalization." On the other hand, indoctrination was not always voluntary. Eck (2010:43) also describes how schoolchildren, businessmen, and Nepali police were forcibly sent to education camps to work the fields and study communism before being released to join the movement. Indoctrination in its various forms proved exceptionally effective, and while mass gatherings raise no obvious objection, re-education is bound to raise some hackles when accompanied by abduction at gunpoint. But substitute social incentives for abduction, and re-education is more benign, particularly if prisoners tender consent when taking up with the guerrillas.

In other movements, however, intimidation and abduction are void of consent and far more malignant. Nelson Kasfir (2005:273) notes how FRELIMO insurgents in Mozambique "sometimes coerced civilians, forcing them to carry supplies, provide information and even shelling them to show that Portuguese forces could not protect them." The Taliban frequently threaten to kill villagers who do not support insurgents. Nor do they hesitate, albeit infrequently, to abduct young men and boys (Qazi 2011). Elsewhere, guerrillas in Africa, South America, Asia, and the Middle East regularly abduct large numbers of girls (McKay and Mazurana 2004). Most, but not all, abductees are adolescents. Few, if any, gave, or could give, their consent. Suggesting that such an atmosphere is morally ambiguous seems disingenuous.

Conscription and Consent

Considering the ugliness of abduction, consent seems central to justify conscription. Some guerrillas take a wage and fight. Others respond freely to the dictates of local authority, while still others skirt the fringes of consent as they respond favorably to re-education. All these recruitment methods depend on some degree of consent. Yet while consent may be sufficient to justify conscription, it need not be necessary. If states do not require consent, they why must guerrillas? If states may conscript, then why is abduction so odious? The answer lies in a system of conscription that ultimately enjoys a measure of support from the people it affects. Returning to the Vietnam War, liberal philosophers recognized the need for a fair system of conscription when necessary to secure basic rights and freedoms of individuals or peoples (Carter 1998). A fair system distributes the burden of war fighting equally along class and ethnic lines, includes provisions for conscientious objection, and imposes a severe but reasonable penalty for draft evasion. Combining these attributes with just cause and legitimate authority offers the prospect of "higher-order" consent. Individuals do not consent to conscription. Instead, the body politic consents to a fair system of conscription and the coercive methods it entails as it confers authority on a state or guerrilla organization through patrimonial representative or quasi-representative institutions.

Under a system of fair but coercive conscription, there is no room for such recruitment methods as abduction, brutal corporeal punishment or the recruitment of children. Abduction violates the conditions necessary for consent and excludes any opt-out option. Violent coercion, that is, physical intimidation and the threat of injury or loss of life, fails for two reasons. First, it precludes higher-order consent on the assumption that rational members of the body politic would not consent to the harsh measures they or their children might suffer should they refuse conscription. Sanctions may be severe but not deadly or crippling. Second, violent coercion is often ineffective. The Kurdistan Workers Party (PKK), for example, quickly grasped the perils of forced conscription. After adopting a policy of abduction and intimidation in 1986, the PKK abandoned the practice four years later after realizing that it rapidly eroded popular support. "You would take the people," observed one guerrilla leader, "and then the village would react, then the people you took would run away, and then you had to kill them" (Marcus 2007:117).

The recruitment of child soldiers violates the protections due minors unable to tender informed consent and presents special problems. On one hand, diminished cognitive capacity weakens a child's "morally responsible agency" to the extent that children are not liable to killing in war (McMahon 2010). If children are not proper targets there can be no grounds to conscript them. In practice, however, child soldiers are valuable assets. They are easy to recruit, consume fewer resources than adults, provide much needed manpower and often enjoy protection when facing uniformed soldiers who are hesitant about shooting children (Bangerter 2011:371). While the predominant response is to unilaterally prohibit the conscription of children under eighteen years of age, there is room for a more nuanced policy. Observing that preventing children from assuming any war related function whatsoever "makes it impossible for members of armed groups to remain together with their families and to be supported by the whole population, on whose behalf they (claim to) fight," Sassòli (2011) suggest more flexibility. One possibility is to allow children under the age of eighteen to assume some war-sustaining activities. Sassòli provides no details but these might include driving supply trucks, delivering messages, or organizing strikes. Such activities reduce liability to the point where children are not liable to lethal harm. Such a policy does not, however, adequately address lack of consent but only sets the consent condition aside while reducing the risk to children in those cases where a child benefits by remaining with her family.

In contrast to abduction, forced conscription, and violent coercion, recruitment strategies that utilize material and social sanctions are critical for guerrilla success and sometimes morally palatable. Consider the following description of Taliban techniques:

> It is reported that in Swat the Taliban had solidified their control by forcing the populace to back their regime politically and supporting it either monetarily or by sending a member of the household to join the rebels. They threatened to evict or chase out those who defied the orders ... During 2006–2007 members of the Shi'a Turi tribe had refused passage to Taliban and al-Qaeda rebels wanting to infiltrate Afghanistan ... As a result of their non-compliance, Taliban rebels in Kurram have long punished members of the Turi tribe by blocking passage of basic commodities to them, making survival extremely dire. (Qazi 2011:594)

Despite the brutality of the Taliban, this passage offers some indication of what legitimate coercion might look like in the hands of guerrillas. Material incentives are negative and include sanctions, blockades, and fines that raise the price of noncooperation. To gain support, the Taliban impose economic hardship on those who refuse to provide political, monetary, or military support. While "dire" hardship is beyond what justifiable measures permit (see Chapters 7 and 8), some level of hardship is necessary to ensure effective sanctions. At the same time, there is a hint of social sanctions as opponents are evicted from their communities. Ostracism, less harsh than banishment, is also a tactic that guerrillas may wield with considerable effectiveness to enforce support and conscription. Social sanctions – ostracism, deprivation of formal and informal social services, and censure by local political or religious leaders – can be devastating for individuals living in small communities. In later chapters, I describe how these sanctions recur in the context of human shielding and economic warfare. They point to a range of coercive but permissible tactics that guerrillas may utilize to mobilize support.

From Conscription to War Fighting

No less than states, guerrillas require conscription to fill their ranks and coffers. To meet this challenge, guerrillas must mix heavy doses of volunteerism with coercive measures to overcome collective inaction. Here negative and positive social incentives play a much larger role than among states. Lacking representative legislative institutions to pass and enforce laws of conscription, guerrillas can only fall back on thick social and kinship networks. While some individuals respond to the ideological fervor of the cause and others simply accept payment, large numbers of individuals respond to local politics: obligations of mutual aid coupled with fear of social sanctions. Guerrillas must thread their way through a number of recruitment methods, adopting the benign while emphatically rejecting abusive and brutal forms of recruitment.

Violent coercion, physical intimidation, and abduction fail because they ultimately repudiate guerrillas' right to fight. Guerrillas resort to brute force when they are unable to secure compliance through consent and legitimate authority. Absent legitimate authority, guerrillas lose their right to wage an armed struggle. The conditions of recruitment for guerrilla organizations, therefore, are not so different from those we impose on liberal nation states. Both are committed to systems that enjoy widespread support among the population they represent, and while both may employ coercion, states and guerrilla alike will find their legitimacy undermined once they resort to ruthless intimidation.

Once equipped to fight, guerrilla organizations will find no respite from humanitarian law as they struggle for self-determination. Nevertheless, the laws and ethics governing the conduct of war must be flexible enough to preserve a guerrilla army's right to a fighting chance. One obstacle turns on insurgents' right to shed their uniforms as they fight. This is a controversial privilege that also makes it possible for guerrillas to hide among the civilian population. The local population, moreover, is not quiescent and assumes a crucial role as civilians provide war-sustaining financial, legal, and logistical aid to insurgents. As a result, there is room to ask whether these civilians deserve blanket and absolute protection from direct harm or only enjoy some lesser protection commensurate with their responsibility for sustaining an armed conflict? This question is particularly acute when it proves impossible to prevail in armed conflict without disabling civilian facilities. The following sections address these questions in greater detail, investigating first the right to shed uniforms, a far-reaching deviation from the law of war but necessitated by the right to fight. Complementing the right to fight without uniforms, is the principle of "participatory liability" that allows belligerents on both sides to take direct aim at various classes of civilians in defiance of traditional norms of noncombatant immunity.

FIGHTING WELL: THE RIGHT TO SHED UNIFORMS

The right of a guerrilla army to fight without uniforms evolved from the same right granted to partisans by the 1949 Geneva conventions. By 1977, partisans were long gone and guerrillas fighting for national liberation earned the right to discard their uniforms in those situations "where, owing to the nature of the hostilities an armed combatant cannot so distinguish himself ... " (API 1977:Article 44[3]). The last sentence is telling: the right or, permission, to shed uniforms is not sweeping; it is only justified when necessary and effective. The framers of Protocol I had colonial occupation in mind where the requirement to wear uniforms would make it nearly impossible for guerrillas to organize, train, collect intelligence, or mount an attack.

The right to fight without uniforms confers protection in two ways. First, it renders guerrillas indistinguishable from noncombatants thereby allowing them to move freely to and from battle. Second, it makes it more difficult for enemy soldiers to target guerrillas for fear of causing disproportionate casualties among the civilian population. Why do guerrillas gain such a right that not only gives them an advantage when they fight uniformed soldiers but also puts their civilian population at great risk? The right to shed uniforms stems directly from a guerrilla army's right to a fighting chance that prohibits laws that would preclude any chance of success in pursuit of a just cause. However, a fighting chance does not demand a level playing field. Uniforms must pose a significant handicap before guerrillas may shed them.

One handicap might be arrest or assassination. Christopher Finlay (2012) argues that shedding uniforms allows guerrillas to retreat safely to their homes or villages in the same way a conventional army retreats behind its lines after the fighting. Without uniforms, guerrillas are difficult to identify, much less arrest or assassinate. While shedding uniforms might make it easier to retreat, lack of uniforms does not necessarily offer protection against assassination and arrest. To the contrary, conventional armies turn to arrest and assassination *because* their non-state adversaries do not wear uniforms. Arrest is necessary because soldiers are uncertain whether they are facing civilians or guerrillas. Assassination is intelligence driven, a process of identifying, tracking, and disabling guerrillas precisely when out of uniform. Once guerrillas don uniforms, arrest and assassination are moot; guerrillas are simply killed or disabled in the normal course of combat. The danger guerrillas face in uniform comes from decimating attacks against which losses from arrest and assassination pale by comparison.

Instead of providing protection from assassination, the right to shed their uniforms affords guerrillas the ability to maneuver among civilians, reconnoiter, move supplies, and establish firing positions. Mufti allows guerrillas to fight better, not retreat. As they fight among civilians, guerrillas also draw their adversaries into attacks that may disproportionally harm the civilian population and thereby give an attacking army cause to desist. The right to shed uniforms cannot confer protection without the shielding that the civilian population provides. The right to shed uniforms and the right to conscript human shields are, as I suggest in Chapter 6, two sides of the same coin linked together by the right to a fighting chance. It is important to keep in mind, however, that the right to shed uniforms is an exception to the general rule to wear uniforms. International law rightly emphasizes how this exception "recognized that situations could occur in occupied territory and in wars of national liberation in which a guerrilla fighter could not distinguish himself [from the civilian population] throughout his military operations and still retain any chance of success" (API Commentary 1977:Article 44, §1698).

The fact that this exception has morphed into a rule that few now question does not grant it sweeping moral legitimacy. To exploit the right without necessity needlessly places noncombatants at risk. To place noncombatants at risk, guerrillas must be otherwise so hamstrung that they cannot fight at all. For this reason, Protocol I correctly restricts this right to guerrillas fighting for national liberation in occupied territory.

Therefore, the argument for shedding uniforms is most persuasive in conflicts like those in East Timor or the Palestinian territories. It is less compelling when guerrillas fight an invading army. Although Hezbollah regulars, for example, sometimes wore uniforms, "village guards" did not. The latter were not the old and infirmed but experienced fighters who remained in their villages after the civilians had fled. They constituted a reserve of battle-ready, albeit part-time, fighters (Cordesman and Sullivan 2007:135–136). Because Hezbollah did not lack the wherewithal to fight pitched battles with Israeli forces and, in some cases, persevere, there is no justification for their shedding insignia. Hezbollah is an outstanding example of a guerrilla organization abusing the right to fight without uniforms while putting their civilian population at considerable and unnecessary risk. Hezbollah is not fighting a war of national self-determination but instead pursuing Lebanese, Syrian, and Iranian interests by armed force. Although an irregular force, there is no reason to automatically assume that "nature of the hostilities" prevents their fighters from wearing uniforms. The right to shed uniforms, therefore, is contingent; guerrillas must establish their right to fight without uniforms by documenting the operational impediments that come with wearing uniforms. Otherwise the international community must call them to task and condition aid and recognition on adherence to the prevailing principle of international law that requires combatants to distinguish themselves from all others.

Just as the right to fight provides grounds to relax the requirement to fight in uniform, the right to fight also reaches into the principle of noncombatant immunity and other basic principles of just war. State armies often complain that they lack the means to disable civilians who aid a guerrilla army. Conventional warfare consists of two clearly designated groups of actors, combatants and noncombatants, and unfolds with the understanding that victory will come when belligerents can disable sufficient numbers of *combatants* (Gross 2010a:56–57). Asymmetric warfare, on the other hand, must contend with small guerrilla armies and much larger political wings that make no small contribution to their war effort. Here, victory comes only when belligerents can disable sufficient numbers of fighters *and* the civilians who aid them. As such, the right to a fighting chance demands some modification of the principle of noncombatant immunity.

FIGHTING WELL: NONCOMBATANT IMMUNITY
AND CIVILIAN LIABILITY

Although voluminous, the principles of just war and the corresponding
law of armed conflict might be boiled down to two principles of humanity, one pertaining to combatants and other to noncombatants:

1. *Combatants may not suffer superfluous injury or unnecessary suffering.*
2. *Noncombatants may not suffer unnecessary, direct, or excessive, that is,
   disproportionate harm.*

Liability and Harm

The stark difference between combatants and noncombatants turns on liability. Combatants are liable to harm; noncombatants are not. As a result,
there is little to protect combatants from lethal harm. With the exception
of weapons like poison gas, hollow-point bullets, blinding lasers, and torture, there are few injuries one cannot inflict on soldiers. Noncombatant
protections are considerably wider. Noncombatants are not liable to any
harm whatsoever. Nevertheless, it is clear that noncombatants will suffer injury or loss of life during armed conflict. Morally, then, the harm
befalling noncombatants must be necessary and unavoidable. For guerrillas, this means "necessary in pursuit of just cause." The requirement
to avoid unnecessary harm is often overlooked and refers to any casualty
that does not accompany a military advantage. Droves of noncombatants
lose their lives through operational errors, faulty intelligence, or futile
military missions. When artillery shells go awry, or the wrong targets are
bombed, or missions end without any discernible benefit, every civilian
who lost his or her life along the way died unnecessarily.

Alongside the prohibition of unnecessary harm stands the ban on
direct harm. The moral basis for prohibiting direct harm against noncombatants is clear: noncombatants have done nothing to forfeit their
right to life. One may therefore not fire upon a civilian with the intent
to injure or kill. The term "intent" is somewhat misleading because it
suggests that an attacker did not actually mean to harm a noncombatant. This is not always true. When a pilot, for example, attacks a military
target nested among civilians, there is no way to say the pilot does not
"mean" to harm civilians. The pilot knows full well that civilians will

die after the bombs are dropped. Rather, civilians are not the object of the attack; attacking and killing them has no military value. After the fact, one may test for intent by looking for any benefit a mission might have gained from harming noncombatants. Any hint that civilian deaths enhanced deterrence, increased political instability, or brought pressure to bear on policy makers to cease fighting points to an express benefit. Such civilian deaths are not incidental or collateral but direct and intentional. There is far more nuance to this prohibition than simply proscribing cold-blooded murder, rape, or torture. The ban on direct harm also prohibits any military action that benefits from noncombatant casualties.

The prohibition against direct harm may, however, leave room for what might be called "therapeutic harm," that is, nonlethal harm inflicted directly on noncombatants to save them from a greater, lethal harm. This can occur when armies use nonlethal weapons to incapacitate a mixed and indistinguishable crowd of combatants and noncombatants with the aim of arresting and disabling combatants while eventually releasing noncombatants. Under these circumstances, the ban on direct harm is not absolute. I return to this point in the context of cyber-warfare in Chapter 7.

While "unnecessary" and various forms of "direct" harm often pose the greatest dangers to noncombatants, most observers focus on disproportionate harm (Gross 2012). The prohibition on disproportionate or excessive harm is a lynchpin of international humanitarian law: "[Belligerents shall] refrain from launching any attack which may be expected to cause incidental loss of civilian life ... which would be excessive in relation to the concrete and military advantage anticipated" (API 1977:Article 57[2][iii]).

The principle of proportionality enshrined by this paragraph is shrouded in ambiguity. The principle governs the relationship between military advantage and civilian deaths. For guerrillas, evaluating proportionality depends on the goal they hope to achieve. Rarely do insurgents aspire to a decisive military victory. More often they are after a symbolic show of strength that will build support at home and abroad, discredit the state they fight, or wear down their enemy. "Victory" may include successfully disabling a single tank, capturing a single soldier, or breaching a high-security installation. The impact of these attacks is materially minimal but symbolically enormous. As a result, it will be very difficult

for insurgents to quantify military advantage and weigh it against harm
to civilians.

Civilian casualties are likewise difficult to determine. The number of
dead civilians is easily quantifiable assuming, of course, one can distinguish civilians from soldiers. This is not always easy. In Operation Cast
Lead (Gaza, 2008–2009), for example, Israelis and Palestinians counted
approximately the same number of Palestinian dead and, in fact, posted
their names in no time. Competing assessments of who among the dead
were noncombatants, however, varied widely from 25 percent to 75 percent of the casualties (Issacharoff 2009; Lappin 2009; Palestinian Centre
for Human Rights [PCHR] 2009). The dead, moreover, are only one cost.
There is no obvious reason to exclude other harms from the proportionality calculation: the injured, diseased, and displaced together with the
short and long damage to transportation, energy, water, or agricultural
infrastructures. But fixing these numbers is only part of the problem.
Assuming one can quantify military advantage and civilian casualties,
what then does "excessive" mean? How many civilian deaths and injuries
are proportionate, and how many are not? There is no easy answer to
this question. Over time, international law has simply taken the advice of
Justice Potter to heart: one knows it [disproportionate force] when one
sees it. The call is largely subjective.

While subjectivity renders the proportionality principle extremely elastic, it is not entirely vacuous. There is a limit beyond which civilian casualties are entirely disproportionate. But whether this can be quantified
by any neat algorithm or is nothing but a gut feeling is never clear. The
Geneva Conventions can only offer advice of questionable utility: "'Some
cases [of excessive harm],' opines the Geneva Convention commentary
'will be clear-cut and the decision easy to take. For example, the presence
of a soldier on leave obviously cannot justify the destruction of a village'"
(API Commentary 1977:Article 57, §2213). This admonishment, however, is so "obvious" that any order to destroy a village to disable one
soldier on leave is manifestly unlawful. Proportionality, it seems, comes
into play only at extremes, when disproportionate harm is synonymous
with flagrantly unlawful death and destruction. As such, proportionality
provides no useful guidance for commanding officers, whether guerrillas
or not, who constantly face situations far more ambiguous than those the
Geneva Conventions describe. Under these circumstances, the best we
might expect is a reasonable and, more importantly, defensible judgment

that articulates the mission's important objectives, its anticipated civilian casualties, an indication that alternative means were not available, and that the civilian deaths could not be avoided or further minimized without great cost to a commander and his soldiers. Despite the hype, the proportionality principle enjoys little reach. Vigilantly avoiding unnecessary and direct harm as well as injury inflicted in error, on the other hand, goes a long way to protecting the lives and property of the civilian population (Gross 2008). Civilians only warrant protection, however, when they steer clear of belligerent behavior. Not all do.

Combatants and Noncombatants: A Sliding Scale of Participation and Liability

Any successful attempt to apply the principles protecting combatants and noncombatants demands a clear understanding of who enjoys which safeguards. Otherwise, belligerents cannot meet their obligation to discriminate between combatants and noncombatants. Traditionally, the categories of combatant and noncombatant are discrete, so that the two rules of armed conflict apply to two distinct groups. Noncombatants are "persons taking no active part in the hostilities" and include civilians as well as prisoners of war and the wounded (Geneva Convention (IV) 1949:Article 3). Combatants, on the other hand, bear arms and take an active or direct part in the hostilities as members of opposing armed forces, whether regular or irregular. While the combatants are liable to harm, noncombatants are protected from all forms of direct harm.

In conventional war, this strict dichotomy between combatants and noncombatants encompasses nearly every actor on or near the battlefield. There is a grey area, but it is very narrow. Persistent questions surround the status of civilian munition workers or truck drivers employed by the military. Because of their close connection with the military, these individuals are liable to harm at least while on duty. Any other civilians who aid and abet their nation's war effort generally retain their immunity. This includes civilian leaders, presidents, prime ministers, journalists, bakers, and boot makers.

This distinction between combatants and noncombatants and the liability each enjoins is at odds with our moral intuitions because it gives overriding weight to one's military affiliation rather than to one's military contribution. As a result, it is reasonable to ask why an army's cooks and cleaners are liable to attack while civilian leaders who run a war and religious leaders and journalists who exhort their people to fight are not. There are many reasons for this. One is mutual self-interest: to help ensure the survival of their political community, states agree to protect an entire class of its citizens, namely those who are not in the army. At the same time, states allow one another to license its soldiers to kill on the battlefield. Only soldiers can kill, so only soldiers may be killed. While this does not address the issue of civilian responsibility, it helps keep armed violence within acceptable limits. Additional reasons speak to the relatively insignificant contribution that most civilians make to war. Although some civilians bear some responsibility for war, it is insufficient to render them liable to death or injury. Moreover, there are practical difficulties when pursuing civilians. All are without uniform, making it impossible to identify the culpable and disable them without inflicting undue harm on the innocent. Under these conditions, legitimate attacks on civilians may spiral into impermissible violations of noncombatant immunity.

Nevertheless, many observers are prepared to permit guerrillas or states to intentionally harm civilians responsible for unjust occupation or grave violations of human rights (Corlett 2003; Steinhoff 2004; Vall 2000; Young 2004). This seems right. Civilians responsible for sustaining war should be legitimate targets when disabling them is necessary and the force employed is proportionate. This can be called the *principle of participatory liability*. Liability does not depend on affiliation but on the actions agents take during war. Participatory liability is a sliding scale that links participation with liability to harm. The greater a person's participation, the greater harm an enemy may inflict when necessary to disable a participant. Participation reflects both a civilian's function within the organization and the magnitude of the threat the civilian poses. Each aspect of participation is usually observable, marked by a person's occupation and the product or service he or she provides. At one end of the scale are *noncombatants* who assume no role in any war related activity. They are not responsible for any threat and, therefore, may suffer no direct harm. At the other end are full-fledged *combatants* who conduct armed campaigns against enemy forces and are liable to lethal (but not inhuman) harm when necessary to disable their person and disrupt their activities. In the vast middle ground are *participating civilians*, analogous to those civilians working for a guerrilla organization's political wing or for a state's bureaucracy or defense industry. Participating civilians provide war-sustaining aid that includes logistics, telecommunications, and financial services. They are only liable to *less-than-lethal* harm. Less-than-lethal harm includes the transitory effects of nonlethal weapons and cyber-attack (Chapter 7), economic sanctions (Chapter 8), or restrictions on liberty, that is, arrest, curfew, confinement, or deportation. In the sections and chapters that follow, the terms "noncombatants," "combatants," and "participating civilians" will be used as just defined.

Participatory Liability, War-Sustaining Aid, and Indirect Participation

The idea of participatory liability never found a firm place in interstate warfare because each side had pragmatic and legal reasons to keep its citizenry out of the fighting entirely. For these reasons alone, Jeff McMahan (2009:235), for example, can assert the force of moral liability but deny it any practical value. On the contrary, he writes, "while absolute noncombatant immunity is false as a moral doctrine, it remains a legal necessity."

While perhaps a necessity of conventional war, absolute noncombatant immunity broke down in asymmetric war as soon as guerrillas stopped wearing uniforms. Once insurgents shed their insignia, state armies faced an amorphous mass of civilians taking a direct or indirect role in the fighting. Gone were the two distinct categories of uniformed combatant and ununiformed noncombatant to which one could easily assign a bundle of rights. This lack of distinction, coupled with the relatively small numbers of actual combatants in most guerrilla organizations, led states to consider the need to target civilian infrastructures. The result was the idea of *associated* targets, that is, civilian targets associated with and supportive of a guerrilla military organization (Gross 2010a:154–162). These were not hard to find; nearly all guerrilla organizations had their political wings. State armies targeted associated targets for two reasons. First, it was difficult, if not impossible, to subdue a guerrilla group without disabling its civilian infrastructure; second, it was reasonable to impute some degree of responsibility to participating civilians, thereby making them liable to harm. Many states, however, adopted only half the principle of participatory liability, extending liability to civilians providing war-sustaining aid but forgetting that a sliding scale of liability requires a sliding scale of force. With the exception of participating civilians who take up arms, lead military operations, or propagandize in favor of terrorism or genocide, most others are only liable to less-than-lethal harm.

With this caveat in mind, there is no reason that guerrillas should also not invoke the principle of participatory liability and target some civilians directly. Like guerrilla organizations, states also have military and political wings. Those employed in the latter may be described as civilians taking an "indirect" role in the fighting or those providing war-sustaining in contrast to war-fighting aid. Unlike direct participation, which signifies acts "likely to adversely affect military operations" and "specifically designed to directly cause ... death, injury or destruction" (Melzer 2009:46), indirect participants embrace the class of participating civilians whose aid is not likely to adversely affect military operations nor specifically designed to cause significant harm directly. For the International Committee of the Red Cross (ICRC) Interpretive Guidelines, indirect participation includes financial and economic services, recruitment and training activities, non-tactical intelligence gathering, propagandizing and diplomacy, assembly and storage of IEDs, and "scientific research and design of weapons and equipment" (Melzer 2009:35, 53–54; also, Schmitt 2004). Other candidates for indirect participation include those who provide legal, media, health, telecommunication, transportation, and police services. While these functionaries do not participate directly in "military operations," they do participate directly in and sustain "the conflict in general" (Akande 2010:188).

For the ICRC's Interpretive Guidelines, the categories of indirect participation reflect a wide range of belligerent activities that civilians may pursue *without* losing their immunity. This is not my purpose. Rather, it is clear that civilians providing war-sustaining services bear no small measure of responsibility for maintaining a war-making capability. If, as Michael Schmitt (2010a:30) claims, direct participation turns on a "clear link between the act and the ensuing harm" and, therefore, warrants liability, then a clear link between an act and a lesser harm also warrants liability. Participating civilians are neither idle nor passive participants; they report for work, draw a pay check, and collaborate with their colleagues. Insurgents fighting for a just cause should be allowed to disable these targets with proportionate, but less-than-lethal, force when necessary. Two notions govern the use of force against participating civilians.

On one hand, participating civilians pose no lethal threat. As such, the military advantage of disabling a war-sustaining facility does not justify killing or seriously injuring civilian employees. On the other hand, there are no grounds to grant participating civilians absolute immunity. Nor are there any grounds to push them into the category of direct participation and the liability to lethal harm this entails. Locating the appropriate middle ground to disable participating civilians effectively remains the challenge.

Disabling Participating Civilians: Liability, Human Rights, and Military Necessity

There is no doubt that military necessity drives participatory liability when neither state armies nor guerrillas can prevail without disabling civilian, war-sustaining facilities. Military necessity, however, does not offer either party unlimited means for this purpose. Rather, the rules that regulate armed conflict constantly juggle the imperatives of military necessity, ethics, and positive law. Definitions of military necessity, such as those appearing in the Geneva Conventions, reference "the means necessary to subdue an enemy and which are not forbidden by international law" (API 1977:Article 35). This formulation does not adequately reflect the tension between law and necessity because the compelling question is: When may military necessity trump the law? The answer, if we follow Protocol I, is never. In reality, the situation is more complex because law and ethics are neither monolithic nor consistent. Some principles lie anchored in explicit agreements between nations, while others draw on natural law or utilitarianism. It is not unreasonable, therefore, to demand that military necessity should defer to some aspects of the law – those that protect a person from murder, torture, slavery, and indignity, for example – but not to every convention or agreement. Balancing necessity and rights is the challenge of participatory liability. The degree of permissible force belligerents may use is a function of the threat individuals pose (liability) and the imperative to choose the most effective and least costly means of disabling participating civilians (necessity). Exercising the right to fight imposes both obligations.

Establishing necessity is the first order of business. This is not always obvious. When civilians are killed collaterally during a military operation, commentators immediately ask: Was the target a legitimate military target? Was the attack legal? However, the first question should be: Was the attack effective? Were there sufficient grounds to believe the attack would have been effective? If the answer is "no," then the discussion stops here. If the answer is "yes," then the second question is: Was the attack necessary or were other means available to achieve the same outcome at a lower cost? Only if the answers are "yes" and "no," respectively, may one ask if the attack violated the principle of participatory liability by subjecting combatants to inhuman harm, participating civilians to lethal harm, or noncombatants to unnecessary, direct or disproportionate harm. This will be the template for analyzing a wide array of military tactics in the following chapters.

Hard problems arise at each stage of this process. Asking about effectiveness and necessity, a guerrilla army may find that it has no effective means at its disposal to fight. Stymied, guerrillas must search for the obstacles that frustrate their right to fight. If treaties and conventions stand in the way, then there may be grounds to demand modifications. Some laws may have to bend to accommodate the right to fight. The right to shed uniforms and the principle of participatory liability meet this demand. If guerrillas, however, find it impossible to wage war without violating the fundamental human rights of combatants, noncombatants, and participating civilians, then insurgents may have to set their arms aside and search for alternative avenues to pursue self-determination. Necessity knows some law but not no law. Will guerrillas respect these constraints no matter how flexible they are?

THE INCENTIVE TO OBEY THE LAW

Common wisdom has it that armed guerrilla groups have no incentive to obey the rule of law during war. "For a predominantly terrorist movement," notes Walter Laqueur (1998:391), "the acceptance of the enemy's rules of war would be a negation of their whole strategy." While state armies scrupulously observe the distinction between military and nonmilitary targets, "terrorism," declares Tamar Meisels (2008:46), "wholly depends upon its opponents upholding moral code which the terrorists themselves reject." In Meisels' view, terrorists free ride and mockingly exploit their enemies' willingness to obey the law.

Free-Riding, Terrorism, and Guerrilla Warfare

One difficulty with claims of free riding is the explicit reference to terrorist organizations. While some modern guerrilla organizations employ terror, none of those pursuing national self-determination is a "predominantly terrorist movement." As subsequent chapters demonstrate, these insurgents mix violent and nonviolent tactics that include not only terrorism but also public diplomacy, economic warfare, and civil disobedience (also, Hicks et al. 2011). On pain of self-contradiction, guerrillas must demonstrate reasonably just behavior to pursue just cause and retain their right to fight. Law-abiding behavior is crucial as guerrillas internationalize their conflict and use the media to push their grievances beyond their borders and enlist international support (Chapter 9). To flout the rules of war flagrantly only undermines a people's goal of international recognition. While the community of states has considerable tolerance for dictators, there is remarkable intolerance for terrorists, war criminals, and rogue states. Insurgent and national liberation organizations will find it much easier to integrate as righteous underdogs.

The incentives that can motivate guerrillas to obey the law are not much different from those that motivate state actors in an anarchic international community. In each instance, the incentives may be normative or material. Normative incentives reach deeply into a moral code that defines a community's identity and its obligation to protect the people it serves. Many communities take pride by upholding principles of honor that forswear intentionally harming the innocent. Observers note the so-called code of early twentieth-century Russian terrorists who struggled with killing children (Walzer 1977:198–199). But one need not look back so far. Catholic guerrillas in East Timor and Moslem guerrillas in the Western Sahara disavowed terrorism as part of their struggle (Stephan 2006; Stephan and Mundy 2006). Although tried for war crimes, Kosovar guerrillas usually targeted spies and collaborators, not noncombatants (Perritt 2008:73–74). The Taliban Code of Conduct calls upon soldiers to minimize civilian casualties and protect "the life and property of the people" (United Nations Assistance Missions in Afghanistan [UNAMA] 2012:12). The Palestine Liberation Organization and the Kurdish Workers Party (PKK) agreed to observe the Geneva Conventions (International Committee of the Red Cross [ICRC] 1994; Yildiz and Breau 2010:15).

Such codes and commitments, common among many guerrilla groups, are not entirely insincere or ineffectual (Bangerter 2011a, 2011b; Sivakumaran 2011). In the Middle East, a loosely structured normative code governs the treatment and exchange of Palestinian, Lebanese, and Israeli prisoners. It is not what the Geneva conventions prescribe, but it is far from the ruthless barbarism characterizing Russia and Chechnya during their wars (Chapter 5). This is not to suggest that all guerrilla groups adhere to the law. Obviously, this is not the case. Nor it is true that those who adhere to some laws of war do not reject others. Kosovar and Aceh guerrillas may have tried to protect those they respected as innocent civilians but their purges of informers, collaborators, and dissidents were far from the norm of international law (see Chapter 5). Elsewhere, the UN Assistance Mission in Afghanistan (UNAMA 2012:15) condemns the Taliban, for example, for failing to make any effort to renounce suicide attacks or refrain from using pressure-plate IEDs, two tactics that violate the Taliban's own rules of engagement. Nevertheless, Giustozzi (2014: 294) describes how the Taliban have strengthened their courts substantially to enforce their code of conduct. Unmoved by international law, the Taliban courts enforce norms of jihad that require fighters to fulfill their religious duties and "avoid excessive harm to non-combatants."

Although moral codes matter, it would be impossible to maintain any legal and moral regime to govern asymmetric war without some measure of reciprocity, fears of a response in kind, or international repercussions. For many guerrillas, just as states, abiding by international law is materially advantageous. Law-abiding behavior garners support from their civilian population and the world community, conserves military resources for use against the enemy, preserves infrastructures for postwar reconstruction, and strengthens morale among soldiers by highlighting courage and self-discipline (Bangerter 2011b). Guerrillas should not target noncombatants for the same reasons that states should not. One reason is surely moral, a deeply ingrained aversion to killing the innocent. The other turns on reciprocity and fear of reprisal: you attack our noncombatants, we will attack yours. "In agreeing not to target noncombatants," writes Yitzhak Benbaji (2013:186) concisely, "each side is able to do less damage to the other, but each side also suffers less at the other's hands."

Because of the vast material asymmetry in asymmetric war and guerrillas' inability to respond in kind, self-interested restraint sometimes breaks down. One option is to free ride and to flout the law and ethics of war deliberately. But flouting the law provokes wrath across the world, no matter how compelling just cause, and, therefore, entails heavy costs for a people hoping for a place in the world community. Witness the continued troubles of the Palestinians. But trouble for the Palestinians is also trouble for the Israelis, and one cannot remain deaf for long to the Palestinian plea of "How then should we fight?" Without an answer to this question, no armed resolve, or in fact any resolve, is possible. So the question remains: What rules can states, non-states, and the international community accept that both allow an aggrieved party a reasonable chance to pursue its right to national self-determination *and* preserve the fundamental human rights of combatants and noncombatants?

While many are reluctant to alter the rules of war, dogmatism thwarts compliance. When the recognized means of warfare do not allow insurgents to press just claims by armed force, there is no incentive to follow any rule of law. If guerrillas don uniforms and attack military targets and still find that they are tortured, tried, sentenced, jailed, or executed upon capture, they have little incentive obey the law. Addressing this dilemma, Marco Sassòli (2011) advocates a "sliding scale" of legal obligations that will release many armed groups fighting for national self-determination from those precepts of international law when they lack the means to comply. Sassòli is careful to point out, and rightly so, that the provisions of the Geneva Conventions' Common Article 3 that protect individuals from murder, mutilation, torture, arbitrary punishment, kidnapping, and "outrages upon personal dignity" are the floor of any change. His list is rather modest and includes suggestions to adjust the age and duties of child soldiers, permit guerrillas to appropriate resources in insurgent controlled territory without violating rules prohibiting pillage and to ignore the complex rules governing prisoners of war as long as captors treat detainees humanely.

Despite such flexibility, one might go considerably further and still retain the principles of humanity enshrined in Common Article 3. As the following chapters demonstrate, permissible tactics push past the limits of international humanitarian law to allow guerrilla fighters to shed their uniforms, conscript human shields, target civilians directly, seize soldiers, assassinate collaborators, manipulate the media, and undertake cyber-warfare. But, as the following chapters also establish, none of these measures is without stringent constraints. Not all guerrillas may shed their uniforms, human shields must tender consent, captured soldiers enjoy freedom from abuse, the liability of participating civilians is limited to nonlethal harm, collaborators enjoy the right of due process, cyber-warfare may not descend into cyberterrorism, and the media may never fan the flames of genocide. Following the war in Kosovo, for example, both Serbians and members of the Kosovo Liberation Army (KLA) found themselves in the dock. The trials offered important lessons for the conduct of guerrilla warfare. Although NATO fought on the side of Kosovar secessionists and ultimately helped secure their autonomy, three members of the KLA were held accountable for their conduct during war. Tried for war crimes and crimes against humanity, two were acquitted and one declared guilty and punished in 2005 (ICCT nd.). As we consider changes to the rules of war, we cannot ignore the necessity of an aggressive legal regime to pursue violators and war criminals who endorse terrorism, shed their uniforms unlawfully, conscript human shields at gunpoint, attack civilian infrastructures disproportionately, encourage genocide, or abuse prisoners of war.

Guerrillas and insurgents have no clear incentive to free ride. On the contrary, their moral and material interests dictate broad compliance insofar as the rules of war accommodate their right to fight. Because many state armies consider insurgents nothing but criminals, they may balk at bending the rules. Before they do, however, they should consider the more radical claim that far from being criminals who flout the law, guerrillas have no obligation to follow the rules of war because their opponents are unjust.

Unjust Combatants and the Rules of War

In contrast to the oft-heard claims of Lacquer and Meisels, consider that guerrillas have no moral obligation to respect the laws that accord their adversaries war rights. Assuming guerrillas must enjoy just cause to fight at all, any state opposing a just guerrilla movement must be unjust. Indeed, a cursory look at the conflicts in places like East Timor, Kosovo, Southern Sudan, or Eritrea bears this out. In each instance, the prevailing state power – Indonesia, Serbia, Sudan, and Ethiopia, respectively – flagrantly violated a people's right to national self-determination and waged a ruthless and brutal counterinsurgency. In these places, at least, there are good grounds to argue that guerrillas need not recognize their opponent's right to fight in self-defense. As such, no captured enemy soldier deserves to be treated as anything but a criminal.

While I acknowledge the importance of moral status, three practical considerations mitigate concern. First, law of armed conflict does not leave a vacuum when no longer germane; human rights law takes its place. Combatants turned criminals, whether Serbs, Ethiopians, or Indonesians, retain the rights that protect them from abuse, murder, rape, torture, and summary execution. They are, of course, liable to arrest, incarceration, and, perhaps, execution. Noncombatants, if anything, come out better. Liable to collateral harm in war, noncombatants remain fully protected during law enforcement, currently the only alternative paradigm to armed conflict. A people fighting an unjust oppressor is little different from the police fighting a criminal gang. This is particularly true if the international community authorizes military intervention on the grounds that an offending state *is* a criminal gang. Nevertheless, no state is going to acknowledge its criminal status and, if so branded, is most likely to scorn the law of war and human rights law as it fights. Assigning criminal status may also reduce the chances of postwar reconciliation. Here, it may be prudent to hold *jus ad bellum* at bay and grant all combatants equal rights in war lest armed conflict descend into mass murder and anarchy.

Second, not all conflicts are as clear-cut as those in East Timor. In some cases, the strength of just cause wanes. While Palestinian claims to national self-determination and statehood are well entrenched, for example, there remain questions whether armed violence has met the last-resort condition, particularly in light of reasonable Israeli peace offers and past Arab aggression. In Indonesia, the Aceh people's right to self-determination stops short of statehood, giving them the right to fight for autonomy but not independence. In these instances, states may reasonably argue that insurgents have turned to armed force before fulfilling the necessary conditions of just guerrilla warfare, thereby permitting states to respond in self-defense. Third, states retain their right to protect their civilians from terrorism. However one understands the Israeli-Palestinian or Chechen-Russian conflict, one cannot deny that Palestinian or Chechen terrorism triggers Israel's or Russia's right to self-defense when noncombatants face attack. Under these circumstances, each side enjoys the right to defend itself.

While it is clear that states act criminally in some asymmetric conflicts, pragmatic concerns support a moral and legal regime that obligates all sides to respect the rights of combatants, noncombatants, and participating civilians. Delineating these rights clearly and in light of each side's right to fight is the compelling exercise. Guerrillas may complain that many principles of just war shut down their war-making ability, while state armies complain that they adhere to a code of conduct that guerrillas regularly flout. Both complaints are frequently valid. Neither enforcing the law of war dogmatically nor disregarding it entirely provides any sort of answer to the challenge of pressing legitimate claims by force of arms without violating the basis rights of combatants and noncombatants. Rather, the challenge is to describe a set of legitimate tactics so that guerrillas retain a fighting chance while conforming to the underlying norms of just warfare.

As the discussion of guerrilla tactics and strategies unfolds, the intimate connection between *jus ad bellum* and *jus in bello* is ever present. It is impossible to consider the ethics of any tactic be it the shedding of uniforms, the placement of IEDs, the capture of prisoners, the conscription of human shields, the use of cyber-warfare or the imposition of economic sanctions in the absence of just cause and legitimate authority. This injunction applies to quotidian issues as well. In Chapter 8, for example, I consider how guerrilla movements may levy taxes to raise the funds necessary to conduct an armed struggle. Aside from the normative guidelines that regulate taxation, such as fairness and nondiscrimination, many would also demand a measure of just cause and legitimate authority. This is correct. Not only must guerrillas establish an equitable system of taxation; they may only raise money, recruit fighters and mobilize civilian support for a just cause and through the exercise of legitimate authority. And, as the following chapters demonstrate, insurgents may only fight while respecting the rights of combatants, noncombatants and participating civilians. Otherwise, they are nothing but gangsters and warlords.

PART II

HARD WAR

4

Large-Scale Conventional Guerrilla Warfare

Improvised Explosive Devices, Rockets, and Missiles

Overwhelmed with images of terrorism, many often forget that guerrillas are fighters first. They shoulder arms, execute battle plans, engage the enemy, and take prisoners. Their military goals and tactics vary as a function of their relative size and power. While some guerrilla groups such as the Eritrean People's Liberation Front (EPLF) fight and win set piece battles, most insurgents set very limited objectives for their military forces. While the EPLF could defeat Ethiopian forces in open warfare, inflict staggering losses, and, most conventionally, capture the Eritrean capital and set up a provisional government (Iyob 1995:108–135; Pateman 1990:88–93), most other guerrillas hope to harry their opponents and make the cost of continued warfare or occupation unbearable. To that end, most employ traditional hit-and-run guerrilla tactics to undermine their enemy's military morale and undercut popular support for continued warfare.

The *military* tactics of guerrilla warfare, that is, operations aimed specifically at unambiguous military targets – troops, convoys, army bases, or any object whose destruction "offers a definite military advantage" (API 1977 Article 52(2); ICRC nd. a) – are not without controversy or moral hazard. This chapter and the following one consider four dimensions of traditional guerrilla warfare: armed attacks utilizing small arms and improvised explosive devices (IEDs), missile strikes, assassination, and prisoner exchange. In many respects, the moral dilemmas accompanying each of these tactics are not so different from those bedeviling state armies. Like any kinetic weapon, IEDs and missiles immediately raise the danger of disproportionate harm when noncombatants are caught in the fighting. Assassination or targeted killings, whether initiated by state or non-state armies, elicit charges of extrajudicial execution, while prisoners of war demand reasonable conditions of incarceration that guerrilla armies, no less than state armies, often struggle to meet.

In addition, guerrilla warfare carries special problems of its own. Often characterized as hit and run or rearguard fighting, guerrilla warfare typically avoids open field confrontation. In theory, this has changed little in contemporary guerrilla warfare. In practice, however, guerrillas utilize an increasingly powerful array of high-explosive and rocket weaponry that are often unsophisticated and inaccurate. Improvised explosive devices (IEDs) are nothing more than hastily buried roadside bombs lying in wait for anyone, soldier or civilian, who may happen to pass by. Missiles, on the other hand, require assembly, targeting, and launch. While the avowed target may be military, most of these missile systems lack the precision necessary to prevent the destruction of civilian infrastructures. As a result, guerrillas face charges of indiscriminate harm each time they bring these weapons to the field. IEDs and missiles are the subject of this chapter; targeted killing and taking prisoners the subject of the next.

GUNS, BOMBS, AND IEDS

In a typical guerrilla, albeit large-scale, operation, Chechen guerrillas defeated superior Russian forces in Grozny in 1994:

On December 11, 1994, the Russian military embarked on what it believed would be a speedy and unproblematic military conquest to wrest Grozny away from Chechen secessionists ... By mid-morning on December 31, the columns of tanks, armored troop carriers (APCs) and self-propelled guns were closing in on the Presidential Palace ... [T]hen the ambush was sprung ... The Chechens had lured the Russian columns deep into the urban maze of Grozny's streets ... They were armed with rifles and with sabers, but they were also equipped with Russian-made rocket-propelled grenades – a deadly weapon against tanks. The Chechens' plan was simple – they would knock out the first and last vehicles in the column. That accomplished, the next step was to destroy the rest of the convoy and then hunt down and kill the fleeing survivors ... As the smoke began to settle, it was clear that the Russians had suffered a devastating defeat. (Shultz and Dew 2006:104)

Like the Chechens, many guerrillas find effective ways to ambush state armies by employing rear-guard or hit-and-run tactics. Camouflage and effective reconnaissance allow guerrillas to lay ambushes, set up snipers, and lay mines (Kilcullen 2009:55–56; Moore 2001:30–31; Perritt 2008:105). But not all guerrillas achieve the success of the Chechens. Perhaps only the ELPF had as much military success against a state army as the Chechens did in their first war with Russia. Hezbollah, too, would probably claim similar but localized successes against Israel in their 2006 war. Most other guerrilla groups, however, can never hope to overpower their adversary in the field. Instead, they undertake small-scale operations to disable convoys, destroy police stations, or demolish checkpoints:

"Most of our attacks were by at most five people," describes one Aceh guerrilla. "We never engaged in frontal warfare. We'd attack, then run. Our aim was to give the following message to the TNI [Indonesian National Armed Forces]: don't sleep, don't eat, don't bathe [we will always be there to attack you]. It was effective. They became afraid of their own shadows ... For a time there were no military posts in the interior of Pase. It was entirely empty." (Aspinall 2009:170–171)

While most guerrilla operations are smaller in scope than the Chechen attack was, they are tactically similar. In each case, a small band of armed guerrillas attacks a military target, either uniformed soldiers or a clearly marked military convoy or installation. If civilians suffer no harm in these attacks, then questions of discrimination and proportionality do not arise. And if prisoners are not murdered or tortured, then violations of combatant's rights are not at issue. As such, these operations are legally and morally problematic only if one finds fault with fighting without uniforms or with fighting in the uniforms of neutral parties or enemy soldiers.

Wearing Uniforms: Yes or No? And Whose?

Fighting without uniforms, as argued in the previous chapter, is a corollary of the right to fight. Fighting in uniform is preferable because it best protects civilians from direct harm by allowing adversaries to distinguish between military and civilian targets. Shedding uniforms is only justified when it is otherwise impossible to fight unjust aggression or occupation. Shedding uniforms opens the door for human shielding and other avenues to conceal the identities and activities of combatants. It therefore reduces the risk to combatants while increasing the risk to noncombatants. Only the impossibility of ever achieving a just victory can vindicate these costs. Uniforms, therefore, are the rule; mufti is the exception.

Even as an exception, the right to shed uniforms says nothing about any right to don the uniforms of neutral or enemy forces as some guerrilla organizations are wont to do. Consider the following examples:

1. "Chechen fighters routinely dressed in Russian uniforms. This simple tactic got them through hostile checkpoints again and again, allowing them to strike repeatedly behind the lines. A variant of this tactic was to ... infiltrate [Russian units] by posing as Red Cross workers" (Arquilla and Karasik 1999:217).

2. "Shooter in Afghan army uniform kills NATO trooper" (Riechmann 2011).

3. "Taliban sleeper agent kills 9 at Afghan military base" (Moore 2011).

4. "According to press reports, twenty-six policemen and two soldiers were driving along the road in Quelicai, Baucau (East Timor) in a Hino truck, when they were stopped by several men wearing Indonesian armed forces uniforms. The truck stopped to pick up the men, when the latter threw a grenade into the truck. An oil drum of gasoline exploded, and in the resulting inferno, thirteen of the people in the truck were burned to death and four were shot as they tried to escape. The dead included sixteen policemen and one soldier" (HRW 1997).

Shedding Uniforms: Right or Wrong?

How do we evaluate the tactics guerrillas use to kill enemy soldiers? One useful tool is the notion of perfidy. International law characterizes perfidy as "an invitation to obtain and then breach the adversary's confidence" (ICRC nd. b). Perfidy is an abuse of the fragile trust and good faith nations need to wage and end wars. The uniform of a neutral medical or aid worker confers protection under international humanitarian law that an enemy soldier could use to cloak the threat he poses. In this sense, it is no different from using a white flag to put an enemy off its guard and then firing upon unsuspecting captors. In all these instances, perfidy brings adverse consequences and violates the rights of noncombatants. Attacking surrendering soldiers (or their potential captors) undermines the institution of surrender necessary to keep battles from degenerating into a fight to the death and violates the rights of soldiers who lay down their arms and are now protected from armed attack. Similarly, masquerading as a medical worker may bring a short-term tactical victory but comes at the price of eroding the conventions that make medical care possible. In an environment of combatants disguised as aid workers, real aid workers will be unable to minister effectively to the civilian population out of fear of attack. This undermines the obligations that armies owe to their people and renders warfare more destructive than it needs to be. Assuming the guise of neutrals also violates the immunity of noncombatant aid workers who now find themselves the objects of mistrust and possible attack. While guerrillas may have the right to put their own people at some degree of risk as they, like states, defend their national interests (a point I pursue when I discuss human shields), they have no right to deliberately endanger neutrals.

Masquerading as an enemy soldier, however, is not perfidious in this sense. The Taliban wearing an Afghan army uniform or the Chechen wearing a Russian uniform is not, like soldiers feigning surrender who then fire on their captors, abusing any special legal protection. Rather, they are exploiting the trust that one accords an ally or compatriot. These "wedge" attacks are devilishly effective. They kill soldiers randomly, significantly undermine morale, and seriously impair cooperation among allies: "Green-on-blue [Afghan on ISAF] attacks drive a wedge between foreign advisers and their Afghan partners. Infiltration breaks down trust. The attacks leave foreign advisers weary and erode the quality of training. Afghan forces wither as commanders wonder who they can trust in their own ranks. The population is left with an impression that their army is a compromised force" (Jensen 2012).

Jensen claims that thirty-one attacks killed forty ISAF troops in the first eight months of 2012. Wedge attacks are successful, cheap, and cost no civilian lives. What, if anything, is the problem? International law, to be sure, frowns upon the practice, and Additional Protocol I (API 1977:Article 39[2]) prohibits "the use of enemy uniforms ... when engaging in attacks or in order to shield, favor, protect or impede military operations." However, the underlying rationale and scope of this prohibition are not clear. While the prohibition forbids using enemy uniforms to stage an attack, it does not exclude donning enemy uniforms to infiltrate enemy positions. Confusing guidelines are the result. The Belgian Law of War Manual states, for example, that "infiltrating enemy lines in order to create panic to the point that the adversary starts firing on its own soldiers ... " is not considered an improper use of an enemy uniform but "opening fire or participating in an attack while wearing an enemy uniform" remains prohibited (ICRC nd. c). In other words, the convention seems to allow combatants to dress in enemy uniforms to disable a troop train but not to attack the fleeing soldiers, or to infiltrate a base but not to attack the soldiers they encounter.

How does all this affect guerrilla operations? The crucial difference between masquerading as a soldier and masquerading as an aid worker turns on the rights of the person who is placed at additional risk for harm. Aid workers are not liable to this risk, so guerrillas who assume their visage violate the rights of aid or medical workers. Soldiers, however, remain liable to harm, so wearing their uniforms does not expose enemy combatants to any unjustifiable risk. For this reason, some states rightly view wearing enemy uniforms as a violation of "good faith" rather than perfidy (ICRC nd. c). In contrast to perfidy, a violation of good faith only brings adverse outcomes, not an intractable violation of the victim's (i.e., the duped party's) rights. Unlike rights, good faith is only worth preserving when it benefits the faithful party. Since, as the Taliban and Chechens believe the benefits of duplicity outweigh its costs, there seems no good reason to adhere to the ban on wearing one's enemy's uniform assuming, as noted, that no third party's rights suffer in the process (as is the case when guerrillas dress as aid workers). This leaves guerrillas to go about their business in uniform or out or masquerading as the enemy. As they do so, IEDs are among their most effective tactics.

IMPROVISED EXPLOSIVE DEVICES (IEDs)

Improvised explosive devices, usually in the form of roadside bombs, are among the most deadly and effective weapons that guerrillas wield. In Gaza and Lebanon, large IEDs disabled sophisticated Israeli tanks and armored personnel carriers (Chadwick 2012; Greenberg and Vaked 2004). Further east, IEDs accounted for 40 to 50 percent of all attacks against Coalition forces in Iraq and roughly 60 percent of all combat-related casualties in Afghanistan (Dietz 2011; Global Security 2003; Wilson 2006). Several factors explain their appeal: IEDs are "inexpensive, simple to construct [and] destructive ... create a sense of insecurity in the populace, retard freedom of movement of coalition and other forces [and] change the political will of coalition partners through coalition casualties" (Dietz 2011:389). Beyond their tactical value, IEDs also generate significant symbolic value by demonstrating insurgents' military prowess, highlighting the failure of the local government to provide security, and sowing discontent and demoralization (Martin 2009). What, if any, ethical proscriptions attend the use of IEDs?

Consider first a critical account of IED deployment by guerrilla forces: "There is an implicit orthodox consensus, or standard model, about the adversary who builds and plants IEDs [that] suggests that these are *ferocious* people ... have the coordination, the resources, and the *lack of scruples* that offer them many strategic advantages, giving them opportunities for real-life experimentation, fine-tuning their tactics and munitions *without much regard for collateral damage*" (Nyce and Dekker 2010:409 emphasis added).

The phrases in italics signify the moral disquiet that accompanies the use of IEDs. When IEDs are delivered by suicide or proxy bombers, hard questions arise if bombers are coerced to risk their lives. When IEDs are rigged to explode a second time as rescuers arrive, then medical workers may suffer direct and intentional harm. When IEDs are left planted at roadside to explode when a convoy passes, they are unjustifiably indiscriminate unless precautions are taken to prevent civilian passersby from triggering an explosion. And, when militants detonate IEDs at busy checkpoints or military sites, they may cause unacceptable collateral harm to the civilians gathered there. None of these defects is without some remedy.

Suicide and Proxy Bombing

"Proxy" bombing marked a short-lived experiment by the IRA in 1990 to force collaborators to drive bomb-laden trucks to army checkpoints and then flee the vehicle moments before detonation. To recruit drivers, the IRA threatened informers with harm to their families. When this proved ineffective, the IRA took to chaining drivers to the seat, creating, in effect, suicide bombers. Unlike modern suicide bombers, however, IRA drivers expressed no interest in martyrdom. And neither did the public. Bowing to public pressure in the face of what one writer called "a public relations disaster," the IRA soon abandoned the practice (Moloney 2003:348). Proxy bombing is marked by significant moral shortcomings that saw guerrillas threatening innocent family members with harm, punishing collaborators without due process, and murdering unwilling drivers. Suicide missions undertaken by militants who willingly give their lives do not carry the same moral baggage. The key, of course, is "willingness." Conventional soldiers are not expected to undertake suicide missions unless they choose to do so. Whether guerrillas observe similar conditions is difficult to say. By many accounts, suicide bombers are not religious zealots, economically deprived, illiterate, or mentally unbalanced. Rather, many are grassroots activists willing to use extreme measures to achieve political and personal ends (Kruglanski et al. 2009; Pape 2005:216; Pedahzur et al. 2003). Under these circumstances there are no obvious objections to training and recruiting suicide bombers to attack *military* targets.

Successive Attacks to Disable Rescuers

Even as they attack military targets, guerrillas sometimes engineer a second explosion to kill and injure rescuers as they arrive. This tactic presents some daunting problems for military ethics. Traditionally, medical personnel are immune from direct harm. This is certainly true of civilian medical teams, but not necessarily true of military medical personnel. It certainly is not true of military rescuers. Take the easy cases first. The immunity of a civilian medical crew is unassailable, not because they are medical personnel but because they are noncombatants. The liability of military rescuers, too, is without question. As combatants, they are always liable to direct harm and knowingly risk their lives if they enter a battleground to rescue wounded comrades. Studying British Army counterinsurgency tactics, the IRA, for example, correctly anticipated the deployment of British troops that would follow an IED attack on a military vehicle. Placing their charges accordingly, their first blast killed six soldiers and the second took the lives of twelve reinforcements arriving in the wake of the initial attack at Warrenpoint, Northern Ireland in 1979 (English 2003:220–221).

The account of the Warrenpoint bombing does not mention military medical personnel whose presence among rescuers poses a difficult problem. Military medics and doctors save warfighters and return them to duty and are, therefore, a military asset. Among state armies, however, military medical personnel enjoy protection through conventions whereby states safeguard the lives of one another's doctors, nurses, and medics. The grounds for protection are pragmatic. Medical personnel are very valuable resources whose singular efforts can save many. But military surgeons enjoy no inherent right to protection conferred by their profession any more than, say, military lawyers. Nor are military medical personnel noncombatants. Rather, they are active military personnel. As such, they are legitimate targets for guerrillas who, given their own lack of medical personnel, have no incentive to agree to protect their adversary's (Gross 2006:175–198; also, Barilan and Zuckerman 2013 for rebuttal). This conclusion is no doubt jarring and clearly at odds with our basic intuition about the immunity due medical workers. But this immunity, unlike noncombatant immunity, is unrelated to moral and material innocence. Rather, the protection of *military* medical personnel is solely prudential and, therefore, only obligates those who find it useful to respect. Lack of any convention to protect military medical workers should not, however, be construed as a permission to attack hospitals or medical facilities. The wounded enjoy the same immunity from direct attack as noncombatants, and any direct attack on hospitals, ambulances, or aid stations flagrantly violates the principle of noncombatant immunity. Ordinary civilian noncombatants enjoy the same protection.

Indiscriminate and Disproportionate Harm to Noncombatants

The principle of noncombatant immunity is of overriding concern when guerrillas set roadside bombs to trigger automatically as vehicles pass. When civilian traffic inadvertently detonates IEDs, the casualties can be catastrophic. The category of harm befalling noncombatants varies with the circumstance. When civilians suffer death and injury in an IED explosion *instead* of soldiers in convoys or armored vehicles, their deaths are unnecessary. They add nothing of military value and might be avoided were guerrillas to take care to detonate IEDs manually when targets are in sight (Kopp 2007). In contrast, *collateral* noncombatant deaths are an unavoidable and necessary by-product of legitimate military attack on a convoy or checkpoint. Collateral harm is permissible unless disproportionate, that is, excessive, with respect to the military advantage guerrillas expect from an armed attack. Consider these attacks on military targets:

- On February 27, 2012, a suicide bomber killed nine and wounded twenty-one in an attack on Jalalabad airfield (a joint Afghan-ISAF airfield). Among the dead were six civilians, one Afghan soldier, and two security guards. Four ISAF troops were wounded (Rubin and Bowley 2012).

- [A] suicide attack in Faryab province [Afghanistan] wounded eighteen people and killed ten, including four U.S. troops, on April 4, when a bomber exploded his suicide vest when foreign troops were about to enter a municipal park.
- A car bomb exploded near a police headquarters in Kandahar City in Afghanistan on February 5 (2012), killing seven and injuring nineteen. Five policemen were killed while six police and thirteen civilians were injured. (Checchia 2012a, 2012b)

Evaluating these attacks requires answers to the following questions: Are these attacks effective? Are the targets military objects? Are civilian casualties proportionate collateral damage? Consider first, the question of targets. In the above examples, the targets include a civilian-military airfield, foreign troops about to enter a park, and a police station. The first target is a dual-use target, whose military function makes it liable to attack (Solis 2010:534–536); the second one is a clear military object. Both are legitimate targets for guerrillas. Police officers present a more complicated issue. Traditionally, police officers are noncombatants immune from attack because they do not fight in the armed forces but provide the law and order necessary to prevent anarchy. In 2008, however, Israel opened its war on Hamas with a devastating attack on a graduating class of police cadets with the claim that they were an auxiliary paramilitary force. Although the UN Human Rights Council vigorously condemned the strike, police officers who unequivocally take up arms in national defense may lose their immunity (Sivakumaran 2011). But are police who fight an insurgency doing nothing more than enforcing the law and so retain their immunity?

States have an obvious interest to affirm the criminal status of insurgents lest they afford guerrillas just cause and undermine the state's right to self-defense. As long as insurgents are criminals, they have no right to fight and remain the responsibility of a police force. So it is not without cause that in Afghanistan, for example, "building a professional and competent Afghan National Police force is key to the war against the Taliban" (Royal United Services Institute [RUSI] 2009:3) or, that in Israel, the Border Police function as a counterterrorism unit. If, on the other hand, insurgents enjoy combatant status and the right to fight then police fighting against guerrillas waging a just armed struggle will find themselves in the crosshairs just as combatants do. The grounds for permissibly targeting police are, then, twofold. First, the police must assume a prominent place in a counterinsurgency force. Second, an insurgency must meet the general demands of just cause and legitimate authority. Otherwise, insurgents are no better than criminals, and the police are not liable to attack.

Weighing the effectiveness and the proportionality of the three attacks cited runs together. Proportionality demands necessary and reasonable (not excessive) civilian casualties relative to the attack's expected military advantage. What were the insurgents after as they struck at an airfield, foreign troops, or the police? The two latter operations killed military and police personnel. The attack on the airfield killed no military personnel, but the airfield was, arguably, a very high-value and visible target. Returning to the list of IED aims, one might reasonably expect an attack on an airport to impede ISAF troop movement. But unlike state armies, guerrilla armies also trade on the symbolic value of their acts by merely demonstrating their military prowess or highlighting the government's inability to provide security. The bar for effectiveness is therefore much lower than the one we impose on state forces and not so difficult to realize in these high-profile strikes.

But if the bar for effectiveness is low, then the bar for permissible civilian harm is high. In other words, symbolic gains cannot entail the same level of collateral harm as material gains that procure weapons or destroy an enemy position. With this very coarse rule of thumb, one can reasonably argue that the collateral harm described was not necessarily excessive if one compares the relatively low number of civilian deaths to the military and symbolic advantages of attacking these military targets. Whether the collateral harm was necessary and unavoidable is another difficult question that cannot be answered definitively. All that insurgents must do is reasonably argue that no alternative scenario would allow them to attack an airport, police station, or parading troops and harm fewer noncombatants. This is the same test that state armies must meet and renders proportionality a very elastic principle that both guerrillas and state armies can usually satisfy as long as there is no disagreement about the *military* nature of the targets. Attacks on nonmilitary targets, however, constitute terrorism and remain prohibited (Chapter 7).

When the use of IEDs is sometimes morally defective and, indeed, reprehensible, the cost of remedying these defects might be feasible and is, indeed, necessary if guerrillas are to wage an armed struggle justly. By shunning pressure sensitive IEDs for remote controlled but riskier detonations, guerrillas can threaten military convoys while reducing noncombatant casualties. By avoiding unnecessary and disproportionate harm, guerrillas can enhance the symbolic value that they hope to gain from successful IED attacks on military targets. With careful resolve and a measure of additional risk to themselves, guerrillas and insurgents can use IEDs effectively and within permissible moral bounds. In some conflicts, however, IEDs are already giving way to missiles. Missiles are deadly, terrifying, less risky to deploy than IEDs are, and capable of inflicting widespread indiscriminate and/or collateral harm. How might guerrillas use these to pursue a just war?

ROCKETS AND MISSILES

Although guerrilla groups regularly deploy mortars, two organizations, most notably Hezbollah and Hamas, have recently turned to short-range and long-range rockets and missiles. In the hands of guerrillas, missiles raise the same questions of effectiveness, discrimination, and proportionality that they do for state armies. However, missiles also raise two special concerns: human shields and terrorism. When Hezbollah positioned missile launchers in front of a UN refugee camp near Qana, Lebanon in 1995, guerrillas made a counterattack exceedingly risky. And, in fact, the Israeli response went awry, killing more than a hundred civilians and leaving Israel no choice but to curtail and soon cease military operations. Such shielding offers a win-win outcome for guerrillas: either civilian shields prevent a counterattack or civilians suffer devastating injuries and hand guerrillas a public relations coup. Second, Hamas used low-grade missiles to terrorize the civilian population of Southern Israel from 2001 through 2008 before turning to more powerful armaments in 2014. Although few civilians died as a result, the attacks inflict significant psychological anxiety, disrupt everyday life, and severely interfere with the operation of municipal institutions. I address the question of human shields in Chapter 6 and that of terrorization in Chapter 7. In this section I focus more narrowly on the hazards of employing missiles during attacks on *military* targets while drawing a parallel with similar questions that arise in the course of interstate warfare.

As Hezbollah and Hamas successfully launched missiles, they found that rocket weaponry can erode their enemy's resolve, afford guerrillas a deterrent capability, disrupt civilian life, and bring their adversary to invest heavily in civil defense (Even 2010:48–49; Lieberman 2012:201). These outcomes are little different and no less legitimate than the goals that states pursue when they deploy missiles. As guerrillas strive and achieve these aims, questions about discrimination and proportionality surface. As with IEDs, discrimination turns on the capability to disable military or dual use installations while steering clear of noncombatants. Two distinct phenomena trouble guerrilla warfare: indiscriminate targeting and indiscriminate weapons. Targeting is indiscriminate when crews fail to aim exclusively at military targets. A weapon is indiscriminate when it is insufficiently accurate or reliable to hit its target consistently without causing disproportionate collateral harm.

Indiscriminate Targeting

To evaluate indiscriminate targeting, outcomes, not intentions, are determinative. In 2006, for example, Hezbollah leader Hassan Nasrallah declared that Hezbollah missiles only targeted military installations so that any subsequent harm to noncombatants was unintentional – that is, collateral – harm. States make this claim all the time, but it requires some proof, typically in the form of some nearby military installation that was the "true" object of attack. Here, Hezbollah sometimes failed. In its 2007 report, Human Rights Watch (HRW 2007:9; also Erlich 2006), describes how "Hezbollah repeatedly fired rockets in the direction of civilian populated areas in which there was no evident military target." Evidence of this sort is sufficient to establish indiscriminate targeting, and HRW's condemnation of such instances should not be controversial. Nor, as HRW (2007:95) emphasizes, is there any room to allow Hezbollah to argue that, as a weaker party, it may target civilians to compel its stronger adversary to discontinue its attacks on Lebanese civilians. Attacks of these sorts violate noncombatant immunity in the most egregious way and repudiate any right to a fighting chance.

HRW, however, overlooks Hezbollah's initial argument. Careful to allude to the presence of military targets, Hezbollah warned Israel of the consequences of *collateral* harm: death, destruction, and displacement. The argument deserves more attention because it is central to conventional deterrence theory. No less than states, non-states hope to deter an enemy by threatening their adversary with devastating attacks and significant collateral harm should it undertake or continue aggression. Is it legitimate to use missiles as a deterrent, and, if so, does discrimination matter?

In conventional war, deterrence is a function of a credible threat to harm civilians. Nuclear deterrence was the subject of much debate and raised the pressing question early on: If it is impermissible to harm civilians directly during war, is it permissible to threaten them with harm? One response relies on collateral harm. Nuclear deterrence does not work by targeting civilians directly. Instead, deterrence gains traction because civilians will die collaterally following a nuclear "second strike" against *military* targets. Another answer pins its hope on outcomes. While deterrence might immorally threaten noncombatants, successful deterrence prevents a greater calamity, enforces the ban on nuclear weapons, and preserves the peace. This forces a tragic choice, to be sure, but a choice superior to nuclear war (Blake and Pole 1984; Coady 1988, 1989; Ethics 1985; McMahan 1985).

These arguments are open to various criticisms. First, it is hard to imagine that nuclear war or limited nuclear war could ever avoid disproportionate casualties. Permissible collateral harm in the context of nuclear war, therefore, is a fiction. Second, and no matter how limited nuclear war may be, one is always faced with the utter immorality of a policy that not only contemplates but also lays plans to obliterate civilian population centers. Although "doing evil to avoid evil" is only a temporary expedient, and one that Walzer (1977:283) uneasily assumes under the rubric of "supreme emergency," it is evil nonetheless.

In the context of conventional war, arguments criticizing deterrence carry less weight. First, collateral harm in conventional war is not a useful fiction, but an integral part of just war. Combatants can and should observe proportionality. As a result, conventional deterrence does not threaten evil to avoid evil. Deterrence works by threatening to destroy military capabilities and inflict insufferable collateral harm. If collateral harm is permissible, then the threat to inflict collateral harm is likewise permissible. These casualties can be devastating, but there is no reason to assume they must be disproportionate to effect credible deterrence.

Second, conventional deterrence, unlike nuclear deterrence, may fail without catastrophic consequences. As credibility weakens, conventional opponents, unlike nuclear adversaries, may realistically call another's bluff and raid, skirmish, or wage war. When deterrence fails, the deterring party may resort to force to restore its deterrent capability. Three important implications for guerrillas follow. First, they must reasonably believe that deterrence will succeed. Second, the response to failed deterrence is not to target civilians indiscriminately, as some guerrillas might suggest, but to wage a self-defensive war against an undeterred aggressor with the ensuing collateral harm this entails. Third, discriminate targeting matters, but discriminate weapons may not. Discriminate targeting matters because only military targets are the proper object of attack. A credible deterrent capability cannot operate otherwise. Discriminate weaponry, on the other hand, is a function of technology. If relatively indiscriminate weapons, such as unguided missiles, can cause significant collateral harm when attacking military targets, they may enhance deterrence.

Indiscriminate Weapons

The principle of distinction requires belligerents to distinguish between combatants and noncombatants as they wage war. Threatening collateral harm to establish a credible deterrent capability is often permissible but will come very close to violating the principle of distinction if belligerents *adopt* collateral harm as a persistent feature of war to demoralize the civilian population. Charges of indiscriminate targeting also fly when belligerents fail to take care that their bombs or missiles land on military targets. These charges are sometimes ill founded because the problem is often technological, one of indiscriminate weapons rather than indiscriminate targeting. Indiscriminate targeting suggests little regard for noncombatant immunity, while indiscriminate weapons are mechanically incapable of fine distinction. Inaccurate and unsophisticated weapons fill the arsenals of some guerrilla organizations.

Compared to Israeli ordnance, for example, Hezbollah's missiles were extremely inaccurate in 2006. They contained no guidance systems and, therefore, require a large target. Of the most common, the 122 mm Grad missiles have a range of 20 km, carry a payload of 7kg, and are accurate within a rectangle of 336 meters by 160 meters (HRW

2007:34). The largest of the missiles, the Fajr 5, has a range of 60–70 km and a 90 kg payload, but it is only accurate within a radius of 1–3 km (Global Security nd. b; Army Recognition nd.). Commenting on future developments, the Jerusalem Post reported: "By 2017, the IDF believes that Hamas and Hezbollah will have a pool of about 1,600 missiles with a level of accuracy of a few hundred meters and 800 with a level of accuracy of just a few dozen meters, giving them the ability to hit what they want" (Katz 2012).

Interpreting this to mean that Hezbollah – arguably the most advanced guerrilla army today – cannot now hit what it wants (nor could hit what it wanted in 2006), we are left with a tough question: Can a guerrilla army use a powerful weapon knowing it cannot always hit what it wants? Is it ever permissible to use an indiscriminate weapon?

Some assume not. Observing Hezbollah tactics in the 2006 war, HRW (2007) writes: "[E]ven assuming Hezbollah had been intending to hit the military target instead of civilians, the unguided rockets it used were incapable of distinguishing between the two" (p. 4). "Because the weapons it used are insufficiently accurate in populated settings, these operations would nonetheless violate the humanitarian law prohibition against indiscriminate attacks" (p. 9).

HRW is unconcerned with Hezbollah's intent, focusing instead on the indiscriminate nature of its weapons. Other commentators suggest that Hezbollah exploited its technological disadvantage to turn collateral harm into a weapon: "Thus long-range artillery rockets cannot be used to pinpoint military targets, which are usually small and focused," wrote Reuven Erlich (2006:141). "They can be used only if the target covers a large area, such as a city or town, effectively and indiscriminately killing, terrorizing and demoralizing the local residents."

These descriptions raise two questions. First, is it permissible to use collateral harm as a weapon? A belligerent may cause collateral harm in the course of armed conflict and, as I argue earlier, threaten military actions that cause collateral harm to build a credible deterrent. However, the oft-used tactic of masking attacks on civilians under the guise of collateral harm is not permissible whether a weapon is discriminating or not. This occurs when the harm befalling noncombatants is no longer collateral but direct. The best evidence for this is the lack of any appreciable destruction of military targets coincident with the attacks. Collateral harm is the by-product of an attack on a military target. Proportionality, the yardstick of permissible collateral harm, is measured against a military advantage. The preponderance of harm in any armed attack must be military. Rockets, therefore, that repeatedly hit a civilian objects without causing any significant damage to nearby military installations no longer convey collateral harm. Repeating such attacks only reveals that the true targets are noncombatants and the true motive is terrorization or demoralization. Such attacks are impermissible whether perpetrated by guerrillas or state armies. Unlike state armies, however, guerrillas may have only inaccurate weapons at their disposal. This raises the second question: Under what conditions, if any, may they use them?

Deploying Indiscriminate Weapons: Discounts for Guerrillas?

Some missile strategies are more problematic than others. I return to Erlich's charge in Chapter 7 when I consider whether Hamas, for example, specifically chose inaccurate weapons to terrorize Israel's border communities from 2001 through 2008. I believe they did, but the rockets they choose were not only inaccurate but also makeshift and weak. Their purpose was not to launch mass casualty terror attacks or destroy military targets but to terrorize the local population, thumb its nose at Israel, and sow a modicum of fear and panic that would *not* provoke a ferocious counterattack.

In contrast, both Hezbollah and Hamas turned to longer-range missiles during sustained combat operations in 2006, 2008–2009, and 2014 knowing their technological specifications but deprived of alternative weaponry. Does the principle of discrimination and proportionality compel belligerents to adjust their targeting strategy to the sophistication of their weapons? This is not a simple question to answer directly. Today, we applaud nations like the United States who turn to smart weaponry to disable insurgents successfully without causing extensive collateral harm to the many noncombatants living next door. The same nation would face considerable condemnation for targeting insurgents with a weapon that could not avoid harming large numbers of bystanders. Nevertheless, one could easily imagine a scenario where a state army might only disable a very valuable target at high cost in innocent lives. Such an attack will survive censure only if the benefits of disabling the target far outweigh harm to noncombatants. This, of course, is the principle of proportionality at work.

Thus the rules for using missiles look like this:

1. Acceptable targets remain military targets. Neither state armies nor insurgents can target nonmilitary targets directly (the principle of discriminating attacks).
2. Because collateral harm is often a function of a weapon's accuracy, belligerents should use the most accurate weapon available to minimize collateral harm (the principle of discriminating weaponry).
3. Collateral harm cannot be excessive relative to a target's expected military value (the principle of proportionality).

These are the same rules that just war theory and LOAC trumpet all along. There is no reason to think that they apply any differently when insurgents go to war. Insurgents must also ask whether their inferior weapons cause collateral harm and, when this harm is excessive, must refrain from attack. However, one might also ask whether inferior weapons lower the bar of permissible proportionate harm. Consider the following case:

At their maximum range of 70 km, at least 50 percent of all Hezbollah missiles will fall within 3 km from their intended target, that is, within an area measuring 28 sq. km. The other 50 percent will land elsewhere. Afula, an Israeli town of 40,000 adjoining biblical Mt. Tabor is 28 sq. km and was the target of a Hezbollah missile attack in 2006. A nearby military base is less than 3 km from the city center.

What may guerrillas do in this situation? They have no obligation to refrain from attacking military targets. They do have an obligation to use the least harmful means available and to avoid disproportionate harm. Two outcomes are possible:

1. The least harmful means available do not cause disproportionate harm because the expected value of disabling a military target exceeds expected noncombatant casualties.
2. The least harmful means available cause disproportionate harm but no other means are available to destroy a military target.

These are familiar scenarios, and one would certainly grant the same latitude to guerrilla armies as to state armies. For both, the expected advantage has to be high and the casualties low. However, there are several intervening factors that may reasonably affect the calculation of proportionality. One is human shields. Some have argued, and not without good cause, that the deliberate use of human shields should permit greater collateral harm than would otherwise be allowed. Otherwise, the principle of proportionality allows the defender an unfair advantage. I address this argument in Chapter 6. Here, I want to consider whether technological asymmetry may similarly affect the principle of proportionality.

Consider two military targets of equal value, each 2 km from an enemy civilian population center. State A has a precision guided weapon that will destroy the target with no collateral harm. Non-state B has a less sophisticated weapon. It will destroy the target but cause what A would regard as disproportionate harm. May B act? Certainly, B may argue that its attack is not objectively disproportionate, that the expected casualties clearly do not outweigh the military advantage that B hopes to gain. Most belligerents take this route and simply exploit the elasticity of the proportionality principle. But let us assume that B's acts would cause disproportionate harm. Can B, nevertheless, receive a proportionality "discount" because it is technologically backward?

The answer is "yes, but not always." Moreover, B does not receive a discount; instead, one recognizes that B's hoped for military advantage is greater than A's in what might seem, at first, to be identical situations. Thus, one can imagine two circumstances. The first is a planned attack on a military target. Given B's inaccurate weaponry the collateral harm will exceed its expected advantage rendering the attack unlawful. B must then go look for a different target. In the second case, B finds that the principle of proportionality prevents it from fighting at all. There are no military targets it can hit without causing disproportionate harm (just as there are no shielded targets that A can destroy without causing disproportionate harm). Under these circumstances, the military value of these targets grows because disabling them becomes the only means B has to win its war. In this situation, rare for states, but probably more common for guerrillas, the threshold of collateral harm increases so some attacks now become possible. At this point, necessity replaces proportionality and in its extreme becomes "supreme necessity" when the threat is so dire (e.g., genocide) that there are no numbers of enemy civilian casualties that are disproportionate.

Of course, the deciding factor turns on the availability of other means of attack and/or other targets. As this chapter and the others demonstrate, guerrillas have a myriad of hard and soft tactics available. To date, only two guerrilla organizations have deployed rockets and missiles.

Some casualties were the result of indiscriminate targeting but others do not necessarily violate the principles of discrimination and proportionality. HRW (1997:49–94), for example, documents thirty-nine of the forty-three noncombatant Israeli deaths in the 2006 war. Among these, approximately fourteen people lost their lives during what appear to be military operations. Assuming that Hezbollah was aiming at high-value military targets, information that neither Israel nor Hezbollah released, these casualties are not obviously disproportionate. As such, proportionate *military* strikes were probably available to Hezbollah gunners. But because Hezbollah also launched missiles into civilian population centers void of any military significance, causing about twenty-five of the thirty-nine deaths HRW documents, guerrillas also stand justifiably accused of terrorism. I consider this more closely in Chapter 7.

Here it is important to point out that while some attacks may be terror attacks, others are legitimate acts of war. Whether initiated by states or insurgents, acts of war must be judged according to prevailing standards of effectiveness, discrimination and proportionality. When guerrilla groups, like Hezbollah, can launch missiles to deter future attacks or erode present capabilities, then there is room to ask about proportionality. As just noted, few civilians lost their lives in attacks on military targets. In these instances, belligerents, whether state or non-state, need only establish the reasonable probability of disabling a military target given the limitations of their weaponry to comply with discrimination, and offer sober estimates of military advantage and relatively light civilian casualties to establish proportionality. Under these conditions, relatively unsophisticated missiles join the arsenal of guerrilla groups waging an armed struggle for self-determination. Such groups do not obviously include Hezbollah, an armed group fighting on behalf of Syria and Iran, but the lessons gleaned from analyzing their experience offer criteria to guide insurgents who meet the conditions of just cause and legitimate authority.

CONVENTIONAL WARFARE AND A FIGHTING CHANCE

Uniforms, fixed piece or rearguard attacks, IEDs, mines, and missiles are the stuff of conventional warfare. To varying degrees and with few exceptions, guerrillas confront their state adversaries with these weapons. Material asymmetry, however, compels guerrillas to exploit whatever strengths they can muster. As they do so, they get some help from international law; permission to wage war without uniforms is a huge concession. More broadly, guerrilla warfare does not merit sweeping condemnation when insurgents plant roadside IEDs or deploy rockets. Each tactic may be used effectively or abused flagrantly. The rules that apply are the same that govern state armies: effectiveness, discrimination, and proportionality. Circumstances, however, vary. State armies measure effectiveness in terms of hard military advantage, while guerrillas are often content with symbolic gains. Effective discrimination requires a careful assessment of the rights of ambiguous actors such as medical personnel or law enforcement officials whose protections are often contingent rather than absolute. Finally, proportionality, an elastic notion to begin with, can make room for the technological disparity between the sides and sometimes permit guerrillas to employ unsophisticated weaponry without facing charges of indiscriminate targeting or disproportionate harm.

In all instances however, the conditions regulating the right to a fighting chance remain of paramount importance. Namely, that guerrilla warfare does not violate the fundamental rights and protections afforded combatants and noncombatants. This caveat creates the tension, common to military ethics and law, between what guerrillas need and those whose rights remain sacrosanct. Within this narrow alleyway, there is space to consider the legitimate use of IEDs, rockets and missiles. Similar constraints govern targeted killing and taking prisoners.

5

Small-Scale Conventional Guerrilla Warfare

Targeted Killing and Taking Prisoners

Setting aside bombs and missiles, guerrillas may eschew military confrontation in favor of targeted killing and taking prisoners. Sometimes
derided as assassination, targeted killing is a common tactic that
often allows insurgents to aim at isolated enemy soldiers and officers,
informants, collaborators, and high-ranking enemy civilians without
exposing themselves to great risk. The idea has also caught on among
state armies who avidly pursue militants in Gaza, Iraq, Pakistan, and
Afghanistan. Both targeted killing and assassination suffer from the
same deficiencies. First, victims may include military and civilian targets
thereby leading to charges of indiscriminate harm. Second, targeted
killings cause collateral harm that may be disproportionate. Third,
these tactics smack of extrajudicial or summary execution, which not
only eliminates enemy assets but conveniently purges informers, collaborators, and rivals.

   While assassination kills, taking and housing prisoners usefully dis-
ables enemy combatants and provides coin to exchange for one's own.
The second half of this chapter considers the ethical challenges that arise
when guerrillas apprehend and detain prisoners. Guerrilla attempts to
take prisoners of war are often met with disdain. Viewing guerrillas and
insurgents as little more than criminals or unlawful belligerents, some
commentators disparage guerrilla tactics as nothing more than kidnapping or hostage taking. But the exigencies of war and the injunction to
minimize harm may sometimes demand prisoner taking. Still, prisoners of war present special problems for non-state military organizations.
Many guerrilla armies have no facilities to house prisoners or to meet
the barest conditions of reasonable treatment. As a result, they usually

take few prisoners, mistreat others, and demand lopsided exchanges, all to the frustration of state armies. But the treatment of prisoners varies enormously among contemporary guerrilla armies, just as it does among the states they fight.

ASSASSINATION AND TARGETED KILLING

Given the prominence of targeted killing today, it is easy to forget that the laws of war traditionally took a dim view of targeted killing. Consider Francis Lieber's 1863 code from the American U.S. Civil War: "The law of war does not allow proclaiming either an individual belonging to the hostile army, or a citizen, or a subject of the hostile government, an outlaw, who may be slain without trial by any captor, any more than the modern law of peace allows such intentional outlawry; on the contrary, it abhors such outrage" (Lieber 1863:Article 148).

Despite Lieber's indignation, the prohibition was always odd. Killing combatants, after all, is the way of war. How is assassination any different? One answer turns on perfidy, and, indeed, it might be simplest to define assassination as perfidious or treacherous killing. Inviting an enemy to peace talks and then striking him down after offering safe passage is one example. Another problem troubling Lieber is the attribution of criminal liability. Combatants are liable to lethal harm because they and the government they represent pose a significant material threat to their enemy. Traditionally, however, soldiers are innocent of any crime unless committing specific crimes of war (murder, rape, pillage, etc.). As such, the law of war forbids criminal trials of prisoners, torture and executions. To identify a target by name, search him out, and kill him is redolent of execution, and Lieber feared that assassination would undermine the rule of law during war. Assassination, on this view, criminalizes combatant participation in war and deprives soldiers of protection should they fall captive. There are also reasonable fears of retaliation in kind. Targets of assassination usually include high-ranking political and military leaders that neither side is anxious to deprive the other of lest (a) the one side expend similar energies liquidating the rival's leader and/or (b) the one side decapitates the other's leadership making it difficult to end the fighting in an orderly manner.

In modern "targeted killings" however, killing by name does the work that uniforms once did (Gross 2010a:100–121). Targeted killings are not extrajudicial executions but strikes against combatants whose identity requires confirmation by assiduous intelligence gathering rather than by uniforms. Because state armies allow guerrillas to fight without uniforms, they must develop some other way to ascertain combatant status. This they accomplish by gathering the information necessary to name members of a guerrilla military organization. Reasonably sure of a person's identity and combatant status, state army operatives may then disable a guerrilla by the same means permitted to disable a uniformed combatant. In contrast to opportunistic strikes by guerrillas on convoys or military installations, assassination turns on killing a particular person whose liability is often predetermined. Targets for assassination include military figures, political leaders, collaborators, informants, and rivals:

1. Chechen insurgents regularly detonated bombs to kill senior Russian military figures. One of their most spectacular operations critically wounded Lieutenant General Anatoly Romanov, deputy commander of the Russian forces in Chechnya, and killed two of his aides in a 1995 attack (Shultz and Dew 2006:128, also 113). No less dramatically, the Eritrean Liberation Front "assassinated the highest Ethiopian commander in Eritrea, General Teshome Ergetu in a daring broad daylight ambush" in November 1970 (Weldemichael 2013:14, n52).

2. Attacking civilian political leaders, Palestinian guerrillas assassinated Israeli minister Rehavam Ze'evi in 2001, while the IRA assassinated the British ambassador to The Hague in 1979, Lord Mountbatten in 1980, and attempted to kill Prime Minister Thatcher in 1984. The IRA also set its sights on civilians who worked for the local government including prison and police officers, judges, census takers, and construction workers arguing, in the latter case, that anyone "involved in building or maintenance work for the security forces would be considered a legitimate target (English 2003:249)." In Indonesia, "GAM [Free Aceh Movement] made it clear that politicians supporting Jakarta are siding with the enemy and are therefore considered legitimate targets" (Schulze 2003:260). The Taliban avidly targeted the local police before realizing "that killing national army and police officers is not well received by the locals" (Thruelsen 2010:269).

3. Many guerrilla groups routinely executed collaborators and informers. Hamas, writes Gunning, "was considered to be more rigorous than other factions in ensuring that only those who were actual security risks were executed" while eschewing the "moral deviants" that other Palestinian factions pursued (Gunning 2009:188–189). One urgent goal of the Kosovo Liberation Army (KLA) was to: "eliminate key members of the Serb police, military, and security apparatus, including ethnic Albanian collaborators and spies" (Perritt 2008:56, also 73–74). Among the Aceh guerrillas, "more than 100 executions of suspected informers or cuak (collaborators) took place in Aceh between late 1998 and mid-1999" (Schulze 2003:257; also Aspinall and Crouch 2003:6). Between 2009 and 2011, the Taliban killed more than a thousand Afghanis suspected of supporting ISAF forces or the Afghan government (UNAMA 2012:11, 18–19). This is hardly a new phenomenon. In October 1942, "the Jewish Combat Organization carried out its first death sentence, assassinating a Jew serving as a policeman in the [Warsaw] ghetto. They had sent a message: there was a price for collaboration" (Shore 2013).

4. Guerrillas also conduct systematic purges.

> Having incarcerated those suspected of involvement in alleged subversion, the nascent EPLF [Eritrean People's Liberation Front] leadership held lengthy deliberations and ... determined that Menka'e [a left-wing splinter group] was a "destructive movement" [and] ... in August 1975, the EPLF executed the Menka'e ringleaders. This "disciplinary conclusion" set a precedent for future purges and sent the clearest of messages to any individuals or groups who would deviate from the prescribed code and pose a threat to the EPLF (Weldemichael 2013:142).

Clearly, there is overlap between categories. National and local police serve military functions and become military targets when they battle insurgents. As is the U.S. president, high-ranking political leaders are often commanders of the armed forces. Civilians working for the occupying forces are not necessarily informers but do collaborate on a professional level. Despite the membership overlap, each category is morally distinct. Assassinating combatants is a military tactic akin to targeted killing, killing civilians verges on terrorism, while executing informants, collaborators, and rivals suggests corrupt judicial proceedings.

Which, if any, are justified in the hands of a guerrilla organization struggling for national self-determination or against unjust occupation and aggression?

Assassinating Military and Political Leaders

Assessing the permissibility of assassination and targeted killing depends on some assessment of their effectiveness and the rights that each victim enjoys. Consider first military targets. Chechen and IRA attacks on military leaders do not demand a very high bar of effectiveness. These strikes constituted part of a campaign to weaken Russian and British resolve respectively and only require discrimination to ensure that the target is properly identified and proportionality to protect noncombatants from excessive harm. Liability to harm is assured by military affiliation, a fact made easy by the uniforms that state soldiers wear.

Among civilians, liability must be tuned more finely. In general, states refrain from assassinating rival state leaders, but opt instead for arrest and trial. This was the case with Saddam Hussein. As things developed, Saddam was the exception to the more general rule permitting assassination when arrest is not feasible. "Not feasible" may mean impossible but more commonly refers to excessive risk of harm to bystanders or to the arresting soldiers. The inability "to arrest first" is a common theme of permissible targeted killings. The Israel Supreme court elucidated this most clearly by demanding that Israeli troops first attempt to arrest Palestinian guerrillas attacking or planning to attack Israeli targets (Israel High Court of Justice [HCJ] 2005). The United States has adopted similar guidelines in the course of high-profile targeted killing whether bin Laden or Anwar al-Awlaki, the American born Al Qaeda agent killed by U.S. forces in Yemen. In each case, arrest was deemed unfeasible leaving assassination the only way to prevent an imminent attack (Amnesty International 2011b; Holder 2012).

In the hands of state agencies, targeted killing often bypasses military personalities to assassinate insurgent leaders with the claim that high-ranking political leaders are nothing but military commanders. Such reasoning suffused Israel's justification for killing Abbas al-Musawi, the former Secretary General of Hezbollah, in 1992 and Ahmed Yassin and Abdel Aziz al-Rantissi in 2004, both political leaders of Hamas. More flagrantly, the Russians had few qualms about assassinating Chechen president Dzhokhar Dudayev in 1996 in a failed attempt to put an end to the First Chechen War. Although the political leaders of states usually retain their civilian status and noncombatant immunity, Insurgent leaders enjoy no such protection. Citing the lethal and overwhelming threat that some guerrilla leaders pose, states will argue that such leaders forfeit their immunity thereby allowing counterinsurgents to arrest, try, and punish their captives or, if need be, shoot them on sight.

How do the arguments that states use to justify assassination and targeted killings play out among guerrillas? Whom may they target? Because the soldiers they face are usually uniformed, naming is not required to establish liability. The leader of a repressive and rapacious regime, like Saddam Hussein, may certainly be a legitimate target for guerrillas but is often beyond their reach. The justification for tyrannicide is the same whether proclaimed by states or insurgents: killing tyrants is wrong but excused by the greater evil that will surely come from letting them live. Guerrillas may also echo government claims of civilian combatancy. When ranking government leaders assume military roles, they lose their immunity from harm.

Each of these arguments is precarious whether in the hands of guerrillas or nation states. There are strong grounds for tyrannicide assuming one can identify a tyrant easily. When severe human rights abuses provide just cause for war, then, the leader of the state guerrillas fight is, by definition, tyrannical. Lacking the means to arrest and try the tyrant, guerrillas seem justified killing him or her just as states might have been justified killing Saddam Hussein or Hitler. Identifying such cases is not simple. Grounds for tyrannicide are considerably stronger for the Kosovo Liberation Army who faced Serbian president Milošević, a ruthless adversary who would eventually stand trial for war crimes, than the for the IRA, who confronted Margaret Thatcher, the elected leader of a liberal democracy. More importantly, however, is the validity of the lesser evil argument. Legitimate tyrannicide does not draw solely from the wickedness of the tyrant, but also depends on the outcome of his or her assassination. In the long run, there must be a reasonable prospect that things will change for the better. This is not assured following the death of the leader of a stable and democratic nation and, in fact, not assured when any government is well organized for a smooth transition following the demise of its leader. There is no reason to assume, for example, that the death of Milošević would have brought leaders any more sympathetic to

Kosovo's cause. On the contrary, assassinations often fuel moral outrage and cement prevailing enmities.

Finally, consider the simple alternative of ignoring the immunity of high-ranking political leaders, whether tyrant or not, because they fulfill a military function. Morally, this is attractive because high-ranking leaders – whether military or civilian personnel, presidents, prime ministers, or insurgents – often direct military policy and the generals who implement it. As such, civilian leaders are liable to lethal, disabling harm in view of the overwhelming threat they pose. On the other hand, state and guerrilla political leaders serve two functions: warriors and diplomats. Which function defines their moral status? True, they pose a threat, but they also hold the key to peace. Keeping leaders alive not only avoids the pitfalls of decapitation that may leave a state or guerrilla organization unable to make war or peace, but reaffirms their overriding moral status as peacemakers. For this they deserve immunity. Those who abjure their duty to make a just peace are no different from tyrants. Morally, they are liable to harm. However, given the adverse effects of high-level assassination, the rapid descent into tit for tat killing that may easily follow, and the inability to assign the label of tyranny with any certitude, the prosecution of tyrants might be better left to the world community following the end of hostilities. The international community is certainly showing growing interest and aptitude for trying war criminals. These remarks, however, do not belie the rare case where tyrannicide might bring a quick cessation to a war and the recognition of just claims.

Informers, Collaborators, and Dissidents

As it stands, most victims of assassination are neither tyrants nor political figures but, instead, civilians who either inform or collaborate. Additionally, guerrillas systematically suppress dissent. Informers are disaffected insurgents or local civilians who supply government forces with sensitive information about guerrilla operations. Observers agree that informers can play a definitive and devastating role. "Nothing is more demoralizing to insurgents than realizing that people inside their movement or trusted supporters among the public are deserting or providing information to government authorities," declares the latest U.S. Counterinsurgency manual (Department of the Army 2006:1.19). "Informers," writes one intelligence scholar "provide almost the only means of penetrating non-state terrorism" (Herman 1996:65 cited in Amble 2012:349). Guerrilla organizations admit this readily. "Except for informers," claim the IRA and its supporters, "the war would have been over long ago" (Sarma 2007:1083).

Informers pose a double threat. Beyond the obvious threat to military operations, informers also undermine the fabric of a society. The Palestinian informer, notes Abdel-Jawad (2001:18), "serves the purpose of creating mistrust, spreading confusion and undermining collective self-confidence within Palestinian society." Spies, informers, and compromised friends and family subvert strongly held beliefs about integrity, trust, honor, and loyalty that hold together traditional societies. And, informers create a cycle of violence within the local community:

Every time a wanted individual was captured, wounded, or killed, the public immediately suspected the work of an informer. It was the beginning of a vicious cycle in which the wanted individuals were hunted by the security forces, while the suspected collaborators were hunted by the wanted, who held them responsible for the death or capture of their comrades. (Be'er and Abdel-Jawad 1994:164; also Rigby 1997)

Unlike informers, *collaborators* do not trade in sensitive information nor do they rend a traditional society as powerfully as spies do. Instead, collaborators are usually civilians who provide services that support the local or occupying government. Collaborators include teachers, judges, police, municipal workers, and private contractors. In this regard, collaboration is a significantly different moral phenomenon than informing. Informing undermines security and the military capability of a guerrilla organization to exercise its right to fight while collaboration only poses a more general threat. Collaborators affect guerrilla struggles by enabling an occupying army or central government to function in territory claimed by guerrillas. Because insurgents disparage local government institutions, many guerrilla organizations will establish parallel institutions to provide welfare, health care, and education (see Chapter 3). Compliance is the key to guerrillas' success, so any compatriots who continue to work in or patronize government facilities may impair non-state institutions of governance. But unlike informers and spies, collaborators do not compromise security.

In contrast to informers and collaborators, *dissidents* pose an internal threat to political leadership and prevailing doctrine. Even among those groups that eschew terrorism, treat prisoners reasonably well, and edge toward openness, few tolerate dissent and they often kill their rivals with alacrity. While collaborators might be bought off or otherwise induced to toe the line, dissenters cannot be easily co-opted. The fear of dissent comes from schism that undermines an insurgent group's authority and erodes its ability to wage war. As rival groups vie for support, schism splits the local population base and the well of international supporters. In the worst case, competing groups battle one another much to the satisfaction of their enemies. Dealing with informers, collaborators, and dissidents is a constant challenge. Just as rights respecting nations may prosecute treachery or treason, guerrillas may also do so assuming that the offences are defined clearly and guerrillas maintain some semblance of due process.

Prosecuting Informers, Collaborators, and Dissidents: Defining the Crime

What loyalties do a people owe insurgents and what crimes do they commit when disloyal? By definition, treason is a crime against the state by those who owe allegiance but wage war against it (United States Code 2006:Title 18 § 2381). Treachery is a more expansive concept, couched more generally in terms of betrayal and disloyalty and may embrace informing (espionage), collaboration, and dissent with little discrimination. As George Fletcher (2003) notes, treason and related ideas of betrayal immediately raise the question of whom is the object of allegiance and who is the agent of betrayal. Among states, loyalty is owed by their citizens; only citizens can betray a state. Underlying the idea of state treason, argues Fletcher, is the idea of a nation or people whom its members might betray. Among insurgents, a guerrilla movement takes the place of the state while the people encompass the aggrieved group fighting for national self-determination. This leads to several complications.

In the hands of non-state actors, crimes of treason and treachery first assume that a people has the moral right to demand loyalty from its members and, second, that guerrillas represent the people so that betraying a guerrilla organization constitutes treachery. Let me first dispense with the second claim. Representation, whether rational or patrimonial, is a condition of legitimate authority and just guerrilla warfare. If a guerrilla organization is waging a just struggle then betraying the movement is to betray the people. But do a people enjoy any right to demand loyalty from its members? States demand allegiance from their citizens as a matter of law. Underlying the right of the state to exact loyalty is a utilitarian claim: the state best protects the interests of its members and without their allegiance the state will fail. Such a failure is doubly harmful because it endangers the private interests of individuals as well as the collective or intrinsic value a people represent. Insurgents waging a just war can make the same claim so that any attempt to undermine a legitimately authorized guerrilla movement will endanger a campaign that serves the individual and collective interests of their people. A just armed struggle requires a critical mass that disloyalty endangers. However, it does not seem that all kinds of disloyalty endanger the struggle or the people equally. Espionage and treason pose significant threats while collaboration and dissent present far less danger.

In contrast to espionage, collaboration poses neither a military nor a security threat. Collaborators often provide a service that supports the apparatus of occupation and government control. Sometimes the support is direct; this was the IRA's argument against construction workers who helped build facilities for the British military. Often though, the support is indirect and includes teachers, judges, police, or telecommunication workers. Government institutions such as schools, judicial systems, or health care facilities allow a regime to entrench itself, impose its judicial and cultural norms, and collect taxes for its further operation. Those working for the occupying forces prevent guerrillas from establishing the parallel institutions they think necessary to build their nascent state. To do so successfully, guerrillas have to both enlist trained personnel and wean their compatriots from their government jobs. It is unlikely, however, that the critical mass necessary to operate military and political operations requires everyone to participate. Americans are fond of recalling that only one third of the colonists supported war with Britain. Under such circumstances, insurgents need to persuade a sufficient number of collaborators to support the movement but usually not all of them. Collaboration, moreover, is not necessarily driven by disloyalty but by apathy and economic distress. Kurds who continued to work as village guards or school teachers and Palestinians who build houses for Israeli settlers may remain unconvinced of the insurgents' chances of success and/or be insufficiently interested in the movement's goals. Inducing compatriots to leave their posts as judges, teachers, construction workers, or secretaries may require such material incentives as financial support or such disincentives as social sanctions or fines. But unlike espionage and treason, there are no grounds to regard collaboration as a criminal offense.

The same is true of dissent. Legitimate guerrilla movements have good cause to *encourage*, rather than suppress, dissent without suffering the risk of schism. Dealing with dissidents depends upon the risk they pose. When rival groups go to war, sympathizers become enemy combatants and may easily find themselves targeted for killing. At other times, the risk that dissidents pose is personal, aimed at a leader and not the movement. Quashing dissent may then become an exercise in autocracy, intolerance, and abusive authority. And while some guerrilla leaders, like the PKK's Abdullah Öcalan, show little tolerance for dissent (Bozarslan 2001), others understand its value. In East Timor, for example, guerrilla commanders who challenged doctrinal decisions were first treated with exceptional brutality. According to one account, one recalcitrant commander was "buried up to his waist in a standing position, without clothes and with his hands tied. Then they burned a car tire, allowing the melting rubber to burn his body." The commanders of the Southern Frontier and East Central sectors similarly disagreed with their respective political commissars on the conduct of the war only to meet their deaths soon afterward." Nevertheless, guerrillas soon understood that bloody purges only weakened their hand significantly leading to them to reorganize, centralize their leadership, and search for more constructive ways to address internal differences (Weldemichael 2013:102–103). As they do, many also retain a measure of due process.

Prosecuting Informers, Collaborators, and Dissidents: Due Process

Clearly, guerrillas have an incentive to prevent treachery and collaboration and to tamp down dissent. In place of assassination, it is noteworthy that some guerrilla organizations uphold a semblance of due process. The IRA maintained court martial procedures that allowed suspected informers (as well as those charged with other crimes) to defend themselves before a panel of three judges, retain counsel, and bring evidence and witnesses. The punishment for informing varied widely from banishment, to knee capping, to execution (Coogan 2002:151, 176, 235–236, 775–776; English 2003:274–755). Among the Aceh, "local GAM commanders, police, or judges would capture the alleged *cuak*, call in witnesses, and pronounce sentence only if there was clear evidence." Punishments ranged from incarceration to reeducation to the execution of those who "whose betrayals had caused the death of a GAM fighter, or something similarly serious" (Aspinall 2009:175). The Taliban maintain a parallel court system to settle disputes, try and punish criminals, identify and prosecute alleged informers and discipline Taliban commanders for breaches of military discipline (Giustozzi 2014). Although many villagers turned to these courts in the absence of a reliable Afghani judicial system, UN observers nevertheless faulted these "public court style hearings" for violations of Afghani and international law (UNAMA 2012:20–21).

Judicial processes that incorporate elemental due process, or "cross checking" of information, are often possible in areas where insurgents exercise control (Kalyvas 2006:187). In general, international law (APII 1977:Article 6) demands independent and impartial courts that respect an accused person's right to know the charges against him, assure a legal defense, presume a person's innocence, protect the right to be present at one's trial, and prohibit any to attempt to compel an accused person to confess or testify against himself. In spite of their reach, the same provisions also permit the death penalty (for treason, for example) and make no mention of any right to appeal. Compliance varies considerably. In some cases, people are fingered and killed based on anonymous tips or accusations intended to settle personal scores (Kalyvas 2003). In many cases, informers, collaborators, and dissidents are summarily executed at the decision of a local commander or functionary (HRW 2010, 2012; King 2007:285–287; H. H. Perritt, pers. comm; Schiff and Ya'ari 1991:214), while those incarcerated often endure abysmal conditions (Aspinall 2009:175; International Criminal Tribunal for the former Yugoslavia [ICCT]:nd.).

On the other hand and as just described, guerrillas sometimes construct a quasi-formal apparatus for arresting, trying, and punishing informers and collaborators. Studies of rebel courts in Sri Lanka, El Salvador, and Nepal demonstrate that guerrilla groups can meet the rudimentary demands of due process and provide impartial judges, advocates for the defense (although not necessarily trained lawyers), and fair punishment (Sivakumaran 2009:490–495). Properly constituted, such courts provide a reasonable and legitimate avenue to try and punish informers and spies.

But while many scholars favorably view insurgents' attempts to preserve "peace and good order among citizens," they look askance at "politically motivated prosecutions" (Sivakumaran 2009:511–512). This is precisely what often occurs when guerrillas go after collaborators and dissidents.

As the treatment of East Timorese dissidents demonstrates, guerrillas often persecute dissidents and collaborators ruthlessly. But vigilante justice and the grotesque punishment it metes out may easily undermine a movement's legitimacy and erode respect for law and order across an entire community. Collaboration and dissent, unlike treason, provide no grounds to deprive a person of the right to life. While states and non-states may criminalize espionage and severely punish spies and traitors, they have no cause to physically intimidate collaborators and dissidents. Persecuting, intimidating, and executing collaborators and dissidents who pose no immediate threat only abuses innocent compatriots. And, apart from terrorism, it is among the most reprehensible of guerrilla activities and the hallmark of an organization that does not enjoy the legitimacy it claims.

While guerrillas will sometimes brutalize compatriots at the beginning of their campaigns, many, like the East Timorese, come to realize that violence is counterproductive. Elsewhere, the PKK eventually realized that attacking schools and health clinics and murdering the Kurdish "village guards" who served the Turkish government only alienated the very population they claimed to represent. In response, the PKK curbed its killing of civilians and offered amnesty to village guards who would join the PKK. Their "tactical versatility," according to Aliza Marcus (2007:118), won the PKK growing respect. Insurgents sometimes, but not always, show similar perspicacity with regard to prisoners of war.

PRISONERS OF WAR

Among modern states, the conditions for taking and holding prisoners of war are among the most legislated and least observed norms of war. Despite a plethora of conventions and agreements to protect prisoners of war, the track record of state armies is none too impressive. In the very best – but rare – cases, prisoners enjoy sufficient food, shelter, and medical care before they are released or exchanged for enemy soldiers who were similarly treated. More commonly, captives receive but minimal sustenance as states exploit prisoners of war for slave labor while parading them publicly to galvanize compatriots and demoralize the enemy. In the worst cases, prisoners suffer abominable conditions: starvation, disease, prolonged captivity, torture, and execution (Scheipers 2010).

The Geneva Conventions guarantee a wide range of rights to prisoners of war. Some of these are fundamental human rights, while others are merely privileges the sides find prudent to accord one another. To accord rights, however defined, one must first ask: who should enjoy prisoner of war rights? For a long time, the answer was obvious: combatants. These included the uniformed soldiers of a party to a conflict but excluded guerrillas and partisans who often suffered torture and execution upon capture. But if partisans and guerrillas were a marginal phenomenon in the World Wars, they moved front and center in the colonial wars that followed. Here, too, there was little controversy as colonial powers across Africa, the Middle East, and Asia refused to treat captured insurgents as anything but rebels, criminals, and brigands. To do otherwise, would grant legitimacy to a national movement that colonial powers were loath to recognize (Bennet 2010; Branche 2010).

Nevertheless, the international community would slowly recognize the rights of guerrillas fighting wars against colonial, alien, and racist regimes and accord them prisoner of war status upon capture. Progress has been slow. Following British military intervention in Northern Ireland in 1969, for example, sweeping arrests, incarceration, and torture of IRA detainees was common. And while conditions improved considerably following a ban on certain interrogation techniques in 1972, it cannot be said that the British ever regarded IRA detainees as anything approaching prisoners of war (Bennet 2010). Today, two of the major state parties still fighting guerrilla wars – the United States and Israel – do not recognize the provisions of Protocol 1. Israel, therefore, has never recognized Palestinian guerrillas as anything but unlawful combatants subject to arrest, trial, and incarceration. The United States similarly classified the Taliban until dropping them from the U.S. list of unlawful combatants in 2009 to pursue peace negotiations (United States Code 2009:Title 10 §948a).

What then about guerrillas? Whom might they capture? The targets of capture are the same as those for killing. If participatory liability permits the use of lethal force to disable a state's military personnel and high-ranking civilians, then it permits capture, a lesser harm. Participatory liability also permits the capture and incarceration of civilians providing war-sustaining aid. Thus, guerrillas have a wide range of potential captives to choose from. How, then, should guerrillas treat them? At one level, certainly, the behavior of state armies should not affect how insurgents treat their prisoners. That Britain, Israel, and the United States tortured detainees in their war against the IRA, Hamas, and the Taliban does not allow these same groups to deprive their detainees of similar rights. But unilateral compliance only pertains to fundamental human rights. Unilateral compliance is not morally obligatory when each side grants the other rights and privileges for pragmatic or self-interested reasons. Thus, if one side deprives its prisoners of family visits, it seems reasonable that the other side might permissibly retaliate in kind. So the first task is to delineate categories of non-derogable rights and abrogable privileges. These categories then make possible a moral and, to a lesser extent, legal evaluation of the conduct of guerrilla organizations toward prisoners of war. Consider a range of examples from the relatively enlightened to the ruthlessly barbaric.

- **Eritrea**: The Eritrean Peoples' Liberation Front (EPLF) held thousands of Ethiopian prisoners of war. "Their living conditions were basic, with meager but adequate food and accommodation – but in this respect they lived little differently from the local population or indeed the members of the rebel fronts. Blankets, clothes, soap and cigarettes were supplied when available, but rarely footwear. Medical care was provided ... There are no reports of physical abuse or execution. The prisoners were sometimes able to correspond with their families. Prisoners were used for manual labor on road construction and other infrastructural projects, but the work was not excessive and discipline was not enforced in a humiliating manner" (HRW 1991:304–305; also Johnson and Johnson 1981).
- **The Middle East**: It was rare for Hamas, Fatah, and Hezbollah to capture Israeli soldiers. In 1982, Fatah captured 6 Israeli soldiers, Hezbollah held a retired Israeli colonel in 2000, and in 2006, Hamas captured a member of an Israeli tank crew. By some accounts, the captives received sustenance level food, shelter, and medical care but were held virtually incommunicado, often in solitary confinement and with little access to Red Cross inspections, visitors, parcels, and letters

(A. Ayalon, pers. comm.; Y. Schweitzer, pers. comm).[1] Later all were exchanged in what many considered lopsided and irresponsible prisoner exchanges.

- **Northern Ireland**: Like the Palestinians, the IRA took few captives from among the British troops in Northern Ireland. In one dramatic episode captured on film, two off duty British corporals wandered into a Catholic funeral in 1988. Immediately set upon by the crowd, the two soon turned up dead. In response, the IRA stated: "Despite media reports, we are satisfied that at no time did our Volunteers physically attack the soldiers. Once we confirmed who they were, they were immediately executed" (Malone 2010).

- **Chechnya 1994–1996; 1999–2009**: [Chechens] "hung Russian wounded and dead upside down in the windows of defensive positions," decapitated prisoners and placed their heads "on stakes beside the roads leading into the city," and "routinely booby-trapped Russian and Chechen bodies" (Arquilla and Karasik 1999:218; also Renz 2010). "Torturing the wounded and desecration of the dead were widespread," writes Tishkov (2004:139). "Captured contract soldiers and air force pilots were nearly always executed. Rank-and-file conscripts were held hostage and used as slave labor for various works, from building fortifications to handling domestic chores."

These accounts raise some obvious as well as subtler questions. First, what conditions do prisoners of war deserve from guerrillas? Second, is there any moral difference between disabling an enemy soldier by capturing him in battle and between capturing an enemy soldier for the express purpose of trading him for imprisoned compatriots or other political benefits? Third, what of exchanges? In Chechnya, a cottage industry took root to facilitate the exchange of prisoners (often through ransom payment or hostage taking), while in Israel many complained bitterly that Hamas or Hezbollah would trade in bodies, live or dead, to gain the release of their comrades held in Israeli jails. Commenting in 2011 on the failure of Hamas to release Corp. Gilad Shalit, Amnesty International (2011c) nicely summarizes the complexities of taking prisoners:

Amnesty International is today calling on the Hamas de facto authorities in Gaza to comply with their obligations under international humanitarian law to ensure

---

[1] Ami Ayalon was director of Israel's General Security Services from 1995–2000; Yoram. Schweitzer is a Senior Research Fellow at Institute of National Security Studies, Israel.

that Gilad Shalit is well treated, held in humane and dignified living conditions, and allowed to communicate with his family, including through sending and receiving letters, and given immediate access to the International Committee of the Red Cross. They must cease treating Gilad Shalit as a hostage, which is a flagrant violation of their obligations under international humanitarian law.

These remarks are interesting in two respects. First, Amnesty International ranks proper treatment, humane and dignified living conditions, in the same category as sending and receiving letters and access to the Red Cross. Second, Amnesty International suggests that Shalit was a hostage rather than a prisoner of war (POW) implying that this depends upon the treatment he receives or on the way Hamas conducts its negotiations. In this vein, Amnesty International adds elsewhere in the report "We deeply believe that neither Gilad, nor the Palestinian prisoners detained by Israel, should ever be used as pawns or bargaining chips." What, exactly, might this mean? After all, belligerents often take prisoners to trade for those the other side holds. Moreover, forcing an exchange is Hamas' and Hezbollah's explicit strategy and often accompanied by calls to seize Israelis for this very purpose.

POW Rights and Privileges

Although the Geneva Conventions offer expansive rights and privileges to captured soldiers, most non-state military organizations do not subscribe to these conventions. Nevertheless, legal scholars demand respect for human rights and humane treatment for prisoners under international law. While the demand for humane treatment is not controversial, some observers wonder whether guerrillas and insurgents have any right whatsoever to take prisoners or if they are nothing but kidnappers. As a matter of law, however, international law does not prohibit armed groups from taking prisoners and, in fact, implies their right to take captives. Otherwise, the provisions safeguarding detained or interned persons in Common Article 3 and APII Articles 5 and 6 are "superfluous" (Tuck 2011:765). Legal scholars also fear that denying guerrillas the right to take captives leaves them no incentive to take prisoners alive (Casalin 2011). Morally, moreover, the right to take prisoners follows from insurgent's right to fight. Indeed, taking prisoners is often necessary to prosecute a just war and is a less violent method of disabling enemy soldiers and removing them from the battlefield than exercising lethal force.

Captives may also demoralize an enemy, provide valuable intelligence, and present a bargaining chip for prisoner exchanges. As a practical matter, guerrillas often take prisoners. Sometimes the numbers are very small. In recent years Hamas, Hezbollah, and the Taliban held only one foreign captive each. The IRA took only a handful, while the Chechens and Eritreans captured many. As a result, one must reasonably ask about what kind of treatment insurgents owe their prisoners.

Common Article 3, which stipulates respect for fundamental human rights in any armed conflict, provides one answer. Humane treatment demands respect for life and dignity, requires adequate medical care, and precludes murder, torture, and degrading treatment. On this basis alone, some guerrilla organizations do not meet the threshold for humane care and severely undermine their right to fight when they deny their prisoners their basic rights. In non-international armed conflict, Article 5 of Additional Protocol II, demands that prisoners receive adequate food, water, shelter, "protection against the rigors of the climate and the dangers of the armed conflict, freedom to practice religion, working conditions and safeguards similar to those enjoyed by the local civilian population," separate quarters for men and women, and the facility to send and receive letters (APII 1977:Article 5). For those nations party to the Geneva Conventions, the requirements are more stringent still and include very specific rules regulating clothing, exercise, periodic visits by the Red Cross, labor conditions, and remuneration. The latter are not human rights, but privileges conferred by treaty and agreements between states. These rights are anchored in reciprocity and in the hopes that by easing the lot of captured enemy prisoners, one's adversary will respond in kind.

Given their complexity and cost, however, some of these rules are difficult for guerrilla organizations to fulfill. Guerrillas lack medical and administrative personnel and the resources necessary to house prisoners. Guerrillas are often mobile so that infrastructures are often makeshift and transient. Most guerrilla movements lack the necessary oversight to assure that prisoners of war are treated properly (Tuck 2011). Guerrilla bases, moreover, are often clandestine thereby forcing captors to restrict or prohibit third party visits. And, given the very few captives some insurgents hold, the rationale for reciprocity may be very weak particularly when states treat insurgents as criminals. Nevertheless, guerrillas and insurgents remain bound by human rights principles akin to those outlined in Common Article 3 and Protocol II that protect a person's physical and mental well-being. Among the more contentious principles is the prohibition on hostage taking, an accusation frequently leveled at guerrilla armies no matter how well they treat their prisoners.

Hostage Taking

Hostage taking is a war crime. What does this mean and how does it apply during guerrilla war? Typically, a hostage is a civilian, and it is not without purpose that the prohibition against hostage taking is enshrined in those Geneva Conventions pertaining specifically to the rights of civilians and noncombatants. Commonly, a belligerent seizes a hostage to force an enemy to refrain from taking a certain action (attacking a troop train or power station, for example) or to compel its enemy to act in a certain way (surrender, give up prisoners of war, etc.). In neither case has the hostage lost his right to life or freedom of movement. Underlying the prohibition against hostage taking is the doctrine of "individual responsibility," "the natural right of man not to be subjected to arbitrary treatment and not to be made responsible for acts he has not committed" (Geneva Conventions (IV) 1949:Commentary, Article 34, 231). There are no grounds then for depriving a person of rights unless he or she commits some crime punishable by detention or execution. Innocent civilians, therefore, may not be held hostage.

But what of captured soldiers? The Elements of Crime under the Rome Statute of the ICC do not limit the definition of hostages to civilians but include anyone *hors de combat* (i.e., captured or wounded soldiers) and, indeed, anyone else enjoying the protection of the Geneva Conventions (ICRC nd.d). On the face of it, then, kidnapping a soldier for the explicit purpose of holding him until the other side meets some demand – for example, releasing captive guerrillas – is a clear example of hostage taking. But on closer inspection, two mitigating factors stand out. First, captivity violates the rights of civilian hostages because it is an "arbitrary deprivation of liberty" (Casalin 2011:744). This is not true of imprisoned soldiers. Their captors may detain them as part of their legitimate efforts to disable an enemy. Second, nations have a duty to secure their soldiers' release to both restore their liberty and bolster military capabilities by returning them to the line. Absent freeing POWs by force, nations have no choice but to bargain for them. For this purpose enemy prisoners are the only currency. Prisoners of war *are* bargaining chips and to detain them for this purpose does not necessarily violate their rights (in contrast to holding noncombatants for ransom, a practice I explore in Chapter 8).

Nevertheless, one might argue that there is a significant difference between capturing a soldier to disable and remove him from combat versus kidnapping a soldier to trade him for captive guerrillas. In the first case, one acts to remove a military threat, but in the second, one consciously exploits a captive as a means. Some interpreters of international law take this position.

Detention by armed opposition groups also remains absolutely prohibited where this amounts to hostage-taking, as per the customary IHL definition, which requires the intent to coerce someone to take action or refrain from doing so. This is entirely different from a situation where the intention of detaining an individual is in order to remove them from hostilities (in the case of a member of the armed forces) ... Should a detention be effected with a coercive *motive*, this would be excluded from the ambit of the permissible detention and would amount to an act of hostage-taking prohibited by both Common Article 3 and customary IHL. (Casalin 2011:753, emphasis added; also Sassòli et al. 2011:chapter 6, p. 10)

This interpretation is not tenable. Regardless of their intention, whether to disable an enemy soldier or use the same soldier to exact a concession from the enemy, insurgents are vying for a military advantage. Disabling one soldier *reduces* enemy capabilities while exchanging that one soldier for hundreds or thousands of guerrillas *increases* guerrilla capabilities. In each case, the motive is the same: to gain a military advantage. Neither tactic is nobler than the other. Nor is it easy to distinguish between the two in practice. Detention may violate a prisoner's rights, however, when detainees are forced to suffer inhuman conditions or ill-treatment until their captors' demands are met. Thus, the conditions under which prisoners are held are important.

Threatening a prisoner with ill-treatment or execution to wrest concessions from his government flagrantly violates a detainee's rights. Curtailing visits or withholding information to gain an upper hand, however, does not. While the Geneva Conventions consider the right to receive letters and packages a crucial component of mental health, periodic Red Cross visits are only a privilege that parties to the Geneva Conventions enjoy. To protect a detainee's rights, belligerents will demand information about their captured service personnel. They will want proof of life and humane treatment. But does denying this information or trading it for some tangible benefit, such as the release of more prisoners, violate the right of the detainee? I cannot see how. If a detainee is enjoying humane treatment, the fact his compatriots are unaware of this does not affect the detainees' rights. It does, of course, erode their government's bargaining position which is precisely the point.

In principle, then, guerrillas do not take military *hostages*. Rather they take prisoners who enjoy the fundamental right of humane treatment but whom guerrillas may hold or exchange as military conditions demand. Depriving detainees of their rights for no other reason than cruelty is as impermissible as depriving them, or threatening to deprive them, of their rights to gain concessions. Fundamental human rights are also at the center of prisoner exchanges.

Prisoner Exchange: Israel-Hamas and Russia-Chechnya

Two cases – Israel-Hamas and Russia-Chechnya – illustrate the legitimate ethical demands incumbent upon guerrilla armies during prisoner exchanges. In each case, guerrillas took prisoners for the express purpose of removing able-bodied soldiers from battle *and* securing an asset to trade for prisoners in the hands of their enemies.

Israel does not accord captured Palestinians the rights of prisoners of war, but treats them instead as criminal detainees. This stems from the nature of the conflict where, for political reasons, the Israeli government is reluctant to recognize the combatant status of Palestinian guerrillas and, therefore, confer legitimacy upon what Israel regards as a terrorist organization. As a result, captive Palestinians face arrest, trial, and imprisonment on charges ranging from aiding a terrorist organization to murder. And while Palestinian detainees largely enjoy the rights stipulated by Common Article 3 and Protocol II, 5 there are recurrent charges of ill-treatment. At times, Israel, for example, severely limits family visits (B'Tselem 2012b). For their part, the Palestinians have had few opportunities to take Israeli soldiers captive. In the past thirty years, two incidents stand out: the capture of Gilad Shalit by Hamas in Gaza, noted by Amnesty International previously, and the capture of six soldiers in the First Lebanon War in 1983. By some prisoner accounts, the prison conditions in both instances were similar. While sheltered and clothed, prisoners were sometimes bound and frequently suffered solitary confinement. Food was minimally sufficient. Medical care was spotty but no prisoner died in captivity. The captives could not send or receive letters at regular intervals nor did they enjoy regular visits from the Red Cross (Aharonovitz 2010; Estrin 2012; Gilboa 2008; Gruf 2003). Conditions such as these do not meet all the criteria outlined, but they do meet most minimally. The prospect of prisoner exchanges gave the Palestinian guerrillas sufficient incentive to treat their prisoners reasonably well.

Over this period, Israelis and Palestinians, as well as Hezbollah, developed a prisoner-of-war regime that adequately, though minimally, respects prisoners' rights. The Israelis continue to arrest, try, and incarcerate thousands of Palestinians on criminal charges (Amnesty International 2011a; B'Tselem 2012a). The Palestinians, on the other hand, take very few prisoners, provide subsistence level living conditions, and maneuver for the release of thousands of Palestinians in Israeli custody. The results are exceptionally lopsided exchanges that predictably infuriate Israelis. But unless Israeli prisoners find their basic rights to humane treatment held hostage to prisoner exchanges, there is no room for charges of kidnapping. Rather, the Palestinians successfully play on the value Israel attaches to its soldiers to demand and receive significant concessions. The Israeli-Palestinian regime is, of course, very unstable and encourages ongoing attempts to take Israeli soldiers captive. In contrast, the United States is steadfast and the exchanges less dramatic. After capturing a single U.S. serviceman in 2009, the Taliban saw exchange talks collapse in 2012 when U.S. negotiators could not gain government approval to swap five Taliban leaders imprisoned in Guantanamo Bay for their captive (Rosenberg and Nordland 2012). It took another two years and the impending departure of U.S. forces from Afghanistan before U.S. officials finally agreed to the deal.

In contrast to the rational and relatively enlightened prisoner exchange norms that emerged during the Israeli-Palestinian conflict, the ruthlessness of the Russian-Chechen regime stands out. As described above, the Chechens thought little of executing their prisoners. Those Russian prisoners that lived became hostages precisely in the sense that Israeli prisoners did not, as the Chechens and their criminal proxies threatened the lives and limbs of the Russians to gain the release of their compatriots or family members held in Russian custody. Phillips (2009:201) describes how "Russian soldiers who were captured in battle were killed immediately only sparing a minority in order to sell for exchange. Hence, the relatives of a detainee could buy a federal soldier from a rebel group and then ransom him for their family member." Chechens and Russians literally bought and sold one another's captives.

How does the Russian-Chechen case compare with the treatment of prisoners by Israel and the Palestinians? On one hand, they seem very far apart as the Israelis and Palestinians cultivated an orderly system to safeguard the lives of their fighters and see to their safe return, while the Chechens acted with abject brutality. On the other hand, both the Chechens and Palestinians invested their prisoners with significant value as instruments of exchange. Because Israel was willing to endure lopsided exchanges, the Palestinians and Hezbollah had a tremendous incentive to treat their few prisoners well. In the Chechen conflict, the market forces were different. The Palestinians hoped to parlay their captives into significant military gains, while the Chechens were only after money or loved ones. As a result, only a few of the many prisoners the Chechens captured had any value. They executed the rest and would easily threaten to kill the living to speed up the negotiations. No less than nation-states, the Chechens and the Palestinians commodified their prisoners. This is the way of war and no great sin; treating them worse than animals, however, is a crime of the highest order.

TARGETED KILLING, PRISONERS OF WAR, AND INTERNATIONAL NORMS OF CONDUCT

In one sense targeted killing and taking prisoners are very different: one kills and the other imprisons. On the other hand, they are both very personal. Unlike IEDs or missiles whose victims are faceless, captives and targets of assassination are often invested with personal liability. As such, guerrillas have to be very careful about whom they target and whom they capture. The right to fight presupposes respect for the rights of participants in war. Only certain groups, namely combatants, are unequivocally liable to opportunistic killing. But this group is usually small. Far more tempting targets include informers, collaborators, and dissidents, the bane of many guerrilla organizations. While some of these agents pose a significant threat, establishing their liability requires due process. However imperfect they may be, judicial institutions and proceedings are essential for guerrillas to establish and maintain legitimacy. At home and abroad, guerrillas face close scrutiny and an ongoing battle to distinguish themselves from criminals and pirates. Guerrillas require good reasons to capture or kill enemy and compatriot civilians.

The fact that many insurgents make an investment in the rudiments of a justice system and strive to maintain some semblance of a normatively compliant prisoner of war regime attests to their position as agents vying for a place in the international community. On one hand, negative incentives are at work. The threat of reciprocal killing and mistreatment of prisoners may restrain the assassination of enemy civilians and heads of state and prevent abuse of prisoners of war. These incentives, however, are often weak. States undertake targeted killing with impunity regardless of what guerrillas do, while many guerrillas take few prisoners. More germane are the ramifications at home and abroad and the harm guerrillas do their domestic and international standing when they run roughshod over fundamental human rights and due process. Compatriots, moreover, are not apathetic about the way they are treated making legitimacy something that guerrillas must constantly earn. As the prior examples attest, compliance with basic norms of justice has its rewards.

The international community plays no small role inculcating respect for norms among guerrilla movements particularly with regard to prisoners of war. Success is only partial. While the International Committee of the Red Cross helped safeguard the rights of prisoners held by the LTTE (Tuck 2011:782), the Red Cross played no significant role in the Palestinian or Chechen conflicts. Here, the parties were left to their own devices. Nevertheless, the international community maintains an active interest in the conditions under which guerrillas hold captives. During war, the ICRC strives to enforce minimal compliance while sometimes adjusting norms to match the capacities of guerrilla armies (Tuck 2011:771–780). After conflicts end, war crimes trials are increasingly common. Following the war in Kosovo, for example, former guerrillas stood trial for abuses in KLA detention camps (ICCT nd.). The efforts of the ICRC and the international justice system are, therefore, important for preserving the rights of prisoners of war and it is encouraging that some guerrilla groups maintain respectable practices. This speaks both to the efficacy of international regimes and the force of mutual self-interest among the warring parties.

Not all guerrilla practices, however, are amenable to international norms. Two ready examples are the human shields and terrorism. Human shields can complement conventional practices of warfare by allowing guerrillas to launch attacks from within populated areas or to position noncombatants to protect military facilities. In each case, noncombatants, particularly compatriot noncombatants, provide defensive cover for military operations. Human shields, much more than terrorism, are a weapon of the weak. Insurgents exploit shields to deter their enemies, and the questions they face are similar to those that state armies encounter when they try to establish credible deterrence: how to put civilians at risk without actually harming them. When shielding works, it is cost-effective and casualty free; when it fails the results can be catastrophic. Finding the route to legitimate shielding is the abiding challenge. Terrorism is an offensive measure that targets enemy noncombatants as an adjunct to conventional military strikes. While terrorism is unequivocally condemnable, there is room to consider strikes against participating civilians. Such attacks, however, require a firm degree of civilian liability and appropriate force. Human shielding and terrorism are at the center of the following chapters.

6

Human Shields

Human shields drive state armies to distraction. Consider the following:

- "There are two things that restrict our movement: Taliban mines and the fear of civilian casualties," said Brig. Gen. Moheedin Ghori, who commands the Afghan brigade ... Taliban fighters have put women and children on rooftops and fired from behind them ... *the Taliban's strategy has slowed the advance.* (USA Today 2010; emphasis added)
- Hamas chose deliberately and systematically to exploit Palestinian civilians as shields for military targets in the IDF's Gaza Operation ... The deliberate strategy of Hamas to blend in with the civilian population *made it difficult for the IDF to achieve the objective of the Gaza Operation –* reducing the threat of deliberate attacks against Israeli civilians – while also avoiding harm to Palestinian civilians. (Israel Ministry of Foreign Affairs [MFA] 2009:8, emphasis added)

Apart from claims that Hamas positioned their fighters and missile launchers in urban areas to shield them from Israeli attack in 2008–2009 (a claim that the report of the United Nations Human Rights Council questions, UNHRC 2009:§ 480), the Israel Defense Forces filmed startling videos depicting how strikes on apartment buildings suspected of housing weapons stores were cancelled when tens of individuals went to their rooftops. Others clips show how armed Hamas fighters called to children to escort them from under fire and, thereby, shield them from attack (Intelligence and Terrorism Information Center [ITIC] 2008, 2009).

The prevailing picture of human shields usually depicts civilians converging around militants or military sites to protect them from attack.

The Atlit Detention Camp, located near Haifa on the Mediterranean coast, was established by the British in 1939 to detain Jews illegally entering Palestine. On the night of October 9, 1945, soldiers from the Palmah [a pre-state paramilitary organization] staged a breakout, freeing more than 200 prisoners. As the fighters and immigrants fled north, members of the Palmah shed their uniforms. Upon reaching the outskirts of Haifa, they were enveloped by hundreds of civilians, thereby, preventing the pursuing British from identifying and capturing the fighters.[1]

In this incident, civilians successfully shielded the fighters from capture and harm. The British, needless to say, were stymied, and the heroes of the Palmah live on in history and myth.

As each of these descriptions highlight, human shields easily frustrate sophisticated military organizations. Bound by international law and their own military ethos, state armies find themselves hamstrung when confronting guerrilla armies willing to draw their own civilians into battle. From the viewpoint of state armies, human shields represent a gross violation of the law of armed conflict. From the perspective of guerrilla organizations, human shields are anything but, and they offer a tempting strategy to offset an organization's military weakness. When effective, human shields harm no one and offer a significant military advantage at no cost.

Human shields are any person who claims immunity from direct attack and whose presence near military operations or infrastructures confers protection against an opposing army. Customary international humanitarian law and numerous treaties, conventions, and military manuals prohibit belligerents from using "the civilian population or individual civilians ... to render certain points or areas immune from military operations, to shield military objectives from attacks or to shield, favour or impede military operations" (API 1977; Article 51[7]; also International Criminal Court [ICC] 2011; 8[2]b[xxiii]; ICRC nd. e).

Using civilians in this way willfully exposes them to bodily harm and grossly violates their right to life and security. As it considers shields, international law narrowly emphasizes the rights of prisoners of war, enemy

---

[1] This is a composite description drawn from the audio-visual presentation at the Atlit Detention Camp Museum, Atlit, Israel, and Diamant (2009).

civilians, or civilians living under occupation (API Commentary 1977; Article 51, §1986–1988). The conventions warn belligerents about abusing its enemies and exploiting an adversary's concern for the welfare of their own soldiers and civilians. Rarely did it occur to anyone that an army might use its own civilians for shields. When one did – Saddam Hussein forced his own citizens to shield military sites – worldwide outrage ensued (ICRC nd. e; n20). This case, like those described by the Geneva Conventions, focuses upon coerced shields; no cases address the legitimacy of consenting, volunteer shields, or the right of a guerrilla organization to conscript shields from among the people they represent. Voluntary compatriot shields, not coerced enemy shields, are a central feature of guerrilla warfare. "War is not fair; it is expedient," writes Anthony Cordesman (2007:43), "and a non-state actor is virtually forced to use human shields as a means of countering its conventional weakness … " Cordesman ponders expediency and pragmatism, not justice. If, however, the non-state actor is "forced" to use shields, wherein lies the moral transgression?

Properly conceived and despite vociferous objections from the international community, there is room for human shields in just guerrilla warfare. Many guerrilla organizations cannot wage war without aid and cover from the civilian population. Military necessity provides an obvious first defense of human shielding, but it is incomplete. Guerrillas must realize they expose their civilian population to significant harm when they employ shields. Proximity to the fighting places all civilians at risk for collateral harm, while civilians taking an active role are at risk for direct harm. When noncombatants find themselves close to the fighting by happenstance, they may become *involuntary* human shields and remain immune from direct attack. Anyone choosing to shield, on the other hand, is a *voluntary* shield and, like any other direct participant, may lose immunity from attack. Contemplating these cases, international legal discourse assumes the perspective of the *attacker* and asks who among the shields is liable to harm and who is not. I turn to these same insights to investigate shielding from the perspective of the *defender* to ask, who is a permissible shield and who is not? Voluntary shields are one candidate, but under certain conditions, guerrillas may also conscript shields.

Regardless of whom insurgents enlist as human shields, they must consider the limits of necessity and proportionality in much the same way a state army does. Guerrillas must ask whether the military actions they undertake cause disproportionate harm to *their* citizens. Not all instances of shielding are just. Using shields to precipitate a disproportionate response is a particularly egregious form of unjust shielding. Subject to these and other caveats, the use of human shields can be part of a rational and just strategy for guerrilla armies. Under these circumstances, responsibility for safeguarding human shields does not shift entirely to those employing shields. Rather, state armies remain liable when they cause disproportionate harm to human shields.

HUMAN SHIELDS: THEORY AND PRACTICE

Can guerrilla fighters wage war without human shields? Cordesman clearly thinks they cannot. Human shields provide at least three benefits. First, they make it difficult to distinguish combatants from noncombatants, thereby increasing freedom of movement and operation for guerrilla forces. Second, shields provide more bodies to fight a numerically superior enemy. Finally, human shields offer significant deterrent effects. Guerrillas, therefore, use human shields because they can provide a significant military advantage at little or no cost and because no other less costly means are available to them. In this sense, the justification for the use of human shields flows directly from a guerrilla organization's right to a fighting chance and, in particular, its right to fight without uniforms described in Chapter 3.

The step, from shedding uniforms to using human shields, is very short. Commenting on a guerrilla's right to fight in mufti, Charles Chaumont links the two:

Surprise tactics, ambushes, sabotage, [or] street fighting takes the place of war conducted in open country and confrontations between comparable military units. In such procedures, the visible carrying of arms and distinguishing signs may ... be incompatible with the practicalities of the action (for example, if the guerrilla fighters use the population for support or are intermingled with it). Because of this, refusing to allow specific procedures would be to refuse guerrilla warfare. (API Commentary 1977:Article 44, n. 40)

Note the parenthetical remark. What difference is there, if any, between using the population for support and intermingling with it? By shedding uniforms, guerrillas hope to confuse the enemy who, unable to recognize militants, will refrain from shooting lest they injure or kill a noncombatant. Shedding uniforms is to intermingle and to intermingle *is* to use the population for support and not necessarily with their consent. If no civilians were present, the right to shed uniforms would be vacuous. If the right to fight without uniforms has any bite at all, it must embrace a guerrilla's right to use human shields. Shields are as necessary as shedding uniforms if guerrillas are going to have any ability to wage war at all.

The right to shed uniforms entails the right to use shields but only under the same conditions that underlie any permissible tactic. An assessment of effectiveness and necessity is critical. Although human shields may prove useful, one can certainly imagine situations where guerrillas put human shields at risk unnecessarily, had recourse to alternative means of defense, and/or only gained a slight military advantage by employing shields. As such, the right to employ shields requires guerrillas to search for alternative measures to protect their assets when feasible. Guerrillas may only turn to shields when it is absolutely necessary for them to fight at all. Under these conditions, to whom may guerrillas turn? Volunteers might be one option; if adults (not minors) consent to shielding, then the guerrillas who employ them are not violating their rights. But consent only trades one problem for another. While consent provides guerrillas with grounds to enlist their compatriots as human shields, consent also undermines the noncombatant immunity that guerrillas expect to exploit. Once one consents to aid guerrillas by acting as a shield, then shields forfeit their immunity and are no longer useful as shields. This is the shielding paradox. The trick then is to navigate between the minimal consent necessary for volunteer shielding and maximal consent that constitutes full-blown participation in the fighting and strips noncombatants of the immunity they need to shield effectively. Doing so requires a closer look at voluntary and involuntary shielding.

Voluntary and Involuntary Human Shields

In light of the growing propensity of insurgents to use human shields, observers are quick to draw a distinction between voluntary and involuntary shields. Typically, voluntary shields are civilians who knowingly and willingly congregate at a military installation to protect it from attack. By doing so, they jeopardize their immunity and become legitimate targets for direct and lethal attack (Schmitt 2010a:31–33). In contrast, there

TABLE 6.1. *Human Shielding, Intent, and Immunity*

Intent to Shield	Noncombatant Immunity
NO (Involuntary shield)	Immune
YES (Voluntary shield)	NOT immune

are two types of involuntary shields: those coerced to shield a military target such as a command post, missile launcher, or arms depot and *passive* shields, those entirely ignorant of the purpose they serve as guerrillas mount attacks from residential areas. Involuntary shields lose none of their immunity and remain protected (Schmitt 2008). It is important to emphasize that noncombatant immunity does not confer blanket protection. Rather, noncombatant immunity protects noncombatants from unnecessary, direct, or excessive harm. Noncombatant shields protecting military sites from attack may still suffer necessary, proportionate, and not inconsiderable *collateral* harm.

Table 6.1 shows how the phenomenon of human shielding is dichotomous according to the view just outlined. Involuntary shields are immune from harm; voluntary shields are not. Such distinctions, based solely on the intentions of the shields, may be difficult to discern in practice and are, therefore, unworkable without unambiguous evidence of voluntary intent or coercion. In some cases, things seem clear: civilians unaware of guerrilla activities in their midst are not voluntary shields. They are passive shields who had no intention of shielding and do nothing to lose their noncombatant immunity. In other cases, intent is not easy to establish. Civilians congregating on the roof of an apartment building that houses a command headquarters may have ascended voluntarily or at gunpoint. Attacking armies cannot know. Attempts by attacking forces to impute intent by warning residents of an impending strike are also fraught with ambiguity. Do those who remain stay to shield or did age, infirmity, or lack of transportation prevent them from leaving?

Intent, however, is only one problem. More significant is the right of any civilian to aid his or her armed forces, particularly when these forces cannot successfully wage war without assistance from the civilian population. As noted in Chapter 3, any attempt to define the scope of permissible aid is elusive. Commentators generally agree that civilians cannot provide direct, war-fighting aid without forfeiting their immunity.

Civilians may, however, participate indirectly and provide war-sustaining aid. Is human shielding a form of direct or indirect participation regardless of intent?

The answer to this question is important for state and guerrilla armies. State armies need to know whom they can attack and who enjoys immunity. Guerrillas need to know the same thing because only protected individuals can shield effectively. While some observers argue any attempt to voluntarily "defend a valid military objective" constitutes direct participation in the hostilities and, therefore, nullifies a shield's immunity (Schmitt 2005:459; also Lyall 2008), others concede that some form of shielding is analogous to indirect civilian participation and, therefore, protected by noncombatant immunity (Melzer 2009:54–57). If shielding is protected, then it is also permissible. Civilians have the right to shield their forces in a way that constitutes indirect participation.

Additional Protocol I (1977:Article 51[3]) defines direct participation as "acts which by their nature and purpose are intended to cause actual harm to the personnel and equipment of the armed forces." Interpreting "actual harm" to mean immediate harm, Human Rights Watch (2003:3) concludes: "[C]ivilians acting as human shields, whether voluntary or not, contribute indirectly to the war capability of a state. Their actions do not pose a direct risk to opposing forces. Because they are not directly engaged in hostilities against an adversary, they retain their civilian immunity from attack."

In its latest guidelines (Melzer 2009:56), the International Committee of the Red Cross (ICRC) concludes that civilians who "attempt to give physical cover to fighting personnel" or "to inhibit the movement of opposing infantry troops" are participating directly in the fighting and lose their immunity. On the other hand, voluntary shields who congregate around a military installation to make it more difficult to destroy are only participating indirectly. Civilians retain their immunity because they are only trying to forestall an attack, not facilitate one. "Although the presence of voluntary human shields may eventually lead to the cancellation or suspension of an operation by the attacker," writes Melzer (2009:56–57), "the causal relation between their conduct and the resulting harm remains indirect." These distinctions between direct and indirect participation only apply to voluntary shields. Coerced, involuntary shields enjoy protection whether they facilitate or forestall an attack. These permutations offer a richer portrait of human shields (Table 6.2).

TABLE 6.2. *The Immunity of Involuntary and Voluntary Shields Indirectly and Directly Participating in the Hostilities*

	Involuntary, Coerced, or Passive Shield	Voluntary Shield
INDIRECT PARTICIPATION	A. Immune	B. Immune
Provide *war-sustaining* aid by: • Shielding military HQ or supply depot • Protecting retreating troops • Political wing activities		
DIRECT PARTICIPATION	C. Immune	D. NOT immune
Provide *war-fighting* aid by: • Shielding a primed missile launcher • Protecting attacking troops		

In this categorization, behavior augments intent. Shields providing the equivalent of war-sustaining aid always retain their immunity regardless of intent (category A, B). Shields coerced to protect attacking soldiers or, more commonly, shields who find themselves employed to protect ongoing military operations in their midst but without their knowledge retain their immunity (category C). On the other hand, those who voluntarily shield primed missile launchers or combatants on their way to fight lose their immunity (Category D). Such activities are tantamount to war fighting.

Examples of human shields voluntarily entering the fray of battle, however, are hard to come by. Hypothetical examples might include " ... large numbers of unarmed civilians who *deliberately* gather on a bridge in order to prevent the passage of governmental ground forces in pursuit of an insurgent group" (Melzer 2009:81, emphasis added). Voluntary indirect shields (category B) include civilians working for the political wing of a guerrilla organization and are particularly important. ICRC guidelines confer full protection on civilians whose war-sustaining activities range from recruitment, training, and the provision of financial, media, and propaganda services to the design, production, shipment, and maintenance of weapons (Melzer 2009:34–35, 51–52, 66–67). These participating civilians will function as voluntary human shields if they find themselves in harm's way when conflict erupts. Their shielding potential is significant. Because they are part of the guerrilla organization, they cannot or will not always flee. But as civilians who provide war-sustaining services, they are immune from lethal attack. As the fighting rages, a state army may find itself stymied by their presence. The number of participating civilians can be high and their proximity to the military wing of the guerrilla organization will shield militants from attack if state armies are wary of harming civilians.

By emphasizing a shield's behavior, this typology properly pays heed to outcomes, that is, the kind of material advantage shields offer. Shields that provide war-sustaining aid are never liable to lethal harm, while those who provide war-fighting aid might be liable to lethal harm. This is the principle of participatory liability applied to shields. Intent, however, cannot be ignored. In all of these cases, the outcome criteria are subordinate to intent: if civilians did not aid militants "deliberately" then their contribution, whether a war-fighting or war-sustaining threat, is irrelevant. Without knowing intent, the ICRC will always err on the side of immunity. It must be exceptionally difficult, if not impossible, however, to know anyone's intent or state of mind in the context of a singular or sporadic event. How might one verify whether civilians *deliberately* block a bridge or a woman *intentionally* shields guerrillas behind her "billowing robes?" (Melzer 2009:56, n139). One might, however, establish intent on the basis of a person's long-term activities and commitment. Employees of a guerrilla political wing who regularly report for duty or receive payment for services are *voluntary* human shields. Nevertheless, these voluntary shields are generally protected from direct harm because they only provide war-sustaining services. In contrast, those who whose participation is opportunistic and confined to a single incident are involuntary human shields simply because there is no way to ascertain otherwise. These civilians have made no long term commitment to provide war-sustaining services. Most just happen to be in the wrong place at the wrong time. Involuntary shields retain their noncombatant immunity.

Human Rights Watch and the Red Cross use these categories to establish noncombatant immunity and emphasize the duty of state armies to observe proportionality and avoid excessive harm as they attack shielded targets. I doubt that they mean to suggest that guerrilla armies and their people have a right to place human shields. If, however, some types of shielding are nothing more than a form of indirect participation that confers immunity from lethal attack, then there is no reason to think that voluntarily serving as a human shield is unlawful or immoral. On the other hand, many civilians do not volunteer. Some are either coerced or uninformed about the dangers they face. Here, one must consider a guerrilla army's right to conscript human shields.

Conscripting Human Shields

Appealing to the same goal of collective self-defense that drives military conscription, Fabre (2012:263) suggests that "noncombatants may well be under a duty to act as shields for the sake of combatants." The grounds for this duty are also the same that proscribe treason (Chapter 5). Animated by just cause and legitimate authority, just guerrilla warfare demands the loyalty and support from the people on whose behalf insurgents fight. Without a critical mass of compatriots to provide combatants, participating civilians, and, when necessary, human shields, guerrillas will fail to achieve the security they seek for their people. What steps, then, may insurgents take to enforce this duty and conscript shields? At first blush, the answer is none; intentionally placing noncombatants at risk without their consent violates their immunity. On the other hand, a people's right to a fighting chance might be irreversibly stymied without recourse to involuntary human shields, hence the duty to shield.

During war, information may be sketchy, time is of the essence, and guerrillas may be more inclined to coerce than convince. Nevertheless, justifying the conscription of shields requires the same kind of public consent that underlies military conscription in state and guerrilla armies more generally. As demonstrated in Chapter 3, military conscription must be nondiscriminatory, distribute risk fairly, include a feasible, albeit onerous, opt-out option, and enjoy the support of both traditional leadership and nascent public bodies. Permissible methods for enforcing conscription include social sanctions, fines, or incarceration but preclude violent coercion, that is, physical intimidation and threat of injury or loss of life. How might these conditions play out among insurgents seeking shields?

Consider two cases of shielding. In the first, unarmed civilians are conscripted to surround an arms depot. In the second, shields are passive and unknowingly protect a missile launching site in the yard next door. Like military conscripts, shields should benefit from an equitable distribution of risk and a costly but practicable opt-out option. Distributing risk might require careful attention to how and in which neighborhoods arms are stored. This is not only necessary to endanger the least number of people, but to endanger the least number of people *equally* by moving stores from place to place when feasible. The opt-out penalty for active shields must be reasonable and come not at the point of a gun but in the form of social or material sanctions. For passive shields there is no opt-out option, and, for this reason, a process of ex post assessment and public consultation is imperative. Recall that legitimate authority is anchored in tacit or active consent of quasi representative or deliberative bodies and/or grounded in the traditional and patrimonial structures of local authority that antedate the rise of a guerrilla movement (Chapter 2). Military conscription comes at the behest of these institutions and so must the employment of shields if the practice is to enjoy any measure of legitimacy.

Gaining legitimacy may mean that following an operation, community institutions and/or community leaders debate the merits, costs, and equity of the shielding tactics insurgents employ. The affected population may demand an end to shielding if ineffective and accompanied by excessive casualties. But the affected population may also agree to shielding thereby allowing the authorities to conscript shields. In this way, support for conscription comes in the form of higher order consent from governing bodies, local leadership, and/or public debate reaffirming the right to resist occupation despite the costs. Suggesting that Gazans see Hamas "as a legitimate resistance force, the closest thing they have to a national army, operating within its community and using what advantages it can against a far more powerful enemy" (Barnard and Rudoren 2014) is an example of such higher order consent. At the same time, individual shields, like military conscripts, submit involuntarily and without explicit consent. Conscription, therefore, exculpates shields and preserves the immunity they need to shield effectively. Higher-order consent combined with first order coercion mitigates the shielding paradox noted at the head of the chapter by providing sufficient consent to legitimate shielding but insufficient consent to render shields liable for direct harm.

Nevertheless shields, like any other noncombatant, remain vulnerable to proportionate, collateral harm. This places an added constraint on the risk that guerrillas may take when they place civilians in harm's way. Ideally (and unlike military conscripts), shields should face no danger at all, and when shielding works, this is precisely what happens. In fact, there is no point to human shielding if it does not work, so that there is no room to permit any harm to befall shields. Because this cannot be guaranteed, the best guerrillas can do is to minimize the risk of harm by competently judging their enemy's willingness to respect shields, bringing sufficient numbers of shields to deter an enemy from attacking a military target, and learning from past mistakes to prevent the needless death and injury to shields in the future.

To ensure a level of permissible risk among conscripted shields, guerrillas must pay particular attention to proportionality. The demand that guerrillas employ sufficient numbers of shields to prevent disproportionate harm understands that an attacking army may cause proportionate collateral harm in pursuit of legitimate military objectives. The more important the guerrilla target, the more civilians may suffer collateral harm as an attacking army seeks to destroy it. At one point, the guerrilla army will find it difficult to bring sufficient numbers of noncombatants to protect the target and shielding will not work. In other words, shields work best when military targets are not extremely valuable. In these circumstances, insurgents may be able to bring enough shields so it becomes obvious that any harm befalling them is disproportionate, thereby deterring an attacking army. The number of shields must be so great as to manifestly overwhelm the value of destroying the military target they are protecting. For this reason, shields are far less useful when trying to protect a war-fighting capability where attacking state armies will have far more leeway to cause incidental civilian casualties before they are deemed excessive.

Permissible shielding utilizes either voluntary shields who provide the equivalent of war-sustaining aid or involuntary shields who provide either war-sustaining or war-fighting capabilities. The former includes the many civilians who work for a movement's political wing while the latter includes both active and passive shields that protect military assets. Constrained by consent, equitable risk, and proportionality, human shields offer an effective means to deter military attacks. Deterrence, in fact, provides the driving force behind shielding and offers a compelling analogy to further explore the ethics of human shielding.

Shielding and Deterrence

In Chapter 4, I noted how successful deterrence turns on a credible threat to harm civilians collaterally. Indeed, nuclear deterrence hinged on the capability to counter a devastating attack on one side's military facilities with an equally destructive strike against the other side's civilian population centers. Because threatening to kill noncombatants is no more permissible than killing them, theorists had to content themselves with the knowledge that (1) the harm befalling civilians is incidental and unintended and (2) nuclear deterrence works so that no noncombatant actually suffers harm. Conventional deterrence, however, is wont to fail occasionally, thereby leaving lawyers and philosophers to settle for deterrence built on the threat of collateral harm. If collateral harm circumscribed by proportionality and necessity is morally permissible, then certainly the threat thereof is equally acceptable.

Human shielding is akin to deterrence in reverse. Just as the USSR was deterred from attacking the United States out of fear that Soviet citizens would bear the awful brunt of an American second strike, the United States is deterred from attacking the Taliban out of fear that Afghani shields will bear the brunt of an *American* first strike. These cases are significantly different, but the underlying factor motivating American or Soviet behavior is still fear of collateral civilian deaths. When the Taliban uses human shields, however, the civilian deaths the United States wants to avoid are Afghani. That is, insurgents like the Taliban put compatriot, not enemy civilians, at risk. The United States will deter a Soviet attack if the expected cost in Soviet lives is excessive. Likewise, the Taliban will successfully deter a U.S. attack if the cost in Afghani noncombatant lives proves excessive.

"Excessive" means something different in each case. In the Soviet–U.S. case, excessive means the unbearable, catastrophic, and overwhelming harm that would come in response to a nuclear attack. Conventional deterrence works on the same principle, although "unbearable" is far more tempered. Israel, for example, hopes to deter Hezbollah or Hamas from future attacks with the threat of air attacks and collateral harm, not with nuclear genocide. In the course of human shielding, "excessive" harm means "disproportionate" harm as attackers turn away when faced with causing more than proportionate casualties among the shields. Unbearable harm, as a function of what is just beyond endurable human suffering, is far greater than disproportionate harm. Significant pain and suffering is somewhere in between. Any nation will be deterred by the prospect of unbearable harm. Most nations will be deterred by the prospect of significant pain and suffering, but only nations with a long history of humanitarian commitment will be deterred by disproportionate harm. Enemies must therefore know one another for deterrence to

TABLE 6.3. *Deterrence and Human Shields: The Moral Distinctions*

	Deterrence	Human Shields
State of war	Future	Present
Population at risk	Enemy	Compatriot
Consent	None	Actual consent or fair conscription
Proposed act	Killing	Letting die

DEFINITIONS

Deterrence: Future victim (endangered nation) threatens enemy civilians to deter aggression.

Shields: Present victim (endangered people) endangers compatriot civilians to deter aggression.

succeed and likewise know one another for human shielding to have a deterrent effect. Guerrillas must know where their enemy will stop.

Human shielding is a powerful weapon. When shielding successfully deters an enemy or mitigates the force an adversary chooses to deploy, shielding achieves a significant military advantage at little cost. When deterrence fails, guerrilla armies and their people persevere nonetheless. Because they have not suffered unbearable harm, they go on. One might ask whether it is fair to put shields in this position, but shielding is superior to conventional deterrence. Table 6.3 shows why.

Reading across the first row of the table, notice how deterrence confronts a future threat. To prevent this threat from materializing, an endangered nation threatens its enemy's civilians with war. Human shields, on the other hand, confront a clear and present danger. Here, the victim (an endangered people) endangers compatriot civilians to deter aggression. Acts of deterrence theoretically gain purchase because they hope to thwart future aggression. They are, however, threats tendered in the absence of any immediate danger and may be abused (to gain economic or territorial concessions) or misguided (mistaking defensive measures for belligerence). Permissible human shielding, in contrast to deterrence, occurs after the outbreak of armed conflict as a party defends itself and its people against clear and present armed attacks.

Moving down the table, note that deterrence threatens to collaterally kill an enemy who would hardly consent to the danger imposed upon him or her. Shields are significantly different in this regard. Some have given explicit consent (those shielding war-sustaining targets), while the others have given their support to a fair system of conscription. Finally, consider the difference between killing and letting die. Deterrence threatens to kill enemy civilians (if only collaterally), while shielding lets them die. The moral differences between killing and letting die open up a host of philosophical questions. In some cases – euthanasia, for example – it is permissible to let patients die (whether by withholding or withdrawing treatment) but largely forbidden to kill them. In other instances, as the countless permutations of runaway trolleys show, it is easy to construct cases to show how a person can be liable for allowing someone to die. In general, the rule of rescue also condemns a person for allowing another to die if the cost of rescue is reasonable. Rarely is letting die morally worse than killing and, in most cases, letting die is the morally superior choice. Shielding, then, is equal to or better than deterrence in this regard.

When deterrence fails, then the side that had hoped to deter its enemy will go to war, and enemy civilians will suffer at their hands. When shielding fails, deterrence fails. The enemy then initiates attack, and shields suffer collateral harm at their hands. Guerrillas who have no means to protect these civilians can only let them die. One might immediately argue that these civilians would not have died had they not acted as shields. But we have already seen that shields find themselves at risk under a fair system of conscription that guerrillas may utilize when necessary to sustain their right to fight. This is little different from states that put their civilians at risk when they go to war. Sometimes states do this to save more civilians, but, no less often, wars take more civilian lives than they save, particularly if a besieged state may easily surrender. Shields are no different. As a result, it is difficult to make the case for shifting responsibility for civilian deaths from one side to the other when nations go to war and civilians find themselves acting as shields to deter enemy attacks.

HARMING HUMAN SHIELDS: WHO IS RESPONSIBLE?

- "The onus for avoiding collateral damage altogether is on the terrorists. They have to stop exploiting their status as civilians, stop using civilians as human shields, and homes – as headquarters, as locations to store ammunition and for snipers to ply their deadly trade" (Etzioni 2010).
- In a hypothetical example, Avishai Margalit and Michael Walzer (2009:21–22) describe how Hezbollah has seized an Israeli kibbutz and drafted civilians as human shields to prevent an Israeli attack.

The identity of the civilians varies from case to case to include Israelis, Lebanese, or foreigners. Regardless, Margalit and Walzer argue that calculations of proportionality and the rules of engagement remain unchanged. And, "if [Israel] observe[s] those rules, and take the morally necessary risks, responsibility for the deaths of Hezbollah's human shields – in all the cases – falls only on Hezbollah."

These cases raise two sets of interrelated questions. First, what is the responsibility of the attacking party when defenders resort to shielding. Second, what is the responsibility of the defending party that utilizes human shields?

Attacker Responsibilities

Observers disagree about an attacker's responsibility when an adversary employs human shields. In its extreme form, Etzioni's argument shifts responsibility entirely to the guerrilla side. This view is echoed in a minority legal opinion that "urges that involuntary shields should be ignored in the proportionality and precautions-in-attack analyses because an enemy violating the law should not be allowed to benefit from its malfeasance" (Schmitt 2008:49). In this view, the attacker has no obligation to protect shields or refrain from disproportionate harm. While such a view would render shielding useless because an attacker has no obligation to protect the lives of shields, it suffers from two defects. First, it is not clear that guerrillas using shields are malfeasant. Guerrillas act wrongly when they violate the rights of shields. While this certainly occurs when guerrilla overtly coerce enemy civilians or foreigners as shields, guerrillas violate no one's rights when they enlist or conscript compatriots while securing consent as outlined earlier. Second, malfeasance on the part of guerrillas should not affect the protections due noncombatants. Guerrillas that coerce enemy civilians or foreigners shoulder responsibility for their actions, but noncombatants retain their immunity. This should compel attacking armies to refrain from disproportionate harm.

Guidelines for disproportionate harm when attacking shielded facilities are a matter of controversy. Margalit and Walzer argue that proportionality is unaffected by the identity of noncombatants. Alternatively, other commentators are prepared to discount the lives of human shields and allow the attacker more leeway. "Even if the principle [of proportionality] endures," writes Yoram Dinstein (2010:155) "the test of what amounts to 'excessive' injury to civilians must be relaxed in the exceptional circumstances of 'human shields' ... [and] must make allowances for the fact that – by dint of the presence (albeit involuntary) of civilians at the site of the military objective – the number of civilian casualties can be foreseen to be higher than usual." Given the elasticity of the proportionality principle it is hard to see how this discount works out in practice. The important point is as follows: attackers are only responsible for the disproportionate harm they cause when attacking a target protected by human shields; defenders who utilize shields, on the other hand, bear responsibility for the *proportionate* harm that attackers inflict. As such, one of the major responsibilities of defenders who utilize shields is to prevent proportionate harm and, indeed, to prevent any harm at all from befalling shields.

Defender Responsibilities

What does it mean to say that responsibility for the deaths of shields falls on insurgents using shields? First, as just noted, it only means that guerrillas are responsible for proportionate harm; the attacker still bears responsibility for disproportionate harm. Second, guerrillas bear responsibility because they place innocent noncombatants in harm's way against their will. But this is not entirely correct. Forcing enemy civilians or prisoners of war to shield military installation is a major concern of international law and rightly proscribed absolutely. Insurgents who abuse *enemy* noncombatants in this way are both responsible and morally culpable. But when shielding is voluntary or backed by a publicly accepted scheme of conscription, guerrillas are neither culpable nor entirely responsible when they enlist *compatriot* shields. Rather, responsibility is diffuse and spreads across leaders and followers alike. Still guerrilla commanders make the operational decisions. What are their responsibilities?

Summarizing earlier discussions, there are five areas of responsibility that guerrillas must observe when they employ human shields.

1. **Necessity and effectiveness**
   The use of shields is permissible when no other less costly and effective means are available to ensure the right to fight.

2. **Fair conscription**

Shield users have an obligation to maintain a fair system of conscription, distribute risk as equitably as feasible, and refrain from imposing brutal penalties on civilians who refuse to cooperate.

3. **Proportionality**

Shield users have an obligation to bring sufficient numbers of shields to ensure that a shield attacker can only cause *disproportionate* harm. Otherwise, attacks on shielded sites are permissible.

4. **Minimize mistakes and failure**

Guerrillas have an obligation to cease using shields if the tactic fails because:

- they miscalculate the number of shields they need,
- they are unable to enlist sufficient numbers of shields, or
- the shield attacker is not deterred by the prospect of causing disproportionate harm.

Shields will suffer harm when guerrillas miscalculate and field either too few shields to guarantee disproportionate harm or field sufficient numbers of shields but underestimate the ruthlessness of their adversary. Miscalculations and mistakes do not necessarily bring moral condemnation. Armies err all the time: intelligence is wrong, civilians pop up unexpectedly, identities are mistaken, equipment fails, and so on. Nevertheless, the requirement to minimize mistakes and failure is complex. There are bound to be failures for the reasons just cited, but they do not always mean that shielding is ineffective. One might readily conceive of instances where shields are tested so that their willingness to take casualties testifies to the shields' credibility and, thereby, guarantees future deterrence. Or, there may be instances where attackers miscalculate and harm shields unintentionally. In each case, casualties can only be one-off and nonrecurring. Shielding, therefore, may remain a viable strategy. Only when shielding indisputably falters and is no longer effective must guerrillas give it up. This is true of any tactic in war.

5. **Prohibition on intentional harm**

Insurgents may not use shields to draw attacks with the expectation that shielding will *fail* so that the resulting causalities bring condemnation down on the shieldattackers. This is no longer shielding but a premeditated attempt to provoke and cause harm. Such a tactic is not a miscalculation but a deliberate violation of the duties incumbent upon guerrillas to minimize risk. Theirs is a calculated attempt to kill their compatriots indirectly in the hopes that resulting international outrage will help their cause. In this case, letting die is as bad as, if not worse than, killing. Guerrillas have deceived and endangered their compatriots and abused their own authority. Under these circumstances, responsibility shifts significantly to the shield users. One might, however, ask the epistemic question: How does one know whether insurgents have honestly miscalculated or dishonestly manipulated their citizenry? A single occurrence offers scant ground to decide; evidence to the contrary only accumulates over time as it becomes clear that mistakes become policy. The evidence is particularly glaring when guerrillas do not desist in the wake of ongoing failure, and calls to mind those instances where conventional forces rely on purported mistakes or collateral damage to attack enemy civilians directly. During shielding, recurring mistakes and miscalculations raise the specter of intentionally using shields to draw attacks.

The Limits of Human Shielding

The caveats just outlined are not intended to prohibit the use of human shields nor offer guerrillas a tactic that entirely hamstrings their enemies. In practice, human shielding cannot easily protect vital military targets and find its highest use when targets provide war-sustaining aid only. Consider each of the following categories:

1. **Involuntary or conscripted shields shielding war-sustaining or war-fighting capabilities**. In this case shields do not lose their immunity. They are, however, vulnerable to proportionate collateral harm. To effectively shield war-sustaining or war-fighting targets, guerrillas must conscript sufficient numbers of shields to make any attack manifestly disproportionate. Because this number is likely to be quite large and requires guerrillas to conscript shields in a relatively short period of time, human shielding is not always a feasible tactic. This is particularly true when guerrillas try to shield

high-value military targets. Consider the difficulties posed by the following case:

On September 8, 2013, a drone strike in an eastern province [of Afghanistan] killed up to 16 people ... Afghan officials said the drone was targeting four insurgents who were picked up along the road by a truck with civilians in it ... Women and children were among the dead, officials said. "The insurgents often force local drivers to give them rides in their trucks," said the Kunar police chief, Abdul Habib Saidkha. (Ahmed and Sahak 2013)

The case highlights the inherent problems of shielding war-fighting targets. On one hand, U.S. forces might reasonably claim that twelve civilian deaths reflect proportionate collateral harm assuming they place a high value on the insurgents killed. On the other, there seems no reasonable number of shields that are manifestly disproportionate, thereby leaving guerrilla open to severe condemnations should they repeat the shielding with similar results. In contrast, the evidence from the 2008–2009 Gaza War suggests that successful shielding occurs when civilians are called to protect an isolated war-sustaining facility (such as an arms depot). In these cases, each side might easily calculate the threshold of disproportionate harm because each knows the value of the military target that is shielded and the number of civilians at risk.

Passive shielding is more problematic for the same reasons. An attacking army confronting guerrillas fighting from within a civilian population faces a diffuse military target. Some targets will be important, others will not. The difficulty of estimating civilian casualties complicates calculations of proportionality. Under these conditions, shield users will find it difficult to bring such large numbers of shields to bear as to make an attack manifestly disproportionate. At the same time, guerrillas must also prove that there were no other venues to effectively engage an attacking force.

2. **Voluntary shields shielding war-sustaining capabilities**. By and large this group embraces the many members of a movement's political wing who work in its associated financial, legal, scientific, and telecommunications facilities. Although participating civilians are not liable to lethal harm, their facilities are subject to ruin. As in case 1, the presence of shields might protect a facility from

physical destruction assuming sufficient numbers are present to guarantee the prospect of disproportionate harm.

3. **Voluntary shields shielding war-fighting facilities**. These shields lose their immunity and are subject to direct harm. However, it is important to keep in mind that in most cases it is very difficult to assess one's intentions and thereby distinguish voluntary from involuntary shielding. For onetime, opportunistic shields, there is no ready way to determine intentionality, so an attacking army must assume they are coerced or conscripted and thereby extend immunity to them. This places most shields in category 1 and largely limits shields to low value war-sustaining targets.

4. **Unjustified shielding: LTTE insurgents in Sri Lanka**.

On three separate occasions the government declared no-fire zones, giving the illusion of safety to hundreds of thousands of terrified civilians who fled into them. The LTTE rebels also went in, set up their heavy weapons among innocent men, women and children and proceeded to attack the military fiercely. The army retaliated and large numbers of civilians were killed. (Wijedasa 2011)

LTTE's use of human shields was unjustified for any number of reasons. First, one might reasonably question the LTTE's legitimacy and cause at this point in their history. While the LTTE enjoyed just cause and considerably legitimacy in its fight for self-determination, one must ask whether the people that the LTTE claim to represent did not withdraw their support for the LTTE and its fight by fleeing to safe zones established by the government. In response, one might concede that the civilians fled as a temporary measure to allow the guerrillas room to fight. This occurred in Uganda, for example, where fleeing civilians gave the guerrillas a free hand to fight without causing civilian casualties. The tactic was designed to minimize noncombatant casualties and instrumental for the guerrillas' success (Kasfir 2005:280). This was obviously not the case in Sri Lanka and highlights a fatal defect in the LTTE's use of human shields: human shields and the counter strikes the ensuing guerrilla attacks generated served no *military* purpose. They were not intended to protect military assets but, at best, only to generate civilian casualties and to either excite world opinion or turn refugees against the government. Such tactics hope shielding will fail, put shields at certain risk, and strive for maximum harm. Finally the LTTE had no way to adequately ensure that counterattacks would be disproportionate, a problem that plagues even the most judicious use of passive human shields.

In short, human shielding may serve three purposes: to protect war-sustaining assets, to protect war-fighting assets, and to provoke disproportionate counterattacks that may generate sympathy for guerrillas. The first is permissible and offers some tactical advantage, the second is permissible but very difficult to implement, and the third is impermissible.

HUMAN SHIELDS AND FREE RIDING

As they ponder human shields, observers remain concerned that human shielding may make it impossible for an adversary facing shields to fight at all, thereby violating a state's right to a fighting chance. Moreover, shielding can only work when one's adversary respects noncombatant immunity and so offers guerrillas "a morally dubious free ride on their adversaries' moral code" (Meisels 2008:46). Neither claim is compelling. First, as the previous discussion demonstrates, shield attackers are not prevented from fighting, that is, their right to fighting chance is not impugned. Shield attackers maintain their right to inflict proportionate harm on involuntary shields protecting war-fighting and war-sustaining targets. They retain the right to target voluntary shields protecting military sites.

Nor are shields effective in all situations. If a military installation is extremely valuable, it is unlikely that sufficient numbers of shields can be conscripted to protect it. It is unlikely that shielding would deter an attack on a rooftop missile launcher poised to fire. Shields might, however, prevent an attack on a missile storage facility. When military targets are diffuse and dispersed among the civilian population it will difficult to determine the number of necessary shields. An attacker may err on the side of caution and refrain from attacking but may also choose to attack while hoping to keep casualties within the boundaries of proportionate harm. Shield attackers do not, as Michael Skerker (2004) argues, always face the prospect of a Pyrrhic victory when attacking shielded facilities. It is Skerker's view that international condemnation in the wake of harming human shields will always disregard appeals to proportionality and thereby negate any military advantage shield attackers might achieve. This claim is neither empirically verifiable nor morally compelling. Shield attackers will not and should not desist when they can attack a valuable military target while causing proportionate harm. While Hamas and Hezbollah may have successfully shielded isolated targets from attack in their wars against Israel, the large numbers of civilian casualties in those wars suggest that the presence of noncombatants will do little to stem the tide of war. Nevertheless, shield attackers should think carefully when targets are less than vital, when the limits of proportionality are not clear, and where they command alternative means to disable guerrilla positions. These rules guide the use of any tactic.

It is interesting that some Islamic discussions of shields mirror these remarks. Traditionally, Islamic law is very reticent about allowing Muslim armies to endanger the lives of innocent Muslims when they attack military targets (Landau-Tassero 2006:3, 11). The dilemma arises when a non-Muslim enemy force holds Muslim prisoners or large numbers of Muslim civilians, for example, traders and their families, reside close to non-Muslim military sites. In recent years, however, some theorists have relaxed these proscriptions to allow insurgents to attack Western military targets knowing that resident Muslims may lose their lives on the condition that strikes are necessary and effective and militants take care to minimize civilian casualties (al-Libi 2008). This says nothing, of course, about permitting Muslims to use their own compatriots as shields. Here, there is far less discussion among scholars. Nevertheless, groups like Al-Qaeda have been criticized for utilizing human shields of their own "without consulting clerics on this matter" (Middle East Media Research Institute [Memri] 2006). "This demonstrates," notes Buchholz (2013:51), "that the human shield is a sensitive, ambiguous topic that requires legitimizing."

In the final analysis, human shields have important but limited use. Compatriot shields can replace active or passive defense measures to protect particular installations, logistical facilities, and arms depots of less than overwhelming military value. As a result, attacking armies may simply find it prudent and feasible to avoid attacking these targets. It may cause inconvenience or bring setbacks but shields alone are not decisive. Instead, they join the arsenal of available violent and nonviolent tactics available to guerrillas fighting a just war.

In the course of asymmetric war, human shielding does not offer an unfair advantage and, in fact, offers no advantage at all except in certain cases. To claim that shield users are exploiting an adversary's willingness to respect humanitarian law ignores the fact that shielding, like deterrence, depends upon the civilian immunity of the potential victims of war to prevent attacks and curtail violence. While successful shielding – no less than successful deterrence – demands a firm belief that one's enemies will respect the principle of proportionality, this demand is no different than law and ethics impose on warfare in general. Just as it is now reasonable to expect that nations will not obliterate cities, it is equally reasonable to expect they will avoid harming human shields disproportionately. As I argued in Chapter 3, guerrillas, like states, have good reason to respect the norms of warfare and the rights of noncombatants. It is for this reason that the case for human shields must be argued so closely. There are no grounds to compel enemy noncombatants to shield; the standing legal prohibition is unaffected by just guerrilla warfare. The right to a fighting chance does, however, permit guerrillas to enlist compatriot shields. Nevertheless, the right of shielding is neither blanket nor sweeping but constrained by the exigencies of war and the rights of its participants.

The exigencies of war and the rights of its participants constrain, but do not prohibit, the many exercises of kinetic force described in the previous chapters. The following chapters turn from kinetic force to nonkinetic force and from hard war to soft war. Here, too, similar caveats apply. Unexpectedly perhaps, these chapters open with a discussion of terrorism before turning to economic warfare, the persuasive power of public diplomacy, and nonviolent resistance.

PART III

SOFT WAR

Soft war describes the use of non-kinetic weapons to achieve the legitimate aims of war. Non-kinetic weapons are any that do not explode or hurl projectiles. Non-kinetic weapons are cybernetic and driven by malicious viruses, economic and driven by sanctions and blockades, or nonviolent and driven by moral shame. Each of these weapons is characterized by *force*. Soft war also embraces public diplomacy and "soft *power*." Soft power does not work by force at all but, instead, invites support through persuasion, high ideals and public works. In this way, soft war is much broader concept than soft power and includes all non-kinetic means whether persuasive or coercive. In practice, soft war is generally less destructive but no less effective than bombs and bullets.

While guerrillas are often saddled with relatively weak kinetic forces, they can often muster strong non-kinetic forces. Looking to cyber-warfare and cyberterrorism, insurgents are rankling state armies who must devote enormous resources to cyber defense. Guerrillas, for their part, will need to distinguish between legitimate and illegitimate targets, the same difficulty they face when setting IEDs or launching rockets. In the context of soft war, however, the challenges posed by the principles of discrimination and proportionality are not obvious. Cyber and economic warfare are largely designed to target noncombatants *directly*. Cyberattacks zero in on computer networks, banking infrastructures, and telecommunications. Blockades and sanctions target many of the same facilities and, like cyber-warfare, hope to make civilian life unbearable to the point of surrender. While noncombatants do not endure physical harm immediately, severe mental suffering does the work of bodily harm, compelling civilians and their government to comply with an adversary's demands. Just war theory has little to say about such weapons, even as they form an increasingly significant part of insurgents' arsenals. Nor does just war theory have much to say about how insurgents respond to the financial pressures of war. All parties to armed conflict need money, but guerrillas face the enormous challenge of raising funds when they cannot impose taxes or issue bonds like states do. Instead, they build rudimentary economies often fueled by crime and coercion. How might these be justified?

No less challenging are public diplomacy, propaganda, and media warfare, soft power *par excellence*. Public diplomacy works through good deeds and good press. Insurgents, no less than states, offer public works – clinics, school, and roads – to win the hearts and minds of their compatriots and use the media to trumpet their accomplishments. Insurgents face two challenges. The first is the truth. When, if ever, may the truth be sacrificed to just cause? As it turns out, lies and half-truths often serve just cause very well. Second, when does soft power become hard? When does persuasion and benign information turn to coercion and bullying? The answer is sometimes, and when the media incites violence or needlessly enflames passions, public diplomacy crosses dangerously into indiscriminate and disproportionate warfare.

None of these problems, it seems, should accompany nonviolent resistance. But they do. Nonviolent resistance is not persuasive but coercive. Some even question whether it is entirely non-kinetic, suggesting that throwing stones at tanks (or Philistines) is just another form of violence. Other activists renounce violence but do not balk at provoking bloodshed to gain a sympathetic hearing in the court of public opinion. This tactic, known as "backfire," is often spectacularly successful, but it comes at a price when innocent demonstrators are killed or injured.

Violence, it appears, is ever present and, accordingly, Chapter 7 opens with a discussion of terrorism. Readers may find this surprising. What, they might ask, could be any more kinetic than terror bombing? This is true, but the transition from hard, kinetic warfare to soft, non-kinetic warfare is not sudden, but gradual. While IEDs and nonviolent resistance bookend this volume, the middle ground – terrorism and economic warfare – offers a mix of kinetic and non-kinetic features. Non-kinetic, however, should not be confused with non-harmful.

7

Terrorism and Cyberterrorism

Terrorism is probably the tactic most closely associated with guerrilla warfare. Drawn from images of Al Qaeda attacks in New York, London, and Madrid or horrific bombings in cafes and bus stations by Palestinians, the predominance of terrorism among guerrillas seems a foregone conclusion. But this conclusion is wrong. While public perceptions are shaped by radical Islamic and Palestinian terrorism, many other insurgent groups either disavow terrorism entirely or use it sparingly and in conjunction with military tactics, economic warfare, and public diplomacy.

It is notable that two of the most successful liberation movements of recent years, those in East Timor and Eritrea, repudiated terrorism entirely, preferring military strikes, public diplomacy, and, in East Timor, nonviolent resistance (Kilcullen 2009:206–210; Weldemichael 2013:3, 13–15). In the Western Sahara, the Islamic Polisario also rejected terrorism (Stephan and Mundy 2006). Among groups that turned to terrorism, some either gave it up in favor of other tactics or, when attacking civilians, usually pursued collaborators and informers. The Provisional Irish Republican Army (PIRA), no less reputed for blood thirsty terrorist attacks than the Palestinians, abandoned terrorism half way through its campaign after realizing that killing civilians only brought them the wrath of the local and international community. By 1989, top Republican officials were already reporting "a greater realization than ever of the need for the PIRA to avoid civilian casualties" (English 2003:260). Instead, the PIRA redoubled its political efforts while restricting bombings to transportation hubs and emptied buildings in business districts (Rogers 2000:13). In Kosovo, too, guerrillas assassinated Albanian informers and

Serbian officials but the LDK, the political wing of Kosovo's national movement, never advocated terrorism (Bekaj 2010; Clark 2000:142; Özerdem 2003).

Assassinating informers, however, is not what we usually mean by terrorism. Instead, terrorism embraces violent acts that intentionally kill or injure noncombatants for political purposes (UN 2004:§164d). On this view, and despite indictments of KLA soldiers for war crimes, there is room to distinguish between assassinating informers and harming innocent civilians. "The former," concludes Henry Perritt (2008:73–74), "occurred, by all reports and evidence [in Kosovo]. The latter almost never occurred." Even Chechen terrorism must be seen in perspective. Despite horrifying attacks against schools, hospitals, and other civilian targets in Budennovsk (1995), Kizlyar (1996), Moscow (2002), and Beslan (2004), civilian casualties of Chechen terrorism accounted for less than 3 percent of those killed in the two Chechen wars (Hughes 2007:150). Moreover, Chechen militants, at least initially, sought hostages to extract political concessions from the Russians, not to mass murder. For these reasons, concludes Hughes (2007:149–150), "this is not a conflict that can be characterized as 'terrorism'" (also Arquilla and Karasik 1999; Galeotti 2002).

Among Palestinians, however, terrorism, not assassination or military confrontation, has been the tactic that they hoped would garner public recognition, mobilize support at home, retaliate for targeted killings, demonstrate Israeli vulnerability, and demoralize Israel's citizenry (Arens and Kaufman, 2012; Gunning 2009:201–216). In contrast to the relatively restrained role that terrorism plays in national struggles in Chechnya or even Afghanistan (UNAMA 2012:16–17)[1], Palestinian terrorism accounted for nearly 70 percent of all Israelis killed by Palestinians between 1987 and 2012.[2] Nevertheless, Palestinian terrorism has declined significantly from an average of nearly 100 Israeli

---

[1] In Afghanistan, assassination accounted for 21% of the approximately 5,900 Taliban-caused civilian deaths between 2009 and 2011, while the rest are attributable to suicide bombings (15%), effects of ground engagements (6%), and IEDs (58%) (UNAMA 2012). During the same period 5,912 American, Coalition, and Afghani soldiers and police lost their lives (Chesser 2012; iCasualties.org 2014).

[2] Between 1987 and 2013, Palestinians killed a total of 1,577 Israelis and foreigners in Israel, the West Bank, and Gaza. Of these, 1,061 (68%) were civilians and 496 were members of the security forces (B'Tselem 2013).

civilian casualties per year during the Second Intifada (2000–2008) to less than ten per year from 2008 to 2012 (B'Tselem 2013). Whether this decline is attributable to aggressive targeted killing by Israel (Kober 2007), or to Hamas' maturation as a political party since taking power in 2006 (Slater 2012), or both (Gunning 2009), what was once a terrorism dominated conflict increasingly now turns on public diplomacy, direct military confrontation, and other tactics.

While many contemporary political and philosophical discussions continue to belabor terrorist attacks against enemy civilians (*enemy* terrorism), two variants of terrorism go virtually unnoticed. The first is *compatriot* terrorism, that is, intentional strikes by insurgents or the state against local civilians. While such violence receives enormous attention from anthropologists and sociologists, few discuss its ethical dimensions. The second is *nonlethal* terrorism. The growing use of sublethal, non-lethal, and cyber weapons offer the prospect destroying or disabling infrastructures in the absence of kinetic force and without significant injury or loss of life. Nevertheless, the psychological effects of nonlethal terrorism – post-traumatic stress, anxiety, depression, and incapacitating fear – are not trivial. Introducing nonlethal weapons into contemporary warfare raises two questions. First, may belligerents, whether state or non-state, use nonlethal and cyber weapons to attack and disable civilians who provide war-sustaining aid? In contrast to emerging guidelines for cyber-warfare, I argue that they may. Second, may belligerents use nonlethal and cyber weapons to terrorize and demoralize noncombatants? I argue that they may not. The psychological harm attending nonlethal or cyber weapons violates noncombatant immunity. This chapter considers each of these issues – enemy, compatriot, and nonlethal terrorism – in turn.

ENEMY AND COMPATRIOT TERRORISM

To define terrorism as "any action intended to cause death or serious bodily harm to civilians or noncombatants ... to intimidate a population or to compel a government or international organization to do or abstain from doing any act" is too broad (United Nations 2004:§164d). Instead, enemy or compatriot terrorism is a phenomenon limited to intentional attacks on noncombatants alone rather than civilians in general. Analyses of terrorism, then, must distinguish among the various actors who populate a disputed region and their liability (Table 7.1).

TABLE 7.1. *The Liability of Civilian Actors in Asymmetric War*

Enemy Agent	Compatriot Agent	Liability	Permissible Harm
High-ranking civilians	Informers, spies	Yes	Lethal
Participating civilians	Collaborators	Yes	Nonlethal, psychological
Noncombatant	Noncombatant	No	Collateral

High-ranking civilians, informers, and spies who provide direct military or war-fighting aid are liable to deadly attack. These civilians provide war-fighting aid and are no more immune from lethal harm than uniformed soldiers or insurgents. Such killing is assassination, not terrorism and, as outlined in Chapter 5, requires firm evidence of liability, and, in the case of informers and spies, due process. Compatriot collaborators (row 2) provide war-sustaining services to the state they are fighting (e.g., Kurds who work for the Turkish government). In contrast to collaborators, participating civilians provide their own state with war-sustaining aid. The principle of participatory liability allows insurgents to disable collaborators, participating civilians and their facilities with nonlethal force. Direct, nonlethal strikes against participating civilians are the subject of the second half of this chapter.

Beyond high-ranking civilians, participating civilians, spies, informers, and collaborators reside noncombatants (row 3). They are not liable to direct attack. Noncombatants pose no threat, and while subject to collateral harm, they remain protected from any direct harm. The prohibition of terrorism should be incontestable. Nevertheless, insurgents sometimes find terrorism irresistible and appeal to a variety of claims to defend their actions.

Enemy Terrorism

Justifying terror attacks against enemy noncombatants often entails an appeal to necessity when a national group faces genocide, mass murder, or ethnic cleansing, and terrorism is the only feasible means to ensure survival. In these circumstances, the moral prohibition of harming innocent civilians is set aside during a "supreme emergency" and overridden by a greater good of saving an entire political community from extermination (Coady 2004; Steinhoff 2004; Walzer 2004). As this book abundantly demonstrates, however, insurgents usually have ample recourse to a wide range of tactics to press just cause. Terrorism is rarely, if ever, necessary. Nor is terrorism effective. On the contrary, terrorism is often a public relations disaster and a cause for brutal crackdowns that generate little sympathy in the international community. While it is reasonable to attribute an abstract right of terrorism to a violently oppressed and endangered people – Jews facing Nazi extermination often come to mind – the political reality is often much different. In contrast to post–World War II apathy, the world community today takes its obligations to forestall grave human rights violations more seriously.

Serbia's campaign of ethnic cleansing in Kosovo and Indonesia's scorched earth policy in East Timor, for example, certainly cross the threshold of an existential threat. Each people faced deportation, displacement, wholesale murder, and widespread destruction of infrastructures at the hands of the state. That these very abuses triggered foreign intervention suggests that a crisis comparable to supreme emergency was at the door. It is noteworthy, however, that terrorism was not a first resort or, indeed, any resort in either instance. This reflects a growing understanding that terrorism diminishes and, perhaps, even repudiates the obligation of the world community to intervene once victims of state violence lose their innocence and partake of aggression. Had either guerrilla group resorted to terrorism, as Palestinian and Chechen guerrillas have, the world community might have justifiably reconsidered armed intervention.

Ineffectiveness is one of terrorism greatest drawbacks and stems directly from the utter wrong of grievously harming noncombatants. Gauging effectiveness demands an understanding of what guerrillas hope to gain. Terrorism fosters "fears of mortal dangers" among enemy citizens that insurgents hope will pressure a state to change policy, withdraw from occupied territories, and grant independence or far reaching autonomy (Meisels 2008:29; also Ganor 2005; Schmid 2005). As a political tool, guerrillas use terrorism to demoralize or intimidate an enemy, to erode their willingness to fight, and to generate support for insurgents at home. Or, guerrillas may be looking for a one-off effect hoping that a terrorist attack will precipitate a grossly disproportionate state response that will evince sympathy for the guerrillas' cause (McCormick and Giordano 2007). Alternatively, terrorists may be after a symbolic gain that spotlights their prowess while displaying their enemy's impotence; instilling pride in the one and wounding it in the other.

Some of these goals are elusive, while those that are feasible cannot justify the murder of noncombatants. Consider an extreme claim: terrorism dramatically benefits terrorists. Pape (2005:62–63) in an oft cited study, for example, links suicide terrorism by Hezbollah, Hamas, and the LTTE to significant territorial concessions by the U.S., Israeli, and Sri Lankan governments. However, the most convincing cases he brings are campaigns against *military* targets – U.S. Marines in Beirut (1983), Israel forces in S. Lebanon (1983–1985), and the Sri Lankan air force (2001) – not attacks on noncombatants. This is consistent with empirical data suggesting that attacks on military targets are a more effective guerrilla tactic than terrorism (Abrahms 2011, for review). Among groups that perpetrate mass casualty terrorist attacks, the record is worse than mixed. While massive terrorist attacks by Chechens may have brought the Russians to the negotiating table to end the first Chechen War, the subsequent peace accords in 1996 provided no lasting settlement (Seely 2001:266–290; Wood 2007:59–79). Indeed, the Second Chechen war proved even more brutal than the first and left Chechnya far from its dream of independence. The Palestinians have fared no better. Although some credit terror for Israel's disengagement from Gaza in 2005, gains were short lived (Peters 2010). Terrorism has done little to advance the Palestinian cause, publicize their grievances, or delegitimize their opponent. Just as civilians in World War II were not demoralized by terror bombing but, in fact, resolved to fight on, Israeli citizens overcome terror attacks just as Palestinians overcome the death and destruction that comes with aerial bombing. Moreover, Israel has developed less destructive ways than massive retaliation to deal with terrorism. These include precision guided munitions, targeted killing, and obstacles that restrict freedom of movement. As a result, Israel has largely avoided disproportionate responses to major terrorist attacks leaving world opinion to side with the terror victims. Similar difficulties plague compatriot terrorism.

Compatriot Terrorism

Many observers understand terrorism solely as direct attacks on enemy noncombatants. Americans and British are victims of Al Qaeda, Israelis are victims of Hamas or Hezbollah, the French were victims of the

Algerian FLN, Russians were victims of Chechens, and so on. However, it is a constant feature of many asymmetric wars that insurgents and guerrillas also target compatriots, execute informers and collaborators, and indiscriminately attack the local population. Murder, torture, kidnapping, and sexual assault are rampant during guerrilla wars.

While scholars remain divided about the reasons for violence perpetrated against the noncombatant population, most agree that it is not wanton but rational. In some cases, guerrilla groups find it easiest to recruit those they can pay with booty and pillage exacted from soft civilian targets. These insurgents are largely undisciplined and quickly destroy the base of popular support necessary for legitimate authority (Weinstein 2007:96–126, 139). In other cases, the relationship is not entirely predatory. Instead, insurgents indiscriminately attack noncombatants to wean compatriots from the government (who can no longer protect them) and strong-arm them into supporting guerrillas (who can harm them) (Hultman 2007). Compatriot terrorism is strategic, and carefully chosen for its political and military benefits. Even well-disciplined guerrilla movements assassinate defectors, spies, collaborators (including police officers), informers, and criminal elements to enforce their authority, ensure compliance, prevent defection, and discourage collaboration (Chapter 5; also Graham 2007).

In contrast to carefully selecting targets among those deemed liable, compatriot terrorism is often indiscriminate and employed by some guerrillas when they lack the information necessary to single out collaborators. Not knowing who among the compatriot population is friendly and who is not, insurgents target noncombatants, informers, and collaborators alike in the hope that "'the innocent' will either force the 'guilty' to alter their behavior or the 'guilty' will change their course of action when they realize its impact upon 'innocent' people they care about – or both" (Kalyvas 2006:150). The result can be widespread indiscriminate violence.

When guerrillas choose to murder noncombatants to prevent defection that might cripple their movement, the question of terrorism is not easily dismissed with a declarative prohibition. At some point in their development, some movements – including those enjoying legitimate authority, just cause, and the right to fight – may find it necessary to kill compatriots even if only to abandon the practice later. Kalyvas (2006:12) makes the point forcefully: "the military resources that are necessary for the imposition of control are staggering and, hence, usually lacking. *The rival actors are therefore left with little choice but to use violence as a means to shape collaboration*" (emphasis added).

Kalyvas' claim is particularly challenging because violence is part of an evolutionary process. When groups are very weak, they frequently turn to indiscriminate violence and target compatriots without distinction. The insurgent position is precarious and violence necessary for its survival. As insurgents gain strength, they replace indiscriminate violence with selective violence and seek out informers, collaborators, and dissidents. If Kalyvas is right, then few, if any, armed groups will ever prevail, no matter how just their cause, unless they employ some sort of violence against noncombatants. This is not just a matter of guerrillas causing collateral harm, or even causing harm because they refuse to discriminate carefully between combatants and noncombatants, but of the intentional use of violence to intimidate noncombatants, elicit information, and root out and eliminate dissidents and defectors. But as the arguments against enemy terrorism demonstrate, harming noncombatants directly is impermissible because terrorism abrogates a noncombatant's right to life and is neither necessary nor effective. How might a theory of just guerrilla warfare respond to these uses of violence?

To address this question, it is first necessary to emphasize the distinction between selective violence and indiscriminate violence. The former targets liable civilians – informers and collaborators – while the latter targets noncombatants. Discussing selective violence or assassination in Chapter 5, I asserted that law enforcement cannot turn into murder. Insurgents have a right to curtail espionage and collaboration but a duty to ensure due process and proportionate punishment. Under these conditions, selective violence does not degenerate into terrorism. Indiscriminate violence, on the other hand, is nothing but terrorism, and, in response, two arguments merit consideration. First, compatriot terrorism is justified if absolutely necessary, and, second, compatriot terrorism is never absolutely necessary.

Compatriot Terrorism Is Justified if Absolutely Necessary
As with enemy terrorism, one might contemplate the claim of necessity in the face of an existential threat. Using this argument, compatriot terrorism, like enemy terrorism, is necessary to forestall defeat. Military defeat alone, however, is insufficient cause. Defeated insurgents must

face genocide, ethnic cleansing, or some grave violation of humanitarian law. This claim might be very compelling. Unlike conventional wars, which are rarely waged to avoid extermination or genocide, guerrilla wars are often fought for precisely that purpose, particularly when states respond with brutal counterinsurgency tactics. Losing such a conflict may have serious repercussions. Or, it may not. Wartime barbarism does not always presage postwar genocide or violence. Isolated human rights abuses ("illegal detention, surveillance, clamp downs on free speech and abductions"), not widespread atrocities, were the norm in postwar Chechnya and Sri Lanka (HRW 2008–2011; HRW 2011–2013). And, indeed, these conflicts, in addition to those in the Palestinian territories, Indonesia, and Turkey, score very low on the factors that predict genocidal outcomes (Harff 2003). In retrospect, the defeated party had little cause to anticipate extermination or ethnic cleansing. On the other hand, it is reasonable for an insurgent group to predict the future based on its enemy's present behavior. Guerrillas will need good reason not to fear defeat. Here, the international community must assume a central role assuring post-conflict peace and, as noted earlier, be all the more inclined to do so if insurgents forswear terrorism. Lacking the prospect of international aid, appeals to necessity may resonate loudly were terrorism the only means feasible to effectively forestall an existential threat. No evidence, however, suggests that this is the case; compatriot terrorism is not absolutely necessary.

Compatriot Terrorism Is Not Absolutely Necessary
While assassination and selective violence may prove necessary, not all guerrilla movements need resort to indiscriminate violence to build support. Among those that do, Kalyvas (2006:12–13) has demonstrated how compatriot terrorism is counterproductive. Certain to provoke vehement outrage, indiscriminate compatriot violence often drives civilians into the hands of rivals. As a result, many insurgents abandon indiscriminate violence for positive incentives and security (Wood 2010). Indeed, in some conflicts, including those where guerrillas enjoy a firm measure of just cause as in East Timor, Eritrea, and the Western Sahara, reports of enemy and compatriot terrorism are rare. Other conflicts drown in enemy terrorism but are light on indiscriminate violence. While Palestinians have eliminated many informers and collaborators and fought fierce internecine battles, reports of compatriot terrorism are not widespread (B'Tselem 2013).

Whether all guerrilla organizations can bypass indiscriminate violence on the way to hegemony and ultimate victory is an empirical question that remains to be answered fully. There is no doubt that many rebel groups mature and turn away from indiscriminate violence and the wanton murder of collaborators, real or imagined. Just cause and legitimate authority should not accommodate any form of terrorism. One would like to speculate that groups that repudiate terrorism mature faster and reach their goals more quickly than those that embrace impermissible violence. If a group does not mature to the point where it can give up indiscriminate violence, then the outrage and dissent this provokes should render insurgents incapable of achieving their goals. And when this happens, insurgents forfeit their right to fight. In warfare, however, progress is not linear. Fortunes wax and wane, and compatriot terrorism may retreat only to resurface as insurgents lose territory and control. Under these circumstances only selective violence, targeted killing, and assassination constrained by proportionality and due process remain permissible options.

The preceding discussion repudiates tactics that permit insurgents to intentionally murder or injure noncombatants who provide no war-sustaining support. Civilians who tender such aid, however (and there are many), present a much thornier problem for these are targets that a guerrilla army may permissibly disable without facing charges of terrorism. Participating civilians, like collaborators and informers, face nonlethal disabling force.

PERMISSIBLE ATTACKS ON LIABLE CIVILIANS

Philosophers (and politicians) who consider the permissibility of terrorism draw two lines. One rules out terrorism, that is, direct lethal attacks on noncombatants, but permits deadly strikes against high-ranking civilians (Chapter 5; also Corlett 2003). The assassination of Hamas leaders noted in Chapter 5 is a case in point. The second line, which I pursue here, also proscribes terrorism but permits direct attacks on civilian structures and property if unaccompanied by injury or loss of life. The first aims to disable civilians who take on a direct, war-fighting role. The second hopes to intimidate noncombatants to force government leaders to alter their policy and/or to disable facilities that provide war-sustaining services. Ascriptions of liability surface in all cases.

The Liability of Noncombatants

International law and orthodox just war theory prohibit attacks on non-combatants and, indeed, on any civilian target, absolutely. This is the principle of noncombatant immunity. Revisionist theories, on the other hand, impute a measure of liability to civilians who help sustain an unjust war. In Jeff McMahan's (2009:219 emphasis added) view, for example, "Civilian complicity in an unjust war may also be relevant to the justification of the *intentional destruction of civilian property* as a means of applying pressure to a government and its civilian supporters." Benbaji (2013:176, emphasis added) presents a more detailed position:

... in order to deter civil society from perpetuating occupation and coloniza-tion, freedom fighters are allowed to target civil society, i.e., violate its public space, by *destroying buildings, streets and public institutions*, in those rare cases in which this is necessary for attaining (what they believe to be) a legitimate goal. Notwithstanding this leniency, militants ought to do whatever they can in order to avoid killing or harming enemy civilians (while attacking the civil society of which these civilians are a part).

Guerrillas and insurgents "deter" civilians from perpetuating injustice by forcing them to pay a high price. The reason for targeting some civil-ian objects directly, a grave violation of humanitarian law, returns to liability, albeit collective liability. Unable to assign individual culpabil-ity to civilians, Benbaji (2013: 195): suggests "treating society as if it is responsible for the evil against which [freedom fighters] fight." Thus the public sphere is fair game while noncombatant lives remain protected. Underlying this argument is a broad view of civilian culpability coupled with a very sharp distinction between death and injury on one hand, and the destruction of property on the other. Both claims are problematic but not without remedy.

Although the destruction of some "public institutions," such as a government building or financial institution that provides war-sustain-ing aid, is often permissible, Benbaji, like McMahan, permits the more general destruction of civilian buildings, roads, and commons because all unjust civilians bear a measure of liability. While this claim has some intuitive basis, it fails for many pragmatic and moral reasons. Putting aside the epistemic difficulties of sometimes determining an unjust war, it remains nonetheless difficult to assign liability with any accuracy or consistency. When civilians join a state's war-sustaining institutions or a guerrilla organization's political wing, they take a clearly definable and recognizable action. They get up in the morning, go to work, and spend their day in activities that support the war effort. Participating civilians are clearly defined by where they work and by the job they do. Civilians who do not participate in this way but only by voting or paying taxes or refusing to emigrate do not provide any clear, war-sustaining support, if, in fact, they support government policy at all. They are noncombatants. Some vote for the opposition. Many pay taxes only to avoid onerous penalties. Most citizens neither possess the means to leave their country nor have any place to go could they afford it. Still others may oppose government policy but lack the means necessary to bring about change. As a result, the liability of ordinary citizens is extremely diffuse and, unlike that of participating civilians, rarely passes any significant threshold of contribution (Fabre 2012:59–60; Primoratz 2013:47–56).

Moreover, the destruction of property is not benign. Apart from immediate harm are the longer term effects that come when vital services collapse following devastating attacks. This is particularly true when health, water, sanitation, manufacturing, and agricultural facilities are destroyed or disabled. The effects of war on public health are far reaching and devastating (Darwish et al. 2009; Fattouh and Kolb 2006; Nuwayhid et al. 2011; Utzinger and Weiss 2007). Apart from disease and hunger, noncombatants are also left to suffer fear, anxiety, depression, and other forms of mental suffering. One is hardly, therefore, able to "violate the public space" or undermine civil and economic life without causing extreme hardship.

Taking into account the limits of liability and the severe ramifications of destroying physical infrastructures, one may nevertheless continue with attacks on the public sphere on two conditions. First, permissible attacks are confined to war-sustaining or war-fighting facilities and cannot spill into either the private domain or into public areas of no military value. Second, the consequences of destroying war-sustaining infrastructures cannot be ignored. When disabling war-sustaining facilities, the rules of proportionality apply and require belligerents to prevent severe humanitarian crises just as they must when imposing economic blockades or sanctions (Chapter 8). These conditions demand a closer look at sublethal, nonlethal, and cyber weapons and offer states and non-states permissible means to disable civilians providing war-sustaining aid.

The Liability of Civilians Providing War-Sustaining Aid

Despite objections by most nations, the United States permits attacks on "economic objects of the enemy that indirectly but effectively support and sustain the enemy's war-fighting capability" (Department of the Navy 2007; also Dinstein 2010:95–96; Schmitt 2010b:717–718). The Bush administration, in turn, targeted "al Qaeda leaders responsible for propaganda, recruitment, [and] religious affairs" (Bush 2010:218). These views recognize that civilians who provide war-sustaining aid are integral to modern warfare. Among insurgents fighting in such places as East Timor, Kosovo, the Palestinian territories, and Southern Lebanon, a political wing provides financial experts to oversee loans, fundraising, and money laundering; legal personnel to file international lawsuits; media specialists, journalists, and diplomats to pursue public diplomacy; and telecommunications technicians to maintain TV, radio, and internet outlets. Participating civilians also include students and workers who liaison between the political and military wings and provide logistical support and passive intelligence. Among nation states, civilian bureaucracies provide the same services.

What is the liability of these agents? How may states or guerrilla armies effectively disable participating civilians? To sort this out, the International Committee of the Red Cross (ICRC) guidelines detailed in Chapter 3 draw a distinction between direct and indirect participation in armed conflict. Direct participation reflects acts of war fighting by civilians as they take up arms against an enemy. Indirect participants, on the other hand, embrace a class of civilians whose aid is unlikely to pose any immediate and significant danger. Direct participants enjoy no immunity, while indirect participants enjoy protection from any sort of intentional harm.

Although the ICRC does not say so explicitly, it is reasonable to assume that indirect participants are immune from physiological *and* psychological harm. As I argued in Chapter 3, however, there are solid grounds to attribute participatory liability to some civilians. The greater a person's contribution to armed conflict, the greater one's liability and, therefore, the harm one may suffer. As such, there is room to consider the degree of force appropriate to disable civilians who provide war-sustaining aid. Moreover, there is a practical urgency to do so. Neither states nor guerrillas can effectively wage war without crippling war-sustaining infrastructures and facilities. The United States clearly understood this as it makes room for such strikes. The same philosophy guided Israeli attacks against similar targets, or what HRW (2007b) called "associated targets" (that is civilian infrastructures associated with guerrilla organizations) in Lebanon in 2006 and Gaza in 2008 and 2009.

Attacks by Israel and the United States against civilian facilities providing war-sustaining aid highlight the conundrum that leaves states to disable such facilities with high-explosive kinetic weapons (and cause substantial destruction) or do nothing. Each choice ignores the limited liability of participating civilians that both protects them from lethal harm but renders them liable to proportionate and disabling harm. The challenge is to define permissible force more closely. In this context, sublethal, nonlethal, and cyber weapons that inflict minimal physical harm can be useful.

Disabling Participating Civilians: Nonlethal, Sublethal, and Cyber Weapons

Modern technology offers several ways to disable war-sustaining civilian facilities with minimal casualties.

1. *Physical destruction of infrastructures (sublethal kinetic weapons).*
   In its least sophisticated form, guerrillas will plant bombs in empty buildings and warn civilians away to minimize casualties. Moving away from terrorism in the late 1990s, the IRA struck civilian targets to inflict economic damage, undermine civilian morale, demonstrate the government's inability to protect its citizens, and publicize their national cause by striking targets in "mainland" England (Rogers 2000). Similarly, Hamas rocket attacks on Southern Israel from 2001 through 2008 caused few fatalities but brought widespread paralysis and economic harm. Warnings prior to bombing may help mitigate injury and loss of life. Nevertheless, some severe injuries are inevitable. Kinetic weapons are only sublethal when they do not cause mass casualties.

2. *Nonlethal warfare.*
   Nonlethal warfare takes advantage of technological advances in biology, chemistry, and physics to develop physiological means to incapacitate civilians who work in war-sustaining facilities without causing permanent injury or significant loss of life. Many forms of nonlethal warfare remain the purview of state armies and include

chemical calmatives, electromagnetic technologies, and low impact kinetic arms (Gross 2010a:77–99; Kaurin 2010; Koplow 2006; Lango 2010). These might allow states, and conceivably non-states, to destroy a physical facility without causalities by forcibly vacating and then arresting the occupants of war-sustaining infrastructures without appreciable injury. Unlike sublethal weapons, which almost certainly injure any person struck, nonlethal weapons carry no more than a one percent chance of killing or severely injuring those affected.

3. *Cyber-warfare.*
Cyber-warfare sabotages computer networks by denying service, stealing and manipulating data, or destroying hardware through the introduction of malevolent viruses. Its consequences may include economic costs following the loss of data or computer facilities, severe psychological suffering as financial, communications, and social networks collapse, or bodily injury and loss of life following the destruction of vital transportation, military, or medical infrastructures. The immediate effects of a cyber-attack, however, are to date neither lethal nor physically injurious.

As these weapons are used, two concerns stand out. First, what harms do participating civilians and noncombatants suffer? Second, are these harms permissible? Answers to these questions will help answer the final question of this chapter: Is it permissible to use these weapons, particularly cyber weapons, to terrorize noncombatants?

SUBLETHAL, NONLETHAL, AND CYBER WEAPONS

Sublethal, nonlethal, and cybernetic technologies offer reasonable means to target war-sustaining civilian infrastructure in way hitherto unavailable. Table 7.2 summarizes the direct and collateral harm that participating civilians and noncombatants may experience when these weapons are used.

Sublethal Kinetic Weapons

As guerrillas utilize sublethal weapons they will cause direct harm and collateral harm. Direct harm reflects the destruction of the infrastructure targeted and injury to the participating civilians working there.

TABLE 7.2. *The Direct and Collateral Effects of Sublethal, Nonlethal, and Cyber Weapons*

	Direct Harm	Collateral Harm
Sublethal kinetic weapons	Destruction of infrastructures Minimal injury and loss of life	Terrorization, severe mental suffering
Nonlethal weapons	0.5% mortality; 0.5% morbidity (Davison 2009:1–8; Fidler 1999)	0.5% mortality; 0.5% morbidity
Cyber weapons	Economic breakdown and military impairment; Mental suffering	Knock on effects, significant but proportionate collateral harm

Collateral harm from such attacks is often psychological. Although relatively few people suffer directly when insurgents bomb buildings, many more civilians suffer significant mental suffering.

Sublethal Kinetic Weapons: Direct Harm
Placing explosive devices in a target facility is, of course, nothing new. Guerrillas, however, must get close enough to place bombs undetected. And, if they are intent on reducing civilian casualties, they will issue warnings. One striking example is the change in IRA strategy. Moving away from terrorist attacks in the mid-1990s, the IRA struck commercial, banking, retail, and transportation sites causing widespread destruction in Manchester and London in 1996. Voicing the IRA's frustration at the lack of political progress following an eighteen-month ceasefire, operatives chose to disrupt commerce and transportation, two critical war-sustaining infrastructures. There tactics were largely nonlethal; warnings kept civilian casualties to two dead in Canary Wharf and 212 mostly lightly injured in the Manchester bombing (Carley and Mackway-Jones 1997). These attacks were economic and, by some accounts, put political negotiations back on the table (Rogers 2000).

Not all guerrillas can engineer similar attacks. Palestinians, Kosovar Albanians, and East Timorese insurgents, for example, have had little chance of getting close to war-sustaining infrastructures in Israel, Serbia, or Indonesia. Of late, however, rocket capabilities have altered this assessment. Some guerrilla groups can now attack civilian targets from afar as did Hamas beginning in 2001. These were not discriminate attacks aimed at military, dual use, or civilian war-sustaining targets. Rather, Hamas bombarded small towns and villages indiscriminately, causing few deaths but significant fear and panic. By severely disrupting civilian life in Israel, Hamas hoped to position itself as a major player in the Middle East, upstage its Fatah rivals, enlist support at home and abroad, trumpet its capability of inflicting harm on Israel, retaliate for targeted killing, and incapacitate many urban centers. In this regard, Hamas was partially successful. Unable to cause large numbers of casualties, Hamas made it difficult for Israel to respond forcefully (as Israel would in 2014 when much more powerful missiles threatened major population centers). The rocket attacks continued for seven years until Israel finally lost patience and launched a large-scale military campaign. Despite Hamas's inability to gain any tactical advantage in the Gaza War (2008–2009) or significantly alter Israeli policy, Israel's heavy-handed reaction brought near-unanimous denunciation from the international community. This, too, was a coup for Hamas.

The lesson of Hamas's use of Qassam rockets is twofold. First, kinetic attacks can be effective if they avoid overwhelming deadly harm. Second, Qassam rocket attacks do not meet the demands of participatory liability, because they did not target war-sustaining targets. But they could serve the purpose of disabling war-sustaining infrastructures assuming three conditions: small warheads, sufficient technological sophistication to allow discriminate targeting, and early warning. How small a warhead depends on a target's vulnerability. As one side hardens and protects its facilities, the other side will respond with more powerful weapons. In general, however, the purpose of attacking war-sustaining sites with relatively weak rockets is to disable, rather than entirely destroy, such a facility. More powerful missiles, capable of significant damage and typical of those fired by Hamas in 2014, are more likely to cause unacceptable lethal harm and trigger a massive response. Discrimination is important to protect noncombatants from the harms that participating civilians might suffer in the course of attacks on their facilities. Because most participating civilians work in public or commercial institutions, ascertaining their location should not require sophisticated intelligence capabilities. Warnings are necessary to further minimize direct and collateral harm. Warnings are themselves useful to disrupt routine operations among participating civilians.

Kinetic weapons are never free of injury or death so that direct but defensible attacks on war-sustaining facilities can only aim to efficiently cripple an infrastructure while employing extraordinary measures to minimize casualties (such as warnings). Despite such measures, kinetic weapons will wreak widespread psychological harm precisely because there is always an attendant risk of severe physical injury. Physical harm may be marginal but psychological injury is an abiding feature of sublethal, nonlethal, and cyber-warfare. Participatory liability excludes severe injury and loss of life but what about mental suffering?

Sublethal Kinetic Weapons: Collateral Harm and Mental Suffering

Psychologists have long known that exposure to a traumatic event increases the risk that a person will suffer from anxiety or depression. While no one died following the massive 1996 IRA bomb attack in Manchester, for example, 18 percent of those injured presented with emotional distress and medical problems (Carley and Mackway-Jones 1997). Early studies following similar attacks revealed significant levels of stress and anxiety among bombing victims but little evidence of widespread or lingering "psychiatric morbidity" (Rapin 2009:168). In the years immediately following 9/11, American researchers searched for clues of mental distress in the incidence of post-traumatic stress disorder (PTSD). PTSD, in fact, became the standard that researchers would measure the psychological effects of war and terrorism for most of that decade.

PTSD is a severe anxiety disorder that occurs following exposure to a traumatic event involving death or serious injury to which individuals respond with "fear, helplessness, or horror" (Yehuda 2002:108). PTSD victims re-experience their trauma through intrusive recollections, dreams, and hallucinations and suffer from insomnia, uncontrollable anger, and difficulty concentrating. PTSD can impair day-to-day functioning and puts patients at increased risk for depression, drug and alcohol abuse, eating disorders, suicidal thoughts and actions, cardiovascular disease, chronic pain, and autoimmune diseases. Estimates of PTSD in the United States prior to 9/11 range from 5 to 6 percent of men and 10 to 14 percent of women (Yehuda 2002). Although studies around the world demonstrate a significant increase of PTSD and other anxiety disorders following terror attacks, these numbers drop off considerably after a few months (Sinclair 2010; Yehuda et al. 2005).

In contrast to the one-off character of 9/11, residents of Southern Israel experienced many years of ongoing rocket strikes. Between 2001 and 2008, 9 civilians lost their lives, 600 were injured, and over 95 percent of the inhabitants of Sderot and the communities near the Gaza border experienced rocket fire firsthand (Gelkopf et al. 2012; Israel Security Agency nd.; Sderot Media Center 2009). As in the United States, PTSD and related anxiety disorder symptoms were common in the aftermath of attacks. Nevertheless, a large percentage (40–78 percent) of victims was symptom free (Zemishlany 2012) and "the emotional impact ... fairly moderate," (Bleich et al. 2003) an outcome that did not change much after forty-four months of intermittent attack. People, it seems, are astonishingly resilient in the face of terrorism, an outcome researchers attribute to "coping mechanisms," "self-efficacy," strong community networks, and social cohesion (Bleich et al. 2006). Closer studies reveal however, two distinct groups of individuals suffering fear-related effects from terrorism. One group exhibits the common PTSD symptoms including psychological distress, insomnia, and exaggerated startle responses. The other group does not re-experience a *past* trauma but suffers instead from "anticipatory anxiety," that is, fear and dread associated with *future* attacks. These fears only grow with time and as long as the threat persists. While relatively few people suffer from PTSD following a terrorist attack, many more suffer from various degrees of debilitating fear (Sinclair and LoCicero 2007; Sinclair 2010). These are the secondary victims of terrorism. They experienced no physical harm whatsoever nor were they necessarily present at the site of a terror attack.

The difference between the incidence of PTSD and anticipatory anxiety is striking. While the incidence of PTSD dropped across a U.S. sample from 17 percent two months after 9/11 to 5.8 percent six months after the attacks, 60–65 percent continued to fear future terrorist attacks and worry about harm befalling their family (Silver et al. 2002). Widespread fear, rather than specific incidence of PTSD, is the more pervasive effect of terrorism and more accurately reflects the psychological malaise that accompanies war and terrorism. Terrorism fears correlate with anxiety, depression, insomnia, and feelings of incapacitating helplessness (Sinclair & Antonius 2012a). Random bombings and rocket attacks lead to fear induced changes in behavior. Secondary victims of terrorism avoid public transportation, public forums, and confined venues such as restaurants, cafes, and theaters, while others often disparage those ethnic

groups they identify with terrorists (Canetti-Nisim et al. 2009). As a result
of increased isolation and ethnocentrism, social intercourse diminishes.
Terrorism brings a constant sense of anxiety, fears about harm to family
members, and heightened vigilance with regard to suspicious packages
and people. Ruminations about recent attacks and fruitless efforts to predict future strikes in an atmosphere of acute fatalism become a constant
preoccupation.

By inflicting such psychological harm, sublethal kinetic weapons can
significantly disrupt and paralyze civil society. Sometimes, strong family and social networks or psychological care can mitigate the fears that
terrorization causes. Civil defense, warning systems, bomb shelters, and
reinforced schools are also essential to maintain a semblance of daily life.
For others it is enough to leave. Unlike PTSD, which persists after the initial trauma and regardless of proximity to the danger zone, anticipatory
anxiety dissipates once one leaves an area threatened with attack.

As a result, some communities, such as Sderot, saw their numbers
decline significantly as 10–25 percent of the population fled (Diamond
et al. 2010). Among those who remain in affected communities, work-
place efficiency deteriorates and turnover and absenteeism increase,
while performance, morale, and motivation suffer (Howie 2007). Civil
society perseveres, but community and economic life clearly suffer
from sublethal and psychological warfare. Sublethal kinetic warfare,
then, affects war-sustaining facilities and the civilian population in several ways. First, explosives and rockets disable physical infrastructures
directly. Second, the ensuing psychological distress severely impairs the
ability of participating civilians to provide war-sustaining aid effectively.
Noncombatants suffer the same distress no less significantly in the form
of collateral harm that can undermine the public's morale. The question
is: Is it permissible to inflict such mental suffering upon participating
civilians and noncombatants?

Inflicting Mental Suffering: No Harm, No Foul?
As the prospect of physical harm wanes with some forms of warfare, concern for their psychological effects rises. Considering cyber-warfare, the
Tallinn Manual on the International Law Applicable to Cyber-Warfare
declares: "While the notion of attack extends to injuries and death
caused to individuals, it is, in light of the law of armed conflict's underlying humanitarian purposes, reasonable to extend the definition to

serious injury and severe mental suffering that are tantamount to injury" (Schmitt 2013:93).

Writers of the Tallinn manual are concerned about the psychological fallout from cyber war but do not define severe mental suffering. In general, international law has also refrained from defining "suffering," and offer little guidance for prohibiting weapons that cause "unnecessary suffering" or for setting the bounds of permissible interrogation techniques before they grow into torture. Nevertheless, there are some insights to glean. Considering reins on prohibited weapons, the ICRC (2006:947) hopes to ban weapons that cause "long term or permanent alteration to the victims' psychology or physiology." Turning to the torture debate and in the absence of firm guidelines, U.S. lawyers tried to define torture in terms of suffering that caused the kind of pain "accompanying serious physical injury" and, later, a level of mental and physical suffering distinguished by its intensity and duration (Levin 2004). In each of these cases, severe mental suffering is a function of intense, prolonged, and irreversible pain.

From the evidence cited, it is far from clear that the mental suffering resulting from sublethal kinetic weapons causes the kind of intense, prolonged, and irreversible pain that comes with torture or inhuman weapons. Rather, those who suffer PTSD and anticipatory anxiety show remarkable resiliency and coping strategies that can diminish or alleviate mental suffering. Given the liability of *participating civilians*, these psychological harms, which do not result in death or permanent injury, are reasonable in the course of discriminate warfare accompanied by warnings and low intensity weapons. Noncombatants, however, are not liable to direct harm. Nevertheless, the mental suffering and temporary incapacitation that will inevitably affect many bystanders can be proportionate *collateral* harm if it is not excessive relative to the military value of disabling war-sustaining facilities. And, in figuring proportionate harm, one must also consider the collateral effects of shutting down certain services such as telecommunications, transportation, higher education, banking, or media. These effects can range from inconvenience to psychological distress to significant injury or loss of life if transportation or security services, for example, are significantly disrupted. Only the latter harms, a concern attending cyber-warfare, may prove disproportionate.

It might be objected that the real problem with sublethal warfare is not the psychological harm it causes but the manner in which it arises.

Inducing mental suffering and terror requires a threat of imminent death or severe injury that is not entirely mitigated by discriminate, low-intensity kinetic warfare. To terrify a community, in other words, some people must die or suffer injury. For this reason, nonlethal and cyber weapons are a superior choice for those with sufficient pick of weaponry.

Nonlethal Weapons: Direct and Collateral Harm

Nonlethal weapons currently under development by state armies include chemical, acoustical, and electromagnetic technologies. However, there are no indications that insurgents possess nonlethal weapons, nor are there much data about their operational effects. Nevertheless, nonlethal weapons deserve brief mention on the likelihood that they, like missiles, may eventually wind up in the hands of non-state actors.

Chemical calmatives render subjects unconscious by depressing neurological functions. In practice they would allow troops to take a war-sustaining or military facility with minimal loss of life. State actors would find this particularly useful when then cannot distinguish between combatants and noncombatants. Although the United States endorses the development of calmative weapons, they have only been used with mixed success by the Russians against Chechen insurgents in Moscow in 2002 and remain largely untested (Fidler 2005; National Research Center [NRC] 2003). Electromagnetic technologies developed by the United States utilize microwaves that create the sensation of intense burning but without causing tissue damage. Such "active denial systems" (ADS) are designed to repel mixed crowds of combatants and noncombatants without causing permanent injury. Acoustical and kinetic nonlethal weapons (e.g., pepper balls, beanbags, or rubber bullets), also repel or subdue crowds by causing a minimal pain and suffering (Koplow 2006). As these weapons are deployed, participating civilians and noncombatants may suffer direct or collateral harm respectively in attacks on war-sustaining facilities. Like sublethal kinetic weapons, nonlethal weapons also carry some risk. While 99 percent of all persons exposed to an incapacitating chemical agent or electromagnetic radiation will experience reversible and transient pain, 1 percent may die or suffer significant injury. As such, their use is governed by the same rules that govern sublethal kinetic weapons: participating civilians are liable to direct but nonlethal harm as belligerents disable war-sustaining facilities, while noncombatants may bear proportionate but collateral casualties.

Cyber Weapons: Direct and Collateral Harm

Unlike nonlethal weapons, cyber weapons target facilities, not people, and unlike kinetic weapons they cause no immediate injuries. In 1999 and 2007, the United States and Israel respectively disrupted Yugoslav and Syrian air defense systems. In 2007 and 2008, Estonia and Georgia suffered widespread denial-of-service attacks, most probably at the hands of Russia. In 2010, the Stuxnet worm infiltrated computers of the Iranian nuclear program and destroyed a large number of centrifuges, while the 2012 Flame virus accessed sensitive information from the same facility (Barzashka 2013; Dinniss 2012:281–292; Nakashima et al. 2012). While attacks by non-states have also disrupted service, stolen information, and destroyed data, the targets they attack are usually weakly protected civilian facilities. In 1998, a twelve-year-old hacked into the Roosevelt Dam in Arizona and, by some accounts, assumed total control of the system. In 2000, a cyber-intruder dumped 250 million tons of sewage into the rivers and parks of Queensland, Australia after hacking into a sewage treatment plant (Dinniss 2012:282–283, 285). In 2012, hackers erased 75 percent of the data on corporate computers of the ARAMCO Oil Company in Saudi Arabia.

The impact of many of these cyber-attacks was, however, short-lived. Many strikes (denial-of-service attacks) only shut down websites and created significant, but temporary, inconvenience. Others successfully disabled strategic facilities for short, but critical, periods of time. No participating civilian or noncombatant suffered any harm whatsoever. Alternative scenarios are not so forgiving. When cyber-attacks impair crucial infrastructures and flood cities with polluted water, disrupt air traffic, derail trains, or jeopardize hospital care, noncombatants can suffer great harm that may easily be disproportionate (Cavelty 2012; Lewis 2010; Parker 2009). Such belated consequences are the "knock-on" effects of a cyber strike and not always easy to evaluate. Lawyers and theorists usually content themselves with the immediate effects of an attack and ignore intermediate and long-term mortality, disease, pain, mental distress, and financial loss that come with armed conflict. In conventional war, this makes some sense. The immediate casualties are so devastating as to dwarf any long-term considerations. In cyber wars, however, the collateral effects may overwhelm a war's immediate casualties, thereby requiring belligerents to make a reasonable effort to weigh the foreseeable harms that come much later. These harms join others as part of the proportionality calculation that assesses the collateral harm that armies may inflict as they wage cyber and other forms of sublethal war.

Nevertheless, no cyber strike to date has brought the devastating knock on effects that some dread. As a result, there is a growing perception that cyber-attacks need not kill or severely injure civilians to foster widespread fear (Ariely 2008; Beggs 2008; Mantel 2009; Mshvidobadze 2011). To date, technicians repair damaged hardware, executives fuss over lost data, and people return to work, no worse for the wear. But this may be a dangerous illusion and one of growing concern for lawmakers and scholars who fear that states or non-states may turn increasingly to cyber weapons because they are effective and, seemingly, victimless (Dormann 2004; Kelsey 2008; Lewis 2011; Schaap 2009; Schmitt 2011). Here, as with other nonlethal weapons, belligerents must be sensitive to the prospect of mental suffering.

Cyber Weapons: Mental Suffering

International law prohibits "acts or threats of violence ... to spread terror among the population" and many worry that cyber threats may do the same (Schmitt 2013:93). This is a significant conceptual leap because nowhere does anyone consider that anything but a *physical* threat can precipitate the fears and mental anguish necessary to terrorize civilians. Can cyber-warfare inflict the requisite level of suffering to either incapacitate participating civilians or terrorize civilians? If so, is it permissible to use cyber weapons for either purpose?

The psychological effects of cyber-warfare are at the heart of the Tallinn's manual's concern but remain mostly uninvestigated. Nonetheless, initial data show a significant increase in the hormone cortisol following simulated cyber-attacks in the laboratory (Canetti et al. in press). Relative levels of cortisol reflect stress and anxiety. Following exposure to danger, cortisol levels jump to improve memory functions and increase resistance to pain. Cortisol levels elevated by acute stress, however, will impair decision making. Presently, these data remain undifferentiated and experiments have yet to gauge the full effects of cyber-warfare on human behavior. To fill this gap, two other sources of data are suggestive. One describes the psychological impact of identity theft, burglary, and cyberbullying (Beaton et al. 2000; Brown and Harris 1989; Hoff and Mitchelle, 2009; Maguire 1980; Roberts et al. 2013), and the other the mental anguish that accompanies kinetic terrorism as described previously.

Extrapolating from these data, one can expect two sorts of psychological suffering in the wake of cyber-warfare. First, individuals will experience the distress and anxiety that come with the disruption of everyday services when people cannot ensure their privacy, access their bank accounts, fill prescriptions timely, travel as necessary, maintain communications, and run their computers. Scenarios depicting the impact of cyber-warfare variously describe denial of service, inability to enter websites, lost or stolen data, the unauthorized disclosure of confidential information, destruction of computer infrastructures, the collapse of social networks, and, in the worst (and, as yet, hypothetical) cases, unknown and perhaps catastrophic effects as critical transportation, security, water, health, and financial systems fail. While in most cases these disruptions are free of the fears of injury or death that accompany kinetic terrorism, they would seriously impair people's ability to function effectively in a modern, technological society. These effects may be particularly severe among vulnerable groups such as the poor and elderly. But ordinary citizens may be no less affected. On January 21, 2014, South Koreans awoke to find that thieves had stolen the credit card numbers, names, and addresses of 40 percent of the population. The immediate result was widespread panic, system crashes, a run on banks to cancel credit cards, and massive lawsuits (Lee 2014). The culprit in Korea was an insider, but the effect upon the citizenry was no different than a cyber strike. Attacks such as these feed a second, but amorphous fear that comes with the constant assault by unknown, malevolent agents whose agenda is neither clear nor predictable. Cyberattacks stoke anxieties about loss of control and unpredictability that might be as inescapable as those accompanying war and kinetic terrorism (S. J. Sinclair and F. Antonius, pers. comm.).

Despite these fears, mental suffering must be seen in perspective. Terrorism and the fear it generates does not usually bring secondary victims the kind of severe, persistent, and irreversible mental harm that Tallinn Manual deems analogous to physical injury and loss of life. While studies just cited detail significantly greater rates of post-traumatic stress disorder (PTSD) and other anxiety disorders among terror victims, they also point to remarkable resilience among many segments of an affected population. Terrorism undermines civilian life and disrupts facilities that employ civilians, but the majority of the population does not suffer long-lasting or irreversible injury. Cyberterrorism would, most likely, cause even less severe mental suffering. Nevertheless, civilians will invariably suffer from cyber-warfare. Some will endure chronic PTSD and other anxiety disorders, while others will suffer from persistent fears that disrupt their lives as parents, students, and workers. As with sublethal kinetic weapons, the harm expected of cyber weapons does not exceed that to which participating civilians are liable. Insofar as cyber-warfare avoids lethal knock-on effects, it may offer an effective and permissible means to target civilian infrastructures, disrupt civil society, and demoralize a population to attain military advantages. Without inflicting severe, persistent, and irreversible suffering, however, cyber-attacks also seem to fall short of causing the harm necessary to violate noncombatant immunity. Under these conditions, cyber weapons may prove useful but much depends on the nature and intensity of knock-on effects. When these effects are severe, then cyber-attacks are no different from kinetic attacks and subject to the same humanitarian constraints. But when the effects fall short of physical harm, the door opens to cyberterrorism, that is, direct cyber-attacks on noncombatants.

Cyberterrorism

The goals of cyberterrorism are little different from those of kinetic terrorism. Cyberterrorism, writes William L. Tafoya (2011) in an FBI Bulletin is "the intimidation of civilian enterprise through the use of high technology to bring about political, religious, or ideological aims, actions that result in disabling or deleting critical infrastructure data or information." "Cyberterrorism," he suggests, "targets civilians – the innocent." Cyberterrorism is, therefore, different from cyber-warfare. Cyber-warfare is a means for disabling war-sustaining facilities not intimidating civilians.

How might cyberterrorism intimidate the innocent? There are, of course, many kinds of intimidation. Many are violent and include threats of death and injury. This is how mass casualty terrorism works. Some forms of cyberterrorism may follow this route by destroying nonmilitary infrastructures such as water, sanitation, civilian air control, public health services, and law enforcement facilities. These might bring severe knock-on effects, death, and injury. Assume, however, that cyberterrorism only brings a degree of psychological harm and mental suffering that will not be severe, prolonged, or irreversible. To do this, terrorists will mount widespread attacks to seize control of computers or phones, disable social networks, steal identities, disrupt telecommunications and transportation facilities, and cripple financial services. Under these circumstances, the effects of cyberterrorism may not rise to a level of suffering, injury, and destruction sufficient to violate noncombatant immunity or, if they do, the psychological harm to noncombatants might be offset by its benefits. While cyberterrorism may undermine civilian welfare, such harm pales before the devastation of kinetic warfare whose collateral effects may far exceed the *direct* harm of cyber terror. In this case, cyberterrorism is the lesser evil, which may "contain war's destructiveness and facilitate restoration of civil society after the conflict" (Shulman 1998:964; also Haslam 2000:173–174).

One should, however, urge caution before embracing cyberterrorism. If the physical effects prove benign, one is still left with a degree of mental suffering. Moreover, there are going to be considerable concerns about escalating violence once noncombatants are in the cross hairs. If social networks are permissible targets, then why not go for medical facilities? If mental harm is permissible, then why not use more lethal means to disable noncombatants? In this vein, there is clearly something aggressive about a cyber-attack on noncombatants.

To get a better handle on this, it is useful to look at the controversy surrounding chemical and electromagnetic nonlethal warfare. Incapacitating chemical or electromagnetic weapons were developed to disable insurgents who fight without uniforms among civilians (Gross 2010b). While live fire might harm disproportionate numbers of noncombatants, nonlethal weapons would not. By design, nonlethal weapons target combatants and noncombatants alike thereby allowing troops to move unhindered, incapacitate everyone in sight, and then sort out civilians from soldiers. The moral and legal difficulty lies in the act of harming noncombatants directly, a gross violation of noncombatant immunity. The solution lies in the intent of the attackers who seek to protect noncombatants from death and injury while they pursue un-uniformed combatants, not to simply inflict a lesser harm. The justification for these sublethal attacks on noncombatants is utilitarian: direct but moderate transient and incapacitating harm causes less suffering than proportionate but deadly and permissible collateral harm. Unlike cyberterrorism, however, nonlethal strikes are not intentional at all. Nonlethal weapons only target combatants and noncombatants as one due to their inability to distinguish between the two groups. Were distinction possible, no argument would permit targeting noncombatants with any weapon.

Targeting the public sphere, however, cyber terrorists knowingly seek out noncombatants to demoralize the civilian population and bring pressure upon a government to meet their demands. Noncombatants, however, cannot be subject to direct harm that significantly impairs physiological or psychological well-being. Not only do noncombatants pose no threat, singling them out for any intentional harm whatsoever constitutes a grave affront to human dignity to which noncombatants are entitled. Noncombatants are not instruments of war and warrant every protection in this respect.

RETHINKING ATTACKS ON CIVILIANS IN MODERN WARFARE

Terrorism, that is, direct attacks on noncombatants, is the scourge of all guerrilla wars. Contrary to popular opinion, many insurgent groups understand this and avoid indiscriminate attacks on the innocent while searching for alternative avenues to fight. Insurgents, like states, however, have good cause to disable anyone, including civilians, who contribute to their adversary's war-sustaining capabilities. As belligerents choose their tactics, they must remember there is symmetry between the threat civilians pose and the harm to which they are liable. The threat participating civilians pose is not great, but their contribution is essential. There are only three options to address this threat: lethal means, sublethal or nonlethal means, or none at all. Granting blanket immunity to participating civilians ignores the military value of their aid. Resorting to lethal force, on the other hand, fails to distinguish between those who fight war and are permissible to kill, and those who sustain war and are forbidden to kill. Sublethal, nonlethal, and, especially, cyber weapons preserve these important moral distinctions. In this context, psychological harm and mental suffering may be appropriate if necessary to disable civilians who provide war-sustaining aid. In other instances, state or non-states may find it necessary to detain, incarcerate, or deport civilians employed by war-sustaining facilities, thereby disabling these operations just as if they destroyed them. Here, too, participating civilians may suffer various degrees of mental suffering.

Consider, for example, five possible less-than-lethal means to disable a financial institution or media outlet: calmative agents, electromagnetic technologies, arrest and deportation, kinetic force, or cyber war. The first three tactics disable participating civilians directly either by rendering them unconscious or forcibly removing them from their offices. The last two destroy or disable infrastructures. Psychological harm plays no apparent role in the initial attack. It is, however, a powerful adjunct that sustains the disabling momentum of the initial strike. The trauma of past episodes together with the ongoing threat to inflict physical discomfort, arrest and deport employees, or constantly evacuate facilities upon warning of impending kinetic attack may very well result in the kind of moderate psychological trauma that comes with reliving past events (PTSD) or ruminating about future calamities. For some individuals, these anxieties may prevent them from doing their work well or returning to work following an attack. While the psychological effects of disabling a facility by cyber-attack alone remain unknown, a combined cyber and kinetic attack may have multiplying effects (Schaap 2009). A cyber and kinetic attack may disable a facility *and* its employees for a long period of time. These attacks, which primarily result in psychological harm, satisfy proportionality, particularly because the effects of anticipatory anxiety are relieved once a person leaves an area under attack. Thus sublethal and cyber-attacks, may effectively drive people away (as happened in Sderot) and disrupt their ability to provide war-sustaining aid without causing severe, long-term or irreversible psychological harm. Participating civilians, not noncombatants, however are the only legitimate targets for such attacks.

Mitigating the harm that befalls civilians is the first rule of war, and the obligation to alleviate psychological harm is no less weighty. Evaluating psychological harm, distinguishing between severe and moderate harm, and understanding the liability of different classes of civilians is a task that is only beginning. It is, moreover, a crucial undertaking in light of the development of cyber-warfare, which promises to secure significant military gains. In contrast to sublethal and nonlethal weapons, cyber weapons are both more discriminating and less lethal. While relatively weak, many sublethal rockets are unsophisticated and imprecise. Nonlethal weapons are similarly indiscriminate. Both sublethal and nonlethal weapons will cause some severe injuries. Cyber weapons are still evolving but appear better controlled. Cyber strikes may be weak or powerful and may target military or war-sustaining facilities to deny service, steal or erase data, or cripple machinery.

With care and foresight, methods of war that minimize physical harm but cause psychological harm can be useful for disabling participating civilians. In contrast to conventional warfare, no side in an asymmetric war will prevail by attacking military targets alone. State armies waging counter insurgencies, like the United States, have long understood the need to undermine support services. Guerrillas enjoy no less a prerogative, and sublethal and cyber weapons may be the weapon of choice. While these technologies allow belligerents to disable those who provide war-sustaining services, terrorist attacks against the public sphere, whether cyber or otherwise, should give us considerable pause. Many are enamored of what appear to be cyber-warfare's very low costs in terms of injury and loss of life, leading some observers to question the cogency of noncombatant immunity during cyber-warfare and others to speculate about legitimate attacks on the public sphere. Existing evidence and emerging trends urge caution. While the devastation of mass casualty terrorism is obvious, it cannot be forgotten that cyber and sublethal terrorism are not casualty free. Sublethal kinetic or cybernetic attacks can set the stage for an uncontrolled spiral of violence, pervasive fear, and an uncertain future that brings on various degrees of mental and physical suffering to which the innocent are not liable.

It is not coincidental that this chapter opened with an account of mass casualty and indiscriminate terrorism and ends with a turn to less violent means of guerrilla warfare. The tactics described in this and previous chapters account for widespread death and devastation. Violent measures, however, are not the only means guerrillas and insurgents have at their disposal. Increasingly, they can turn to economic warfare, public diplomacy and media warfare, and, in a number of surprising instances, nonviolent resistance to accomplish their aims. Insurgency and guerrilla war are not entirely about armed resistance. Alternative and often no less successful tactics of irregular war occupy the remainder of this book.

8

Economic Warfare and the Economy of War

Economic warfare and the structure of a guerrilla organization's war economy represent two distinct ethical dimensions of insurgency. Economic warfare, the subject of the first part of this chapter, designates the use of military power to disrupt the local economy. For states, the prime tactics are economic sanctions and blockades. For guerrillas who lack the means to impose sanctions or blockades, the tactics of choice are terror attacks to disrupt tourism and other economic activity. Economic warfare is a coercive tactic that often precedes military force in the hopes of preventing or shortening a war. Like public diplomacy, economic warfare is virtually unregulated by international law and just war theory. Yet the simple fact that economic warfare targets civilians directly creates a host of ethical difficulties for its practitioners.

The economy of war is the subject of the second part of this chapter and designates the methods insurgents use to raise funds for weapons and recruits. Unlike their adversaries, guerrilla organizations lack coercive state institutions to collect taxes or the wherewithal to issue bonds. As a result, insurgents resort to various schemes to raise the money they need to wage war. These range from quasi-voluntary taxes, which guerrilla organizations levy upon their diaspora community, to drug trafficking and money laundering. Important ethical issues remain unaddressed by just war theory: What economic measures may insurgents employ as they pursue just war? To what extent may guerrilla organizations coerce constituents to contribute funds to the cause? Are unlawful activities morally permissible to raise the funds necessary to dislodge an occupying army? Akin to a state's right to levy taxes, guerrillas may also look beyond voluntary appeals to raise money for men and materiel.

ECONOMIC WARFARE: SANCTIONS, SIEGE, AND BLOCKADE

With the exception of laws that require occupying armies to meet the basic needs of the civilian population, international law does not regulate economic warfare (Cohen 2009; Neff 1988). States and international organizations like the United Nations may impose economic sanctions to prevent trade in certain commodities such as oil, timber, or diamonds or institute blockades to prevent war materiel from reaching an enemy state or non-state. In recent years, there are also growing discussions of "smart sanctions" or "financial warfare." Designed to avoid the overwhelming hardship to the civilian population that sometimes accompanies sweeping economic sanctions, smart sanctions target the bank accounts and assets of terrorist organizations or heads of rogue states.

Three features characterize economic warfare. First, many believe it is effective. Sanctions work and, therefore, bear repeating. Second, economic warfare is less destructive than armed conflict. As such, sanctions and blockades are a necessary step that nations must take before they wage war. If armed force is the last resort, then economic warfare is the penultimate resort. Finally, economic warfare places noncombatants squarely in the cross hairs. Lawyers dodge the clear moral implications of attacking noncombatants by noting that international humanitarian law only regulates armed conflict, not economic warfare. Philosophers have to do better, but they, too, struggle to explain why an army may impose severe economic measures that target the innocent. The following sections address these questions. Assuming that economic warfare meets the criteria of effectiveness, lesser harm, and noncombatant immunity, then one more question remains: If states may impose sanctions and blockades, then why not allow insurgents the same latitude to conduct economic warfare commensurate with their more meager means?

Economic and Financial Warfare: Definitions

Economic warfare aims to prevent an enemy's access to military resources and to create hardship for the civilian population by disrupting the supply of consumer goods (Bracken 2007; Lowe and Tzanakopoulos 2012). The twin pressures of insufficient war materiel and civilian discontent should, if economic warfare is successful, cripple an adversary's ability to fight and resolve conflicts at far less cost than armed force. Sanctions block other nations from selling to or buying from the sanctioned state or guerrilla group. Blockades, whether on sea or land, forcibly prevent the physical passage of goods into enemy territory and are, as such, an act of war rarely available to insurgents. Recent examples of blockades and sanctions include those imposed on Hussein era Iraq, Qaddafi era Libya, Hamas dominated Gaza, and Ahmadinejad era Iran.

If the blockaded goods were unequivocally military, there would be little debate about economic warfare. Controversy erupts, however, because sanctions and blockades target "dual-use" goods to disrupt a nation's war-making capabilities. In Iraq, UN sanctions blocked trade in nearly every product but the medical supplies or foodstuffs necessary to forestall a humanitarian crisis (United Nations Security Council [UNSC] 1990). Israel imposed similarly sweeping sanctions on Gaza from 2007 to 2010 (UN 2012). Sanctions by various nations against Iran, on the other hand, are limited to weaponry, nuclear technology, oil, and aviation parts (Johnson and Bruno 2012). In all these cases, financial sanctions accompany other economic measures as nations around the world freeze bank accounts, seize assets, and prevent travel by government officials.

Economic and Financial Warfare: Effectiveness

The effectiveness of economic and financial warfare remains the focus of intense scrutiny. Part of the debate is theoretical: What constitutes effectiveness? And part is empirical: Are sanctions effective by the relevant standard? The most obvious criterion of effectiveness is compliance: Did the targeted state comply with the wishes of the party imposing sanctions? Judgment has been elusive because the data show no unambiguous correlation between sanctions and compliance (Pape 1997; Peksen 2009). While proponents point to guarded indicators of success, opponents of sanctions argue that coercive economic measures only strengthen a targeted nation's resolve to resist and/or allow dictators to exploit sanctions to solidify their hold on power by blaming their opponents for economic hardship (Erdbrink 2012; Weiss 1999). Despite these adverse assessments, the use of economic sanctions has increased dramatically since 1990 (Gordon 2011). Now, the international community is working to make sanctions smarter by concentrating on arms embargoes, travel restrictions, aviation bans, and financial warfare in an effort to target those responsible for threatening international peace and security.

Armed force complements smart sanctions as states assassinate or detain guerrilla leaders fingered through records of their financial transactions. Despite these successes, scholars continue to debate the effectiveness of smart sanctions (Gordon 2011; Lopez 2012). The costs of sanctions are no less contentious.

The Cost of Economic Warfare

Following sweeping sanctions against Saddam Hussein, it was soon clear that the civilian population was bearing the brunt of an economic disaster of catastrophic proportions. Writing in 1999, Gottstein describes how "About 50,000 children under the age of five are dying every year due to the sanctions. A quarter of all emergency patients in the hospitals cannot be saved because of missing medicines. About 40 percent of the Iraqi people are hungry, as the food ration is enough for only 25 percent of their vital needs" (Gottstein 1999:282). Estimates of the number of dead children range from 170,000 immediately following the First Gulf War to 300,000 excess deaths (beyond ordinary mortality) of children under five (Garfield 1999; Gottstein 1999).

In Gaza, on the other hand, the losses were primarily economic. The blockade severely reduced the output of the agricultural and fishing sectors, while the lack of building materials decimated the construction industry. Water supply, sewage facilities, and utility infrastructures broke down with sufficient frequency to raise alarms about potable water for Gaza's inhabitants. Due to economic sanctions, unemployment reached 34 percent in the general population and 50 percent among the youth (United Nations Office for the Coordination of Humanitarian Affairs [UNOCHA] 2012; ICRC 2010). Food insecurity, defined as "lack of access to sufficient, safe, and nutritious food, which meets dietary needs and food preferences for an active and healthy life" rose from 52 percent of the population in 2006 to 61 percent in 2009 (PHR 2010:8). As a result of malnutrition, wasting, stunted growth, and anemia rose during the years of the blockade (Physicians for Human Rights, Israel [PHR] 2010:9, 30–32). On the other hand, infant mortality *decreased* by nearly 20 percent between 2007 and 2010, while life expectancy *climbed* from 72 years to 73.7 years (Central Intelligence Agency [CIA] 2007, 2012). These data, however, offers no evidence that economic warfare is benign. International relief organizations and smuggling prevented a humanitarian crisis by providing basic nutritional and medical needs, so much so that by 2010 nearly 80 percent of Gazans were aid dependent (World Health Organization [WHO] 2010). Gaza is a very far cry from Iraq. Following attempts by an international flotilla to break the blockade in 2010 and the resulting deaths of nine activists, Israel relaxed its blockade without ever achieving its goal of regime change (Chapter 10).

Data from Iraq and Gaza belie the assertion that sanctions will succeed when civilians can no longer adjust to the inconveniences of basic shortages. In Gaza, international organizations and smuggling prevented severe shortages. In Iraq, sanctions failed because the government was adept at shifting the costs on to an impotent civilian population. Sanctions fail, therefore, whether the costs are moderate or, conversely, border on the catastrophic. With that lesson in mind, smart sanctions might be a better tool for conducting economic warfare (Lopez and Cortright 1997). But smart sanctions are not without criticism: arms embargoes lead to thriving black markets in small arms, plane crashes and transportation disruptions follow restrictions on the sale of aviation parts, and the expropriation of assets brings charges that targets were denied due process (Gordon 2011). The high costs of sanctions, whether sweeping or smart, demand more stringent legal and moral criteria to regulate economic warfare beyond those currently provided by international law.

Moral and Legal Criteria for Economic Warfare

All would agree that economic warfare must be cost-effective, but the devil is in the details. Effectiveness may be measured by compliance (as states often demand) or by symbolic acts of defiance that resonate in public's mind (as insurgents often hope). Significant costs befall the civilian population. Which of these are permissible to impose on noncombatants? In response to this question, international law takes a broad view and permits any sanction that does not cause a grave humanitarian crisis. As such, states are not motivated by respect for proportionality but only emboldened by the legal lacunae that allow them to impose sanctions indiscriminately. To remedy lack of distinction, economic warfare must take the principle of participatory liability seriously.

Participatory liability reflects the material threat a person poses and leaves room for direct attacks against certain classes of civilians. The greater the threat, the greater the force to which one is liable. Combatants and other war-fighting assets are, therefore, liable to lethal force while non-combatants remain protected from direct harm. War-sustaining facilities occupy the middle ground and remain liable to nonlethal force, including crippling economic sanctions. But economic sanctions also bring collateral harm and with it the demand that warring parties observe the principles of necessity, discrimination, and proportionality (Christiansen and Powers 1995; Gordon 1999a, 1999b). Coercive economic measures must be necessary, that is, the least harmful means available to achieve a desired military end. They must only target military or war-sustaining objectives and bring harm that is not excessive relative to the military advantage an attacker expects to gain.

Indeed, this is the moral logic behind smart sanctions that impose arms embargoes against rogue states or seize guerrilla leaders' assets rather than impose sweeping sanctions on an entire civilian population. Nevertheless, noncombatants will inevitably suffer collateral harm. Unregulated and illicit arms flood a conflict zone in the wake of embargoes, airplane crashes may follow aviation sanctions (Dadpay 2010; Denselow 2010), and financial sanctions prevent hospitals from purchasing lifesaving drugs (Namazi 2013). When military targets are the object of economic sanctions, this harm, although severe, may be proportionate if the military advantage sanctions provide is significant. When war-sustaining targets are the object of economic sanction, collateral harm cannot be so acute. Permissible sanctions may, therefore, block funds and supplies for war-sustaining financial, legal, and telecommunications infrastructures whether operated by a state or non-state. While this will make it more difficult for a state or non-state to wage war, noncombatants are also adversely affected. Nonfunctioning banks, courts, security forces, and telecommunications networks disrupt civil society in a very fundamental way. The ensuing psychological harm may be little different than that which comes in the wake of a successful cyber-attack. Noncombatants may be cut adrift without services or lose their sources of livelihood. Such collateral harm is only justifiable if proportionate and incurred in the context of necessary and effective economic warfare. Were it possible disable war-sustaining infrastructures by economic means without bringing noncombatants to grief, then such an outcome is morally preferable. Of course, the argument is not usually framed in this way. Many would argue that targeting noncombatants is precisely the aim of embargoes and sanctions. But attacking noncombatants is not necessary, and, ironically, we can see this more clearly by examining economic warfare in the hands of insurgents.

Insurgent and Guerrilla Economic Warfare

"From a moral point of view," writes Amichai Cohen (2009:145), "the limited nature of economic sanctions justifies imposing them on one party who refuses to respect the claims of the other." Although Cohen may not intend the ambiguity, the focus on "respect" rather than compliance or submission sets the stage for the way guerrillas use economic warfare. Insurgents cannot hope to force compliance, but they can earn respect and attention from their adversaries and, more importantly, from the international community.

Insurgents have few means to mount a blockade or prevent the passage of goods. In the very least, insurgents might hope to send a symbolic message when they attack economic targets. At best, they may strive to raise their adversary's costs of war or convince the international community to impose effective sanctions. Such was the case in South Africa. Widely acknowledged as one of the more successful instances of sanctions, economic pressure eventually brought regime change and the end of apartheid in South Africa. Measures included trade sanctions, boycotts, an arms embargo, and divestiture. Sanctions partially contributed to lower GDP, higher unemployment, and, perhaps, greater health problems for many South Africans (Coovadia 1999; Garfield et al. 1995; Levy 1999). Nevertheless, South Africa was the rare case where the civilian community (at least the black community) was not the intended target and where the active consent of guerrillas and the tacit consent of their constituents helped offset concerns about excessive harm to noncombatants.

The asymmetry between the gain and pain of economic warfare is no trivial matter. Noting that sanctions rarely force compliance and nearly always bring suffering to the civilian population, Gordon (1999b:138) turns her attention to the symbolic goals that states sometimes pursue through sanctions. "'Sending a message,' while ordinarily a legitimate undertaking for a state," she writes, "becomes ethically problematic if the means of communication consist of depriving vulnerable sectors of a foreign population of basic necessities." This is true. Under these circumstances, sanctions cause harm incommensurate with their expected military and political benefits and violate both proportionality and the fundamental right to subsistence. Guerrillas pursuing similarly symbolic goals do not usually face this problem because they can inflict little harm. On the other hand, insurgents must often resort to kinetic force to wage economic warfare. Does this cross a red line?

Economic Warfare and Kinetic Force
In the hands of insurgents, economic warfare includes attacks or threats of attacks against tourists, aid workers, foreign businesses, and financial institutions. One notices immediately that none of these tactics is obstructive in the way that blockades or sanctions are. All require destructive armed force. They do not target civilians directly nor bring widespread harm but zero in on those guerrillas believe are agents of oppression. The attacks are, therefore, "smart," but they are not sanctions. Like sanctions, however, guerrilla tactics aim to disrupt the local economy as part of a campaign to press their claims. Unlike sanctions, guerrilla economic warfare is not a penultimate resort but an adjunct of armed struggle. Economic tactics and strategies vary considerably in their destructive force and run the gamut from egregious terrorism to more subtle and permissible means of economic warfare.

Terrorism serves many goals, and one of its most effective is disrupting tourism. In 2002, terrorist attacks in Bali killed 202 people, wounded 240, and decimated the tourist industry. The stock exchange in Jakarta lost 10 percent of its value and economic losses reached $5 billion (2–3 percent of the GDP). In Egypt, attacks on tourists in Luxor killed sixty-two people in 1997, and tourist revenues fell an estimated 53 percent. In Israel, periodic drops in tourism follow terrorist attacks by Palestinians (Lutz and Lutz 2006). Tourism is a very soft and responsive target and the results are immediate. Tourists leave, cancel reservations, defer future visits, or consider alternative vacations sites.

Nevertheless, not all attacks on the tourist economy are violent or accompanied by appalling casualties. Consumer boycotts of South Africa were certainly less violent than terrorism, but probably less effective. Modeled after the South African effort, the Palestinians can boast modest success of their boycott, divestiture, and sanction (BDS) campaign against Israel (Awwad 2012; Erakat 2012; also Chapter 10). In other cases, guerrillas employ armed force but without targeting tourists or causing many, if any, civilian casualties. In the late 1970s,

Basque guerrillas wrecked the tourist economy by detonating bombs at tourist sites, but only after first issuing warnings to prevent causalities. The threat of violence may wreak similar damage and does not necessarily require a track record of prior deadly terrorist attacks to prove credible. Tourists, unlike local citizens or foreign businessmen, have no vested interest in the countries they visit. It takes little for them to pack their bags. Nor are tourist sites a necessary target if the goal is to cripple tourism. Attacking the Colombo international airport in 2001, Tamil Tigers destroyed military aircraft and half the nation's commercial aircraft (Lutz and Lutz 2006). No civilians were killed in the airport attack. Avoiding tourists and tourist sites, the IRA blew up buildings in London's financial district between 1992 and 1997 causing small numbers of civilian casualties. Tourism fell off significantly with each attack (Rogers 2000).

Economic Warfare and Civilian Casualties
The IRA attacks deserve another look because they point to the moral bounds of guerrilla economic warfare. Aware that constituents were growing increasingly resentful of terrorism, kidnapping, and assassination, the IRA turned to economic warfare between 1992 and 1997, and targeted financial institutions and transportation infrastructures. These attacks were kinetic – banks and bridges were blown up – but the goal was economic. Paul Rogers (2000) estimates direct damage at 2 billion GBP. Indirectly, the IRA's campaign weakened morale, raised the cost of security and insurance, led to an exodus of foreign companies, and diminished the image of London as a city capable of protecting its citizens. Four civilians lost their lives in the 1992–1993 Baltic Exchange and Bishopsgate bombings (English 2003:278–279).

As most states, guerrillas employing economic warfare must satisfy the criteria of effectiveness, necessity, and proportionality. Economic warfare must accomplish its stated aims (in this case often symbolic) while causing no more than proportionate noncombatant causalities. Unlike economic sanctions, attacks on economic targets frequently employ kinetic force that may bring *collateral* mental suffering, injury, and loss of life. None of these harms is necessarily disproportionate assuming that economic warfare is necessary and effective and no other less destructive means are available to achieve similar military and political objectives. Proportionate, economic warfare demands minimal injuries, an objective that some groups can achieve by attacking facilities when people are not present or by issuing warnings.

As with cyber and nonlethal technologies, there is a clear distinction between economic warfare and economic terrorism. Attacking foreign workers in Sudan, insurgents from Southern Sudan successfully halted production and disrupted oil sales, making it difficult for Sudan to purchase war materiel (Lutz and Lutz 2006:6). In 2008, for example, five Chinese workers were abducted and killed. This is nothing but economic terrorism, a grave violation of noncombatant rights and no more permissible than other forms of terrorism. Elsewhere, insurgents employ mixed tactics. The Aceh, for example, conducted economic warfare against Indonesia by destroying pipelines and burning refineries and transportation vehicles. There, civilian casualties played no part (Aspinall 2009:66). But insurgents also turned to intimidation, kidnapping, and murder to oust foreign companies from Aceh. Both tactics were effective. Schulze documents how Exxon Mobile closed down operations following guerrilla attacks in 2001 and how GAM guerrillas forced logging companies out of Aceh. In Nepal, Maoist insurgents targeted political and economic targets including the government owned Agricultural Development Bank, land registration offices, and factories owned by Coca Cola, Colgate Palmolive, and Unilever (Bray et al. 2003). With the intention of destroying loan records and property titles, guerrillas justified attacks on banks and land offices in the name of aiding the poor and undermining what they saw as rapacious government policy. As in Aceh, guerrillas accused foreign companies of collaboration and exploiting local resources. Although the Maoists in Nepal struck primarily at military and police targets, they also attacked local government offices and terrorized their staff (Marks and Palmer 2005).

The mix of economic warfare and terrorism is troubling. To justify terrorism, guerrillas might argue that economic warfare will not succeed without terror attacks. Or, they may claim that company employees are collaborators, and company officials rob the local inhabitants of their resources (Schulze 2004). Both arguments are specious. First, there is no proof that casualty free tactics are entirely ineffective or that employees cannot be dissuaded from collaborating by social sanctions, imprisonment, or fines rather than by physical threats. Second, neither collaborators nor company officials are liable to deadly harm but, at best, may only suffer nonlethal harm that is not severe, intense, or prolonged but proportionate and discriminating.

Whether undertaken by states or insurgents, economic warfare aims to disrupt the local economy, impair war-making capabilities, and demonstrate the attackers' resolve. Permissible economic warfare does not target the entire civilian population but specific economic targets. In the hands of guerrillas, these attacks resemble smart sanctions in their specificity. Unlike smart sanctions, however, guerrilla attacks are almost always kinetic and sometimes cause death and injury no different than may result from blockades and other forms of state-sponsored economic warfare. Assuming states and insurgents prevent disproportionate collateral harm, economic warfare conducted by kinetic means is no less permissible than that conducted by blockades and sanctions. Both destroy economic assets, both bring pain and suffering in their wake, and both are condemnable when noncombatants suffering excessive injury, loss of life, or widespread destruction of property.

Economic Warfare: Who is Responsible?

At this point it is natural to ask: Who is responsible for the effects of economic warfare? Targeted governments seem doubly guilty. On one hand, the citizens of Iraq and Gaza, for example, would not have suffered had Saddam Hussein capitulated or Hamas renounced its control of the government. On the other, the same citizens would not have suffered had Saddam or Hamas not squandered funds for their own selfish purposes and left their people destitute. Guerrillas, too, may claim that only the recalcitrance of foreign oil companies endangered their employees. These arguments are not new. Calling for greater sanctions against Cuba, Senator Jesse Helms and Representative Dan Burton blamed Castro alone for the island's troubles: "The economy of Cuba has experienced a decline of at least 60% in the last five years ... At the same time, the welfare and health of the Cuban people have substantially deteriorated as a result of this economic decline and the refusal of the Castro regime to permit free and fair democratic elections" (United States Congress 1996 cited in Gottstein 1999:275).

This is a peculiar argument. What, one might ask, must a government do when subjected to sanctions? It sounds like it must surrender and acquiesce; otherwise its leaders bear responsibility for the harm that befalls their people. Perhaps, but it is odd that this demand is rarely made of countries that go to war. There is no consensus that a nation that fails to surrender bears responsibility for the evils of armed conflict. The same is true of economic war. Two factors repudiate the duty to capitulate. First, nations do not surrender because they hope to prevail at a reasonable cost. When wars begin, the human and material costs are not well understood. They only become clear as conflict progresses. The effects of sanctions are less well appreciated: if nations do not surrender at the first sound of gunfire, they certainly will not lay down their arms when their adversaries cut off oil or withhold airplane parts. Second, advocates of capitulation take a flat, one-dimensional view of harm as death and injury. There is no moral imperative that wars must save more lives than they take. Rather, nations and their citizens routinely and justifiably risk life and limb for freedom and honor. Responsibility for the devastation of kinetic war or economic sanctions, therefore, lies with the aggressor. When aggression is hard to pinpoint, then responsibility for death and injury lie with the proximate cause. Nations and insurgents resorting to economic warfare, no less than those who turn to armed conflict, incur responsibility when they impose sanctions or exercise armed force.

The preceding discussion may give the impression that economic warfare and armed conflict are discrete and mutually exclusive events. Sometimes this is true and, when it is, the demand that economic warfare precede armed conflict is particularly compelling. As the penultimate resort, economic warfare should not compare to armed conflict in the destructive force it unleashes and should make room for full-scale war only when economic measures fail to keep the peace. This is not always the case. While casualties in the 2008–2009 Gaza War (Cast Lead) far exceeded the deaths attributable to the Israeli blockade, sanctions in Iraq brought far more deaths and injuries than the 15,400 civilian deaths inflicted by the Coalition's brief bombing campaign (Iraq Body Count 2012). More often than not, economic warfare is an adjunct to, not a substitute for, armed conflict. This is especially true in guerrilla warfare where insurgents rarely have the facilities to conduct an economic struggle alone. Economic warfare must only precede armed conflict if both feasible and effective. Guerrillas, on the other hand, face the dilemma of finding any means that are feasible and effective. Usually these accumulate slowly, a combination of hard power, soft power, kinetic, and other means to conduct their struggle. Regardless of the measures they choose, insurgents will always need money and a viable war economy to sustain its flow.

THE ECONOMY OF GUERRILLA WAR

Questions swirling around war economies resemble those that bedevil economic warfare. When states resort to coercive means such as taxation or mandatory bond purchases to raise funds during war, few raise any more moral qualms than they do when states inflict punishing sanctions. But when insurgents and guerrillas resort to similarly coercive means, it raises a hue and cry little different from that meeting their efforts at economic warfare. In the following discussion, a brief overview describes state economies during war and the measures they take to finance them. Building on the legitimate coercive measures that states employ to raise funds during wartime, the subsequent sections examine guerrilla war economies and the legitimacy of the methods they utilize to finance their struggle for national self-determination.

Financing Armed Conflict

To finance the crushing costs of war, states turn to taxation, domestic borrowing (war bonds or forced savings accounts), and monetization (printing currency) (Rockoff 2012). Guerrillas and insurgents also rely on taxation but cannot usually borrow from their constituency or mint money. Therefore, they depend on foreign aid or counterfeiting and other unlawful activities.

State Revenue Measures

Commenting on taxation, Rawls (1971:332) reiterates one underlying normative principle: "Taxation ... can be justified only as promoting directly or indirectly the social conditions that secure the equal liberties and as advancing in an appropriate way the long-term interests of the least advantaged."

Tying taxation to the two principles of justice obscures the difference between raising funds for public goods (such as defense and security) and redistributing resources in the name of social justice (Buchanan 1984). These are two different goals, and during war there is often a tendency to defer income redistribution in favor of raising money for national defense. As a result, social welfare programs, health care, and education systems often suffer during wartime (Russett 1969a; 1969b). In contrast, Keynes (1940) was adamant that wartime offered a

compelling opportunity to attend to social justice. As such, his proposal
to finance World War II emphasized large doses of mandatory savings
and war bond purchases to soak up surplus income, reduce inflation,
and offer the working class resources to weather the inevitable postwar
economic downturn (Haensel 1941; Maital 1972). Mandatory savings
combined with tax increases, deficit spending (government borrowing),
and price and wage controls (the latter to Keynes' chagrin) are the staple policies of modern war financing (Studenski and Krooss 1963:280–
301, 459–518).

Schemes of fair taxation vary widely. They may be proportionate (flat)
or progressive and may tax consumption or income. Advocates for taxes
on consumption claim that those who consume more should pay more.
In contrast, progressive income taxes assume that the wealthy are better
able to afford the burden that taxes impose. Progressive income taxes
are also among the quickest methods to redistribute income and raise
revenues. This is especially true in wartime when personal consumption
drops as durable goods feed the war machine. Despite the urgency of
national security and other needs, most individuals are not rationally
inclined to pay taxes voluntarily. To prevent disaster, especially during
war, states require coercive institutions and proportionate penalties to
prevent free riding and to guarantee a functioning system of taxation.
"By enforcing a public system of penalties," writes Rawls (1971:240–241),
"the government removes the grounds for thinking that others are not
complying with the rules. For this reason alone, a coercive sovereign is
presumably always necessary ... " Enforcement measures include fines,
garnishment, property seizure, and imprisonment.

Non-state Revenue Measures
Arguments that justify wartime economic policy among liberal states are
not easy for insurgents to appropriate. Among states, national defense
is a public good that most citizens accept without controversy. Security,
together with the limits it imposes on absolute freedom (e.g., conscription and taxation) is the outcome of agreement among free agents. This
is the mainstay of modern contract theory. Guerrillas face different problems. First, they must prove to their constituents that the good they seek,
that is, self-determination, will effectively alleviate injustice. This is the
problem of just cause. Second, they must establish their legitimacy to
carry out the struggle for national self-determination. States take their

legitimacy for granted and democratic states stake their right to tax on representation. Insurgents, on the other hand, cannot assume that everyone values the new public institutions they propose or consents to taxation to finance the struggle. Third, guerrilla organizations have no ready access to the coercive measures that states command. There is no institutional infrastructure for imposing and collecting taxes or punishing evaders. All this must be built with some measure of consent from their constituents and respectful of the same basic rights and liberties guerrillas fight to protect.

In the best case, guerrillas turn to taxation to finance armed struggle; in the worst instances, they turn to criminal activities. In between are a host of measures that states can rarely countenance. Suitable measures depend on the goals and relative success of a guerrilla organization. At the early stages of armed struggle, guerrillas resort to "predatory" tactics to obtain the funds they need for arms, food, and transportation. Such tactics prey on the livelihoods of others and include blatant criminal acts: robbery, kidnapping, and hijacking (McCormick and Fritz 2009). In the long run predatory tactics destabilize and, eventually undermine, the local economy. Once guerrillas post moderate successes, therefore, they often turn to "parasitical" economic practices to solidify their control of captured territory, raise funds for war materiel, and provide the resources to maintain nascent social, educational, medical, legal, and welfare infrastructures. A parasitical war economy feeds off and profits from the legitimate and illegitimate economic activities that take root during armed conflict. Guerrillas do not interfere with this economy as much as live off it through taxation, toll collection, extortion, and smuggling. Along the way, guerrillas manipulate foreign aid while dipping into drug production, drug trafficking, and counterfeiting.

Finally, some guerrilla organizations may achieve substantial success and begin to operate like a government. Their war economy becomes "symbiotic" as it must now provide extensive public services, including competent law enforcement and well-functioning courts, ports, health care organizations, and government offices. An increasingly sophisticated war economy turns to taxation, import duties, tolls, and fees for government services, as well as foreign aid and economic investments (Ballentine and Nitzschke 2005; Naylor 1993; Qazi 2011; Williams 2007). At these later stages, the unlawful aspects of a parasitical war economy subside but do not disappear entirely. Obviously, these forms of economic activity differ greatly from what states use to finance their wars. In the next sections, I consider the moral implications of engaging in what is often unlawful activity to wage a just war of national self-determination.

To evaluate guerrilla war economies it is useful to begin with lawful and unlawful practices. In the former, we find international aid, taxation, charitable fundraising, and legitimate business interests. Among the unlawful we find extortion, theft, drug trafficking, money laundering, looting of natural resources, kidnapping, smuggling, and counterfeiting. None in either category is without its unsavory, or redeeming, qualities.

Lawful Aspects of Guerrilla War Economies

International Aid

International aid for guerrilla organizations often comes from rogue regimes. The IRA took money and arms from Qaddafi; Iran provides hundreds of millions of dollars to Hezbollah and tens of millions to Hamas, while the Taliban remains partially funded by Pakistani intelligence (Becker 2011; Goodhand 2004; Moloney 2003:326; Rudner 2010). Rogue regimes are not the only source of guerrilla financing. Palestinians receive aid from the EU (Hollis 1997), and Kosovo received humanitarian and military aid from the United States and NATO. While one may question guerrilla movements taking money from dictatorial regimes, any blanket condemnation would have seriously undermined the efforts of the many anti-colonial movements in Angola, Namibia, South Africa, and Mozambique that accepted Soviet aid in the 1960s and 1970s (Thom 1974). The more interesting moral question turns on how these funds are used, because there is no *prima facie* reason to reject funds from any nation state unless they accrue as the direct result of aggression, genocide, or crimes against humanity. The legitimate use of funds, on the other hand, brings us back to just cause. Only guerrilla groups struggling for just cause may enjoy the fruits of international aid. There are then two criteria governing permissible sources of *lawful* revenue. The first is procedural, that is, how the money is raised; and the second is consequentialist, that is, how the money is used.

The second criterion seems less problematic than the first. In fact, to demand that guerrilla use their funds to pursue just cause is somewhat beside the point. The permissibility of all the tactics and strategies described in this book assume that guerrillas seek just cause. Just cause, however, does not offer blanket permission to use funds for any purpose whatsoever. As a practical matter, guerrillas and insurgents use funds like they raise them: morally and immorally. Moral uses should dominate so that guerrillas utilize funds to purchase arms and supplies, pay wages, and fund social services. These are no different than legitimate state expenditures. On the other hand, guerrillas, no less than states, may not utilize funds to conduct terrorism, foment genocide, or otherwise violate human rights egregiously. Interesting questions arise about borderline practices such as bribery, whether it is used to induce government soldiers to defect, pay informants, or corrupt local officials.

To succeed, guerrilla movements must gain support and curb defection. Knowing whom to favor and whom to threaten depends crucially on information. While the human costs of utilizing physical threats and violence to maintain support or acquire intelligence raise severe concerns about human rights violations, such concerns are largely absent when inducements are financial. This is not to say that bribery is without consequences. In some cases, material incentives are accompanied by physical threats. I noted disturbing carrot and stick methods of conscription in Chapter 3. As evidenced in the Palestinian territories, moreover, bribery-induced treachery and collaboration will exacerbate social tensions (Chapter 5). Bribing local officials may also impair local services. Nevertheless, and recognizing insurgents right to fight, such grey uses of funds represents a viable and far less costly alternative to maintain support than violence or terrorism. And while such practices would clearly undermine civil society in peace time, bribery and other financial inducements to defect may constitute an acceptable tactic of just guerrilla warfare if benefits outweigh costs and costs do not ride roughshod over human rights.

Evaluating how guerrillas raise funds by examining where these monies come from is morally more vexing than their use. Revenue sources require proper provenance, supervision, and administration. International aid enjoys little oversight. All kinds of states give aid to whomever they please, the only restriction being aid so tainted with blood that no rights respecting guerrilla movement would touch it. Greater oversight governs taxation and tolls, charitable fundraising, legitimate businesses, and investment management.

Taxation

Guerrilla movements, particularly those with a substantial diaspora community, have developed sophisticated systems of taxation and coercive compliance. In the 1990s, the Kosovo government in exile enacted a voluntary 5 percent income tax on employed residents, 8–10 percent on businesses, and 3 percent on Kosovars living abroad (Clark 2000:103; Perritt 2008:88–89). By publishing names of tax evaders, the Kosovar authorities could apply considerable social pressure to force compliance (March and Sil 1999; IISS 1999). During the same period, the LTTE maintained taxpayer rolls and collected taxes from its local and diaspora community. At home, tax evaders faced imprisonment, travel restrictions, or confiscation of property. To enforce compliance abroad, Tamil representatives would "remind" their diaspora compatriots of their relatives and property still in Sri Lanka. These methods were effective and compliance was very high (Gunaratna 2003). In addition to income taxes, guerrillas collected a myriad of other taxes. Aceh guerrillas, for example, imposed taxes on property, local and foreign businesses, trade in coffee, timber and palm oil, vehicles, and inheritances. Violence and threats of violence including "destruction of businesses, abduction, and even murder" motivated compatriots to comply (Aspinall 2009:180). The Taliban, for their part, effectively taxed land, opium sales, and agricultural produce and collected tolls on shipments between Pakistan and Afghanistan (Goodhand 2004; Pugh et al. 2004:34). In the same vein, the PKK taxed local companies, domestic shippers, and smugglers operating between Turkey, Iran, and Iraq (Kocher 2007; Marcus 2007:182–184).

Several characteristics of these schemes are noteworthy. First, as the Kosovar case demonstrates, the tax rates vary relative to morally relevant groups. Individuals pay the same 5 percent tax, while businesses pay more and expatriates pay less. A variable flat tax reflects a concern for equity, while acknowledging that each group receives different benefits and assumes different burdens. While every Kosovar will benefit from independence and national self-determination, those at home gain political freedom, government services, economic stability, and opportunity, while Kosovar independence provides expatriates with a source of pride and affiliation and the means to strengthen their ethnic identity. Second, the purpose of taxation was not redistributive but aimed solely to maintain the institutions necessary for independence. Third, just taxation requires permissible, that is, proportionate and effective means of enforcement. Acceptable sanctions include the same avenues of enforcement open to the state: incarceration under reasonable conditions, social sanctions, and confiscation of property and exclude execution, bodily harm, collective punishment, or, more commonly, threats thereof. Violent means of enforcement, whether to maintain local support, conscript civilians or soldiers, prevent defection, or ensure tax compliance all suffer from the same defects. They both violate the rights of noncombatants and undermine legitimate authority.

Legitimate Charitable Organizations and Business Operations
Charitable fundraising abroad provides a significant source of revenue for many guerrilla organizations. Like the IRA, Hezbollah and Hamas maintain a chain of charitable organizations to raise funds for the welfare services they provide their constituencies (English 2003:344; Horgan and Taylor 1999, 2003). Hezbollah, for example, spent $100 million in 2004 to run local hospitals, political campaigns, administrative offices, a militia, religious institutions, housing, rural electrification, water and sewage infrastructures, and a martyrs fund (Cirino et al. 2004; Rudner 2010). Hamas allocated $70 million in 2011 for similar services taking direct aid from Iran ($20–30 million), while donations from diaspora Palestinians and other wealthy Arab contributors provided much of the rest (Global Security nd. a; Council on Foreign Relations [CFR] 2011). Many Islamic banks also permit a charitable surcharge to allow clients to funnel significant funds to insurgents or terrorists. Such banks also launder money from smuggling and the drug trade (Looney 2005; Napoleoni 2004; Navias 2002).

Two aspects of charitable giving are morally problematic. First, consider fundraising appeals. While some guerrillas often appeal to duty and solidarity, others turn to coercion. Hamas fundraisers, for example, explain how those who cannot participate in the armed struggle can, nonetheless, fulfill their religious duties of charity and jihad by giving to Hamas (Levitt 2006:66). Hezbollah, on the other hand, does not shy from strong-arming potential donors by exploiting the vulnerability of those with family or assets in Southern Lebanon leading Rudner (2010:705) to describe these as "exactions" and "tax-like levies" thereby blurring the line between taxation and charitable fund raising in Southern Lebanon. While publicizing the names of non-donors to impugn their reputation is an acceptable method to induce the recalcitrant to donate (Bekkers and Wiepking 2007), threats of violence, whether against body or property, violate the protections due taxpayers.

Second, funds earmarked for charitable purposes often find their way to an organization's military wing to buy arms or pay salaries. Organizations like the IRA, Hezbollah, and the LTTE simply siphon off funds with little compunction (Gunaratna 2003; Horgan and Taylor 2003; Rudner 2010) or, quite possibly, without the consent of donors. Were some donors to object, lying may merit consideration in extreme cases. Consider an insurgency so strapped for funds that it must lie to donors who might withdraw their support if they discovered how they were deceived. While none of the cases surveyed in this book fits this case, one might surmise that guerrillas might justifiably lie and then apologize, rather than ask for permission in advance, if this was necessary to wage a just struggle for national self-determination. Other tactics are more subtle. Some guerrillas exploit the gratitude and dependence of their aid recipients. "Recipients of Hamas financial aid or social services," notes Levitt (2006:120), "are less likely to turn down requests from the organization such as allowing their homes to serve as safe house for Hamas fugitives, ferrying fugitives, couriering funds or weapons, storing and maintaining explosives, and more." In the very least, "Hamas' simultaneous engagement in charitable work legitimised suicide operations by increasing the overall legitimacy of the organization" (Gunning 2009:116–117). Charitable aid, therefore, need not be used to buy arms directly to provide guerrillas with significant military advantages.

In addition to charitable donations, legitimate businesses provide guerrillas with substantial revenue. Tamil Tigers, for example, loaned money and operated successful video and film ventures that brought them tens of millions of dollars and roughly half of their revenue needs (Gunaratna 2003). R. T. Naylor (1993) estimates that income from investments funded half the PLO budget in the 1980s. In some cases, legitimate businesses operated under a shroud of coercion. Guerrilla run construction and security companies in Northern Ireland and Southern Lebanon monopolized reconstruction (Horgan and Taylor, 1999, 2003; Rudner 2010). Given the mammoth scope of reconstruction and huge influx of foreign and local funds for rebuilding, there was considerable room for abuse, intimidation, violence, and corruption. These are the same coercive measures that guerrillas must avoid as they raise money through taxation and charitable donations. To behave otherwise risks the loss of local and international support.

The moral restrictions imposed on international aid, charitable donations, and legitimate businesses are not onerous. While these restrictions scrutinize the source of international aid and curb the coercive methods guerrilla may employ, they work to ensure popular support and continued legitimacy by guaranteeing that guerrilla organizations do not violate the rights they must respect to maintain their authority. Mining *unlawful* revenue sources are, however, more problematic. While such activities may abuse the rights of compatriots, others prominently violate the rights of some third party that guerrillas and their compatriots denigrate. These may be the institutions of the state they fight or some of its citizens. Unlawful activities include smuggling, looting natural resources, counterfeiting, kidnapping, and drug trafficking. None immediately violates the rights of compatriots. Rather, the first three – smuggling, looting, and counterfeiting – generally injure the enemy state (or perhaps some neighboring state). Kidnapping aims at specific individuals, families, or companies while drug trafficking speaks to ramifications at home and abroad.

Unlawful Aspects of Guerrilla War Economies

Smuggling, Looting, and Counterfeiting

Smuggling arms and goods are a staple of guerrilla war economies. The Kosovar shadow government and The Kosovo Liberation Army exploited black markets and bartering networks to smuggle arms, aliens, and cigarettes (Ballentine and Nitzschke 2005; Radu 1999; Reitan 2000). Hezbollah smuggled weapons, drugs, diamonds, software, DVDs, and cigarettes through the lawless tri-state border district of Argentina, Brazil, and Paraguay to fund up to 10 percent of its annual budget (Cirino et al. 2004; Rudner 2010). Organized looting pillages government institutions, local resources, and third parties. The IRA and Maoist guerrillas in Nepal routinely robbed local banks (Bray et al. 2003; Horgan and Taylor 2003). The Taliban took control of emerald mines and timber forests after occupying Swat on the Pakistani-Afghan border in 2008 (Qazi 2011). Nor were the Taliban averse to looting relief supplies (Goodhand 2004). Counterfeiting was neither prevalent nor particularly successful.

There are few known attempts to counterfeit U.S. dollars, airline tickets, or traveler's checks (Naylor 1993).

States that violate a peoples' right to self-determination and, with it, a host of fundamental human rights are liable to harm in the course of a just insurgency. Economic warfare can be devastating, but, in the hands of guerrillas, its effects are limited. When guerrillas smuggle goods across the border, loot timber or gems, and counterfeit currency, they gain for themselves and impose economic costs on their adversaries. Consider smuggling, the import and export of contraband goods without paying taxes. Some smuggled goods may be necessary to wage war, some are essential commodities for the local population, and still others are local goods sold abroad. The first casualty is the revenue of the state government guerrillas are fighting. Assuming that guerrillas enjoy just cause and legitimate authority, it would be indefensible for a persecuted people to pay taxes to the same government that denies them a dignified life. Smuggling, then, functions like conventional economic sanctions that disrupt vital enemy industries.

In contrast to interstate sanctions, however, two other parties are affected by smuggling. Locals may bear the cost of an unregulated black market that may drive the price of commodities beyond reach or distribute them inequitably. Here, the responsibility for equity and fairness lies with guerrilla authorities. Guerrillas, no less than a state government, cannot ignore gross inequities lest they lose their authority. Additionally, one must consider the neighboring state whose economy is severely disrupted by smuggling. This is difficult to evaluate because smuggling is part of a much larger problem of lawlessness. Hezbollah, for example, currently operates with a semi-failed state on one side and a failing state on the other. Hamas smuggles weapons and goods through unpoliced Sinai, and the Pakistan-Afghan border is far from the central government of either nation. Under these circumstances, local warlords, chieftains, and brigands probably benefit from smuggling as much as guerrillas do. The people in Egypt, Lebanon, Pakistan, and Afghanistan may suffer, but smuggling is surely the least of their problems, which might only be solved by the end of the fighting and the post-conflict restoration of law and order.

The responsibilities of guerrilla leaders also constrain oil, timber, gas, or mineral looting. Aceh guerrillas, for example, claim local resources for their own and argue that looting (or "reclamation") restores the revenue stream to its proper owners while denying revenues to a repressive regime. Claiming resources may be permissible in the context of a just armed struggle assuming both a firm entitlement to the disputed resources (or more commonly, a fair portion of the disputed resources) and an equitable distribution of the proceeds and benefits among compatriots. Claims of the first kind are common causes of conventional war. Saddam Hussein invaded Kuwait claiming theft of its oil, and oil continues to grease brutal warfare between Sudan and Southern Sudan. Guerrilla demands, as least in Aceh, were relatively restrained and only sought some formula for retaining more revenue locally. Equitable distribution is the obligation of any legitimate authority, requiring a consensus or shared norm about how best to feed the hungry and fund the war effort. Accepting these conditions – just war, valid claims to resources, and distributive justice – belies any attempt to consider the merits of resource looting by warlords in places like the Congo or Sierra Leone.

Apart from their state enemies, insurgents often look to the foreign companies with stakes in natural resources. Respecting these third-party interests may be difficult if they are entwined with those of the state. In many cases, guerrillas are unable to extract resources themselves. As a result, they cannot simply commandeer a facility. Rather, insurgents might resort to extortion and kidnapping. Michael Ross describes how the Aceh attempted to raise money from the Lhokseeumawe natural gas facility by extorting money from its operator, ExxonMobil. Aceh guerrillas hijacked trucks, attacked company airplanes and buses, kidnapped executives, bombed facilities, and killed employees. In justification, guerrilla commanders declared "We expect them [ExxonMobil] to pay income tax to Aceh. We're only talking about a few percent of the enormous profit they have made from drilling under the earth of Aceh" (Ross 2005:48).

How compelling is this argument? While, it obviously leaves no room to kill or terrorize foreign workers, what might be said about extortion or kidnapping? The liability of a foreign company that helps sustain a war is an intriguing question. Recall that enemy civilians who provide war-sustaining aid are liable to disabling force that includes the destruction of infrastructures and nonlethal harms. The facilities of foreign companies that provide a regime with war-sustaining aid are similarly liable to disabling attack and destruction. If it is permissible to destroy an infrastructure, then extorting money *not* to destroy a facility is a lesser harm, particularly if guerrillas have a claim against revenues. Ultimately the decision is based on costs and benefits. While guerrillas might have the wherewithal to shut down a facility entirely (or destroy a single pipeline), it might be to their benefit to let it run if the revenues they gain through extortion and their compatriots gain through employment offset whatever war-sustaining benefits the government derives from continuing operations.

None of these calculations apply to looting or destroying relief supplies or killing and injuring aid workers, a growing concern in Afghanistan, for example (Nordland 2013). Relief workers do not provide war-sustaining aid and, therefore, are not subject to disabling measures. Matters on the ground, however, are not always clear cut. Following an EPLF raid on a UN/U.S. relief convoy that destroyed 23 trucks and 450 tons of food supplies destined for famine struck Ethiopia in 1987, guerrillas claimed, "three vehicles were carrying 'bombs, ammunition and fuel oil' while the others carried food" (Rule 1987). While the military targets were certainly liable to destruction, the others were not. Whether the destruction of the other twenty trucks is permissible collateral harm remains an open question. Permissible collateral harm, however, offers no excuse or justification for intentional attacks on relief agencies and their staff.

Kidnapping

Guerrillas kidnap for ransom and publicity. Rebels in Sudan, for example, kidnapped foreign oil workers to publicize their cause (Janes 2007). In Chechnya, on the other hand, motives were largely material. While guerrillas exchanged some hostages for captive Chechens, many were ransomed. In the fall of 1999, at the beginning of the Second Chechen war, Russian authorities recorded 1,843 abductions. Kidnappers released 851 hostages reaping ransoms estimated at $200 million (Peimani 2004; Phillips 2010; Tishkov 2004:114). For Aceh guerrillas, on the other hand, kidnapping was a form of punishment, justified with the claim that victims either owed money to guerrillas (back taxes, for example) or betrayed the movement (Aspinall 2009:181).

In some of these cases, an appeal to the culpability and liability of the victim can lend support to kidnapping. Among civilians, executives of foreign companies are a common target. The Aceh, for example, received $500,000 after kidnapping a senior oil executive in 2001. Like war-sustaining facilities, war-sustaining agents are subject to disabling force. For a key employee, this might mean incarceration or expulsion. And, if a senior executive might be incarcerated, then he might also be incarcerated and released for ransom, a lesser harm. However, costs, benefits, and rights matter significantly. First, the authorities and the public often react forcefully to kidnapping foreigners, and this may significantly increase the costs of holding hostages. Second, the slippery slope is ever present. Senior executives lead to junior executives lead to the mail boy, not all of whom are liable to disabling force. Incarceration may slide into torture or execution, particularly if ransom demands are not met. The Chechens, for example, executed many hostages, including foreigners (Tishkov 2004:111–115; Napoleoni 2004). Moreover, releasing a hostage belies the reason for capture in the first place. Once released, it is back to work. As a result, capture with the intent to ransom and release, that is, kidnapping, cannot ride on the coattails of permissibly disabling war-sustaining agents unless the ransom far exceeds the marginal benefit the senior executive will provide to his employer upon reemployment or there is some assurance that the same executive will not continue his war-sustaining activities and leave the country. Otherwise, kidnapping is nothing but a criminal act of intimidation or worse, particularly if captives are harmed or threatened with harm to speed up negotiations. Such campaigns quickly become counterproductive. Commenting on the brutal IRA kidnappings of business executives and their families in the early 1980s, Naylor describes "a public relations disaster" that eventually led the IRA to end kidnap operations (Naylor 1993:28; also Coogan 2002:522–524).

The arguments that swirl around kidnapping combatants track similar concerns about rights and efficacy that surround taxation and economic warfare. The Chechens regularly captured Russian soldiers to later exchange them for ransom (and vice versa) (Politkovskaya 2007:49, 66–68). Demanding ransom is not intrinsically evil. One might imagine the Taliban, for example, demanding ransom for the one U.S. prisoner it held from 2009 to 2014 knowing that the United States would not then release Taliban detainees. Severe violations of human rights arise, however, when kidnapping turns combatants (or noncombatants) into a means of raising money. Ordinarily, capture serves to disable enemy combatants. Prisoner exchanges are a byproduct useful to the opposing sides when particular prisoners may be valuable or the cost of maintaining large numbers of prisoners become onerous. To turn capture into a revenue-producing mechanism invites abuse as human prisoners retain no more dignity than slaves. Shorn of human dignity, prisoner hostages will suffer nothing but degradation. This is the lesson of the Chechen case and the source of public indignation and denunciation in the Irish case.

Drug Trafficking

Drug trafficking provides income for some insurgent groups as well as a groundswell of popular support when guerrillas can protect local producers. The IRA, Tamil Tigers, KLA, Hezbollah, Taliban, PKK, and Kurdish guerrillas all traffic in or tax heroin, marijuana, or opium (Dishman 2001; Gunaratna 2003; Naylor 1993; Özdemir 2012; United nations Office on Drugs and Crime [UNODC] 2012:84–85). On one hand, it would be easy enough to dismiss drug trafficking by asking: Where is the foul? – particularly with regard to marijuana, the most heavily trafficked drug in the world (UNODC 2012:97). On the other hand, one cannot ignore the vast criminal network that surrounds the drug trade nor its adverse social ramifications. This is particularly true of heroin and opium. The 2012 World Drug Report (UNODC 2012), for example, cites the grim statistics of drug use: mortality, morbidity, HIV and its associated costs, lost productivity, crime, and social disintegration.

Nevertheless, one must be careful to distinguish among the effects of drug use, drug trafficking, and drug production. Regarding Afghanistan, for example, the World Bank describes the "pluses and minuses" of an opium economy. To its advantage "the opium economy has boosted rural incomes," but at the cost of "price and quantity volatility [and] opium-related debt ... which may precipitate drastic measures like mortgaging and losing land, or giving up daughters in marriage to pay off opium debt ... and drug addiction" (Byrd and Ward 2004). In the final analysis, there is no doubt that Western policy makers see Afghanistan's future in agriculture and "broad-based rural development, with expanded markets and livelihood opportunities for households ... " (Byrd and Mansfield 2012:4), rather than in poppy cultivation. Until then, guerrilla fighters in Afghanistan or elsewhere will need to balance the demands of their constituency with the financial requirement of an armed struggle. While their concerns about drug addiction in Chicago will be secondary, insurgents will need to weigh the local costs and benefits of continued drug production. The evidence suggests overriding costs, but eliminating them entirely requires alternative economic arrangements not readily available in wartime. If wartime poppy production is the best plan available and serves local needs better than alternative economic schemes, then guerrillas may have no choice but to protect the local drug trade. Protection may demand regulation, taxes on production and trade, and controlled smuggling. Western resistance to drug trafficking draws its force from opposition at home, criminality abroad, and the money it brings insurgents. From the insurgents' perspective, foreign opposition to drug trafficking is of secondary importance in a war torn economy (unless it affects aid); criminality is a cost to weigh against local benefits while financial benefits are precisely the point.

THE ECONOMIC DIMENSIONS OF JUST GUERRILLA WARFARE

Economic warfare and the economy of war are crucial components of just guerrilla warfare. In the hands of states, permissible economic warfare depends on avoiding humanitarian crises while a legitimate war economy depends on sometimes severe, but never unlawful, methods to finance armed conflict. The situation is radically different from the perspective of insurgents.

Insurgents cannot mount blockades, impose sieges or remotely threaten their enemies with penury. At best, they conduct kinetic attacks that, by their nature, induce fears of injury and loss of life. While one wonders about international norms that allow states such wide latitude while condemning guerrillas for lesser harms, this bias offers guerrillas no grounds for terrorism. Economic warfare in the hands of insurgents is bound by the same rules of effectiveness, necessity, and proportionality that bind states in the conduct of war. There is no doubt that economic warfare in the hands of insurgents inflicts some harm on civilians. Guerrillas, however, can keep harm within permissible bounds by avoiding terrorism, warning away civilians, and carefully choosing targets whose destruction impairs the state they fight. This is a far cry from state capabilities in whose hands economic warfare is far more deadly. Obliged by these constraints, guerrillas will find economic warfare a useful, but not a decisive, tactic for waging asymmetric war. Far more crucial is their economy of war and the avenues they develop to finance their struggle.

Like state revenue sources, the avenues available for financing an insurgency vary in effectiveness. Wennmann (2009) suggests that

mineral exploitation, drug trafficking, and foreign aid are easier revenue sources for guerrillas to control and generate higher revenues
than taxation, investments, or kidnapping. The latter sources, while
not pivotal, remain important supplementary sources of funds. Unlike
states, guerrillas adopt economic measures that are not morally neutral. But nor are they uniformly condemnable. Describing the guerrilla
war economy of the Taliban, Goodhand (2004:158, 165) notes "disruption to markets and destruction of asset bases, violent redistribution of
resources and entitlements, impoverishment of politically vulnerable
groups, out-migration of educated, political instability in neighboring
countries [and] circulation of small arms." On the other hand, he writes
"the shadow economy may well promote processes of development
[and] link remote rural areas to major commercial centers, both region-
ally and globally." The result is "commercial networks ... governed by
rules of exchange, codes of conduct and hierarchies of deference and
power ... They are not anarchic and do not depend purely on coercion.
Trust and social cohesion are critical."

Trust and social cohesion, not coercion, define the hallmarks of permissible conscription whether of soldiers, civilians, human shields, or
resources and constitute the building blocks of a successful guerrilla
movement and the authority it enjoys. On this basis and cognizant of the
costs and benefits of a war economy, one may cogently evaluate guerrilla
war economies. Here, the simple distinction between lawful and unlawful revenue sources does not suffice because some practices offer benefits independent of their legal standing. Nevertheless, lawful sources of
revenues – international aid, taxation, and charitable fundraising – draw
our attention initially. These finance state efforts, and there is no reason to deny insurgents the same resource. Yet there must be latitude for
implementation. Lacking coercive state institutions to enforce compliance, insurgents may turn to social pressure, fines, incarceration, and
confiscation of assets but must vigorously shun threats to life and limb,
wanton destruction of property, and the punishment of innocent third
parties. And, as the prior examples attest, many movements successfully
regulate taxation without egregious abuse.

Unlawful revenue sources are also tightly constrained. Smuggling, looting, and counterfeiting though unlawful are, sometimes, victimless. Or
more precisely, the victim is often the same repressive state that is liable
to armed attack. Under these circumstances, one would be hard-pressed

to condemn smuggling, looting, and counterfeiting that denies revenue and resources to a state on the wrong end of a just conflict. Smuggling, looting, and counterfeiting are condemnable when innocent victims suffer harm and, thus, exclude violence, murder, intimidation, and attacks on relief agencies. While there are good moral and pragmatic reasons to vigorously condemn kidnapping, there is greater latitude for extortion when liable civilians are threatened solely with economic losses. Extortion may, in some instances, bring greater benefits and fewer costs than destroying infrastructures. Similarly there is no prima facie reason to prohibit drug trafficking simply because marijuana is illegal in New York. Rather, a complex case-by-case analysis that weighs local social costs and benefits will resolve the issue for guerrillas. In most cases, long-term costs are harsh but short-term benefits persist as long as armed conflict continues. These benefits may legitimatize some commercial aspects of drug trafficking, for example, taxing drug proceeds. As in all cases of fundraising, however, permissible drug trafficking excludes violence, murder, and intimidation. Herein lies the rub and presents a challenge that some guerrilla organizations are not always willing to meet.

Economic warfare and the economy of war exemplify two kinds of evolution that push beyond images of mindless guerrilla violence. The first highlights the movement from predatory to symbiotic forms of war economies as insurgents establish hegemony. By their nature the latter are less violent and lawless. Second, the economic dimensions of asymmetric warfare push guerrillas to consider means that are generally less casualty ridden than kinetic warfare. Like cyber-warfare, economic warfare articulates a range of options that point toward softer power. Public diplomacy, the archetype of soft power, is the subject of the following chapter.

9

Public Diplomacy, Propaganda, and Media Warfare

In language not far from that used to express their exasperation over guerrilla's use of human shields, sympathetic supporters of Western armies bemoan their enemy's uncanny ability to manipulate the media. Turning to the media fiasco accompanying Israel's foray into Lebanon in 2006, Kalb and Saivetz (2007:43) lament "how an open society, Israel, is victimized by its own openness and how a closed sect, Hezbollah, can retain almost total control of the daily message of journalism and propaganda." No more successful than the Israelis, the International Security Assistance Force (ISAF) in Afghanistan complains how Western forces are "constrained by legal, political and ethical considerations in getting its messages across which often means that it is unable to effectively rebut or counter Taliban propaganda" (Nissen 2007:11).

Such impressions leave us to wonder how state armies will ever prevail against their media savvy, manipulative, and underhanded opponents. This question aside (for it is not the subject of this book), I want to ask instead about the ethics of media warfare. Are their clear moral guidelines for media warfare that should bind all sides to a conflict? And, if so, are guerrillas acting without care for law and ethics when they wage media warfare against their enemies?

The short answer is "yes" to the first question and "no" to the second. We can formulate guidelines for just media warfare; but they do not necessarily condemn deception, fabrication, propaganda, staged news events, media manipulation, censorship, or the attacks on broadcasting installations. To flesh out this point, the following sections surveys several key components of contemporary media warfare ranging from public diplomacy to military attacks on journalists and media infrastructures. Two key ethical guidelines emerge from the discussion that follows. First, the truth is a legitimate casualty of just war. But killing the truth, like any other harm of war, must be weighed against its effectiveness, necessity, costs, and benefits. The means of manipulating information are likewise the subject of scrutiny: staged media events and censorship are one thing; intimidation to silence witnesses quite another. Second, journalists are among the primary agents of media warfare. Like armed combatants, they pose a significant threat; unlike armed combatants, the threat they pose is usually not lethal. As such, journalists incur participatory liability during war that exposes them to disabling but not lethal force. While their lives are not forfeit, journalists are not immune from capture, incarceration, expulsion, or the destruction or confiscation of their equipment.

PUBLIC DIPLOMACY, PROPAGANDA, AND PUBLIC WORKS

"The problem of the propagandist," wrote Harold Lasswell (1927:627), "is to intensify the attitudes favorable to his purpose, to reverse the attitudes hostile to it, and to attract the indifferent, or, at the worst, to prevent them from assuming a hostile bent." Lasswell's definition resists a pejorative depiction of propaganda and usefully outlines the goals of modern public diplomacy. Public diplomacy like other forms of what Nye (2008) terms "soft power," aims to secure a political goal through attraction and persuasion rather than through threats or coercion. Successful public diplomacy and public works projects provide the means for states and non-states to gain crucial support at home and abroad. For non-state actors, the benefits of successful public diplomacy are striking, allowing such groups as the Taliban, Chechens, Kosovo Liberation Army, FALINTIL (East Timor), Palestinians, and Hezbollah to expand their power base, form alliances, exert diplomatic pressure, recruit supporters, and strengthen morale. Public diplomacy is surprisingly cheap and cost-effective and comes without the destruction and loss of life that attend kinetic warfare. In some cases, public diplomacy may replace military force and hard power and successfully achieve broad national goals. In other cases, public diplomacy augments hard war by winning tactical victories no less impressive than those gained at gunpoint.

When two or more nation's or groups compete for the same audience, media warfare ensues. Two tactics dominate: information operations

(IO) and public works. IO augments public diplomacy with objective commentary as well as staged media events, censorship, and information manipulation. Public diplomacy also leans heavily on public works projects to win the hearts and minds of the local population.

The Structure of Information Operations

While public diplomacy appeals to three distinct audiences – compatriots, enemies, and neutrals – and its message is tailored to fit each, the basic structure of each appeal is similar. Ben Mor (2007) describes four distinct communication strategies that reflect a state or non-state's desire to (1) avoid blame for bad acts (e.g., attacks that killed civilians), (2) take credit for beneficial acts (e.g., ridding an area of warlord crime), (3) discredit its enemies by blaming them for bad acts, and (4) denying an enemy credit for good acts. Pursuing its strategy, notes Mor, each side employs factual and normative arguments. To avoid blame an actor may deny an event or reframe it ("X did not occur as reported.") or disassociate itself from a bad outcome ("We did not do it."). These are empirical arguments suggesting that the facts are otherwise than initially thought. Alternatively, an actor may accept the factual description of events but give them a different normative spin. Such strategies include "justification" ("The act was legitimate.") or "excuses" ("We had no choice."). Information operations can reinforce a praiseworthy action as actors take direct responsibility ("We certainly did do X.") and enhance an act's value ("X was better than you think."). IO may also impugn an enemy's image by assigning blame and/or denying an enemy credit for its accomplishments. Assigning blame can be factual ("It was them, not us.") or normative ("Their actions were wrong, not right.").

During armed conflict, informational operations form part of a comprehensive campaign to secure a military advantage for one side and impose a disadvantage on the other. States do this all the time (Pratkanis 2009), and there is no reason to deny guerrillas similar means of media warfare, public diplomacy, and propaganda.

Public Diplomacy and Propaganda

Some observers are harshly critical of any attempt to identify public diplomacy with propaganda. While the former emphasizes dialogue, objectivity, and understanding, the latter is coercive, deceitful, and manipulative (Brown 2008). Public diplomacy, moreover, focuses narrowly on turning the foreign policy of one nation to the advantage of another (Manheim 1990:4). Both characterizations are unnecessarily restrictive. If public diplomacy is limited to nations, then it has no place in guerrilla warfare. Thus, the scope of public diplomacy must be broadened to include non-state organizations who hope to persuade other nations and *compatriots* of the justice of their cause. Second, propaganda need not be malevolent but turns to discourse or deceit as security interests warrant. Propaganda is the informational arm of public diplomacy and, for my purposes, the same as information operations as just described.

Successful propaganda may be white, gray, or black and corresponds to the truths, half-truths, and lies that make their way into commercial or social media (Betz 2006; Guth 2009). White propaganda is truthful reporting grounded in accuracy and backed by authoritative sources. As the Cold War unfolded, President Truman (1950, cited in Guth 2009:313) trumpeted the virtues of truth telling, declaring, "communist ... propaganda, can be overcome by truth – plain, simple, unvarnished truth – presented by newspapers, radio, and other sources that the people trust." The Communists were, of course, masters of black propaganda (Martin 1982; MacDonald 2006:140) that the Americans publically condemned but unabashedly employed as the CIA overthrew governments in Iran and Guatemala and later disseminated blatant misinformation about Saddam's weapons of mass destruction (Hiebert 2003; Theoharis 2006). Between the unvarnished truth and bald-faced fabrications lies the broad expanse of gray propaganda that colors most of public diplomacy in modern armed conflict.

Gray propaganda embraces what some theorists call quasi-rational and emotional messages. Such messages trade on thinly veiled and generalized appeals to authority figures coupled with half-truths and the manipulation, distortion, oversimplification, and omission of facts that make gray propaganda difficult to dislodge. Gray propaganda deliberately conflates facts with inferences and values, fails to encourage recourse to alternative sources of information and appeals to emotive language to beguile the unsuspecting (Black 2001; Plaisance 2005). Particularly important is the use of "templates," that is, pre-existing historical, linguistic, and cognitive frameworks that propagandists draw upon to weave together a number of related, readily accessible, and historically grounded accusations, injustices, heroics, and stereotypes to justify armed violence. Consider this description of one Taliban video:

One begins with clattering Chinooks disgorging American soldiers into the desert. Then we see the new Afghan government onstage, focusing in on the Northern Alliance warlords ... next it shows bombs the size of bathtubs dropping from planes and missiles emblazoned with "Royal Navy" rocketing through the sky; then it moves to hospital beds and wounded children. Message: *America and Britain brought back the warlords and bombed your children.* (Rubin 2006)

The clips in this film were not fabricated but loaded with half-truths and emotional appeals. Children did die but were not the target. NATO did not bring back the warlords; they were a permanent fixture of the Afghan landscape. But these omissions matter little. In this Taliban propaganda film, the producers successfully invoked the "occupier template" or the "narrative of foreign occupation," to sway the audience (Rid and Hecker 2009:176). Given Afghanistan's long history of fighting foreign occupation, this occupier template is particularly salient and short-circuits rational deliberation. Images like those in the clip are burnt into the national psyche and propagandists need only invoke a few key terms to conjure up an entire range of related concepts – mujahidin, freedom fighter, colonialist, crusader, and infidel – to justify their continued attacks on Western forces. Propaganda is a powerful tool, but no less significant are public works.

Public Diplomacy and Public Works

Public works projects are a key feature of public diplomacy, and leave adversaries to compete intensely for the support of the local population. When an ISAF officer says of local projects, "We tie everything back to the Afghan government [and] we want the people to know that their government and local governor care about them," the underlying military and political goals are not subtle (Johnson 2010). Such posturing is a far cry from the civic-minded duty one expects from public officials in a well ordered democracy. No one would expect a public official to openly admit that he or she is out to curry favor as he or she builds roads, hospitals, or schools.

Among the most significant state-sponsored public works projects designed to win local support are medical civic action programs

(MEDCAP). Prompted by strategic interests, MEDCAPs in Vietnam provided high-profile, short-term medical care. Impressed by the dramatic results of orthopedic or plastic surgery (cleft palate repairs were a particular favorite), Vietnamese villagers were expected to support the U.S. troops and Vietnamese government that brought them these benefits. The results were mixed or at best difficult to evaluate. Military medical personnel routinely report the goodwill that comes with lifesaving medical care, yet it was difficult to translate this goodwill into long-standing political support. Nor were MEDCAPs without criticism. In the absence of efficient follow-up care, critics charged the U.S. army medical corps with pursuing showmanship rather than practicing good medicine or, in more serious breaches of medical ethics, coercing patients to collaborate or spy in exchange for medical care. Cognizant of these difficulties, subsequent MEDCAP programs focus more closely on competent, basic medical care. Under these circumstances, medical and other public welfare projects foster loyalty and support among the target population and can bring significant military and political benefits (Gross 2006:199–210).

Nevertheless, public diplomacy is not just about building roads, schools, and clinics. Such efforts will fail dramatically, as they did in Vietnam, if only the product of propping up a corrupt regime. Alongside public works, then, must come strenuous efforts to fight corruption, institute land reform, afford good governance, and provide public security. All are crucial to establish credibility and legitimate authority. Public security is the linchpin but it vexes policymakers. Absent security, there can be no public works, but without a vigorous public works agenda, intervening troops will lack the legitimacy necessary to maintain security. Achieving the two in tandem presents a constant challenge to intervening state forces.

The benefits (and challenges) of public diplomacy do not pass unnoticed among guerrillas, nor do the efforts of state armies go unchallenged. There is fierce competition for the loyalties of the local population and the sympathies of the world community. Unlike national governments, many guerrilla organizations must constantly win the loyalty and trust of their own people, impugn the efforts of their enemy, and work to convince a skeptical international audience of the justice of their cause and their right to fight. "As the insurgents and counter-insurgents play mind games to gain local support," explains Lawrence Freedman (2006:85–86), "they may be as anxious to create impressions of strength as of kindness, to demonstrate a likely victory, as well as largesse." Pursuing these aims, guerrillas also use information operations and public works projects to win support among compatriots and foreigners.

GUERRILLA PUBLIC DIPLOMACY

Addressing compatriots, insurgents employ a mix of propaganda, education, and public works. As noted in previous chapters, guerrillas face a steep learning curve. To gain traction they must enlist significant support early on. Supplementing the social networks they employ to secure backing and the violent methods they utilize to deter defection, guerrillas routinely turn to educational efforts to pave the way for national self-determination. Numerous national movements placed special importance on comprehensive propaganda efforts to prepare their people for national liberation by invigorating, instilling, and, in the case of the Aceh, for example, inventing the national identity necessary to establish a historic homeland (Bekaj 2010; Schulze 2003, 2004; Wright and Bryett 1991). Similar efforts appeal to the diaspora community. Palestinian activists in the United States and the UK, Kurds in the Netherlands, Tamil in Canada, Kosovars in Western Europe, and Irish Catholics in the United States have all used their presence to influence their home government's foreign policy toward their ancestral land (Demmers 2002; Hercheui 2011; Kaldor 1997; Tugwell 1981; Vidanage 2004). With the rapid expansion of social media, virtual communities of activists use cyber forums to strengthen their identity, preserve their culture, and find like-minded compatriots without having to set foot in the mother country. Their virtual participation is significant. They debate policy, raise funds, monitor opponent's websites, and conduct cyber-warfare (hacktivism) against their enemy's virtual communities and computer networks. Supporters of the Palestinians and Chechens, for example, have been especially active in this regard (Bakker 2001; Estes 2013; Khoury-Machool 2007; Petit 2003; Thomas 2000).

Public Diplomacy and Public Works

Public works are a staple of guerrilla organizations trying to build support for armed resistance among their prospective constituents. Often lacking legitimate authority at their inception, guerrilla groups must provide human security and social services. Such services, provided by well-funded groups like the Kosovo government in exile, Taliban, Hamas, and the Hezbollah, maintain sophisticated police, judicial, welfare, financial, health, and educational institutions (Chapters 1 and 8). The impact of these services is not lost upon Western forces who often find themselves competing with insurgents to provide similar services at the behest of the local government. ISAF forces in Afghanistan, for example, make assiduous efforts to repair roads, maintain educational institutions, provide micro-grants, and fund agricultural assistance programs (International Security Assistance Force [ISAF] 2011; Johnson 2010; Roling nd.). To impugn a state's image, insurgents will also exacerbate crises of human security. Aceh guerrillas, for example, engineered an acute refugee problem by paying or otherwise encouraging secure villagers to leave their homes for refugee camps and urging displaced villagers to remain in camps rather than join relatives at home. Without some measure of consent of the refugees and due care to protect their health, these tactics are extremely problematic. Nevertheless, they formed part of a propaganda program to gain international support for independence (Schulze 2004:42). In such ways, then, each side uses public diplomacy to trumpet its accomplishments and enhance its credibility while denigrating the enemy and blaming it for the misfortunes befalling the local population.

Because public work projects are relatively inexpensive and effective, the competition between ISAF forces and the Taliban, for example, is far from friendly. Each is out to prove it can deliver more comprehensive services than its adversary. To this end, coercion and intimidation are not uncommon. The Taliban, for example, sabotage ISAF efforts by attacking schools, destroying roads, or threatening compatriots who use ISAF services. Ominous threats often come in the form of "night letters" that the Taliban post to warn local villagers against collaborating with the ISAF forces or with the local government on pain of death. Quite often, night letters name specific individuals. The threats are not idle: named villagers are often beaten or killed and their property destroyed while schools are bombed and gutted. The current night letter campaign, like that of the previous round against the Russians, is largely successful, intimidating villagers and scaring off foreign nongovernmental organizations trying to implement government welfare policies (Johnson 2007; Rid and Hecker 2009:172–174; Thruelsen 2010).

Like the Taliban, the IRA also posted death warnings to deter collaborators (Horgan and Taylor, 2003). But the IRA also orchestrated a public relations campaign to accuse political opponents of drug dealing, treason, and other undesirable social behavior before killing them. Coining these campaigns "defensive propaganda," Kirin Sarma (2007) describes how they preempt any moral criticism of killing and assassination by utilizing powerful collaborator or informer templates to denigrate their victims. Bystanders come to believe that victims deserve their fate and that their killing is just. Propaganda of this sort raises intense moral difficulties. While they might be conducive to successful armed struggle, night letters or IRA diplomacy undermine the most basic conditions for balanced moral deliberation and lead compatriots to condone the most heinous of crimes.

Threats and intimidation are no small part of the propaganda campaign that guerrillas employ as they compete with the local government for the loyalty and support of the people. Nor are threats and intimidation foreign to state armies. Human rights organizations have accused the Israeli government, for example, of withholding travel permits or blocking access to medical care to induce Palestinians to collaborate (PHR 2008). Less severe, but no less compelling, were the criticisms surrounding MEDCAPs. Providing rudimentary medical care first in Vietnam and later in strategic areas worldwide, MEDCAPs stood accused of exploiting poor health conditions by exchanging medical care for political support (Gross 2006:199–205; Olsthoorn and Bollen 2013). In all these cases, access to social service can significantly channel support to the government or its adversary. This makes health and welfare projects a potent tool of public diplomacy. Appeals to the media are equally effective.

Public Diplomacy and Propaganda

Foreign support for guerrillas comes from two sources: enemy civilians and sympathetic nations. During armed conflict, insurgents sometimes appeal to an adversary's citizens, often through third parties. Organizations like Hamas, Hezbollah, and Taliban regularly turn to Al Jazeera to bring their message home to English speaking audiences. Local NGOs are another important vehicle for information operations. Palestinian and East Timorese activists, for example, hoped to sway the citizens of Israel and Indonesia respectively through human rights organizations active in those countries (Steinberg 2006). Theses appeals stress human rights violations – torture, indiscriminate killing, unlawful property seizure, restrictions on freedom of movement, lack of access to medical care, and so on – perpetrated by the occupying power. Sometimes these appeals prove effective. The struggle in East Timor, for example, resonated loudly among a growing number of Indonesian human rights groups in the 1990s who also joined a worldwide campaign on their behalf (Hill 2002; Weldemichael 2013:210–215). While Israeli organizations often find themselves preaching to the converted, they have cultivated strong ties with international NGOs and foreign governments that are sufficiently worrying to goad the Israeli government into, as yet, unsuccessful attempts to hamper their activities. For the most part, though, guerrilla organizations appeal directly to the world community: media outlets, the United Nations, the International Committee of the Red Cross, human rights organizations, and bloggers of all stripes.

Pleading their cause directly to the international community, guerrilla organizations employ print media, internet, text messages, videos and DVDs, and face-to-face encounters as they extol their friends and disparage their enemies (Mor 2007). In the process, adversaries utilize a range of arguments, from the rational to the oversimplified to the *ad hominem*, to build a factual and normative narrative about what transpired and who was responsible for praiseworthy acts of resistance and damnable acts of terrorism and murder. While appeals to compatriots emphasize patriotism, religious duty, and historical injustices, appeals to the world community utilize templates that juxtapose democracy and the right of self-determination with brutal violations of humanitarian law, war crimes, oppression, and occupation (Iyob 1995:123–135; Perritt 2008:57; Thomas 2000; Tugwell 1981; Wright and Bryett 1991). East Timor offers an outstanding example:

Dili, East Timor (1991): When Indonesian troops opened fire on unarmed East Timorese demonstrators and killed 275 people in 1991 in what became known as the "Santa Cruz massacre," activists used phone, fax, and email to bring details of the carnage to the immediate attention of international aid and human rights organizations. With dramatic film and still images provided by foreign journalists, the Santa Cruz massacre grew into a central theme of East Timor's information campaign and a turning point in their efforts to gain statehood. For the first time, the world community could see the brutality that met popular demands to end Indonesian rule (Kingsbury 2000).

The campaign bore fruit. After East Timor finally gained independence in 2002 one observer enthused:

"East Timor was the first country born of the Internet Age, thanks to the sophisticated information bombardment of its committed supporters ... Part of the reason for the turn of events in September 1999 when President Clinton and Prime Minister Howard relented to armed peacekeepers going to East Timor was the level of Internet outrage on embassy systems at the White House, in Portugal, and the Australian Parliament" (Connole 2000 quoted in Hill 2002:46).

East Timor affords an example of an information campaign that is potent and truthful. The depiction of the massacre drew from independently supplied raw footage that spoke for itself. The impact and importance of such images are not lost on any guerrilla group. Scenes of massacre, mangled corpses, dead and starving children, and acres of rubble speak volumes for a guerrilla cause. These scenes reframe the narrative, lay damning blame at the feet of the enemy, paint guerrillas as saviors, and offer convincing excuses for their sometimes violent response to occupation. Given sufficient force and duration, information operations that exploit an adversary's killing of civilians may force an enemy to do what shells and rockets cannot: desist and stand down at little or no cost to guerrillas. As media campaigns make an increasingly important contribution to military success the question then arises: Why not exploit it to the full? Why not stage events or otherwise manipulate the media? Consider the following case:

Drone Warfare in Afghanistan (2007): Following a U.S. aerial attack in Afghanistan' Baghni Valley that killed 154 Taliban fighters in 2007, a local Afghan news agency claimed that nearly 200 *noncombatants* died after coalition forces bombed civilians assembled for a public event. In the aftermath, a number of foreign news agencies uncritically circulated the same account with some reports still surfacing several years later (Leigh 2010). Analyzing the Taliban's information coup, Rid and Hecker (2009:181–182) cite a U.S. intelligence report that hardly conceals its admiration:

"'Mullah Ihklas coordinated the movement of media personnel to this remote valley ... and ensured they filmed what the Taliban wanted them to film ... 'The Taliban commander allegedly directed his men to get a group of 50 to 100 locals and instructed them to tell the media representatives that the bombs had hit a civilian picnic area. The U.S. report describes the incident as 'the best manipulation of the international media using video of the 'locals' telling the prefabricated Taliban story in a multimedia interview.'"

In this case, as in many others, the Taliban tie their information operations to images that depict American and other Western forces as foreign occupiers, a resonant and deeply rooted historical theme among the Afghan people. Exploiting the overwhelming hostility these images evoke, the Taliban frequently prevail upon villagers to fight the invaders, resist the occupation, oppose the regime in Kabul, pursue martyrdom, and sacrifice themselves for Islam and Afghanistan.

The Taliban campaign moves from manipulating information to its fabrication and raises ethical concerns, which I address later in this chapter. Here, however, three things should be kept in mind. First, skillful manipulation of templates characterized by national self-determination, historical oppression, and human rights abuses feed successful communications campaigns. In Qana, Lebanon, for example, Israeli strikes killed twenty-eight during the 2006 Lebanon War (Reuters 2007). Initial but erroneous claims of 56 civilian deaths resonated due to an unfortunate coincidence: Qana was the scene of a tragic Israeli artillery strike that killed more than 100 civilians taking shelter in a UN compound in 1996. "Qana I" thereby afforded a ready-made template that predictably boosted charges of gross violations of international law from around the world (Asser 2006; HRW 2006) *and* discredited Israel's attempts to defend its actions. The uproar, successfully stoked by Hezbollah despite accusations of staging photo ops, misusing ambulances to simulate civilian casualties, and moving the same bodies from site to site, brought an immediate cease-fire (Peskowitz 2010; Rid and Hecker 2009:141–161). Israel found itself in a public relations "trap" (Pratkanis 2009:125), forced to face international rebuke or give up bombing. Israel's choice to stand down was reinforced by the diminished credibility imposed by Qana I. Furthermore, the Taliban, like the Hezbollah in Qana, do not cut their story from whole cloth. Rather, they usually employ "gray propaganda," a judicious mix of truths and untruths that exploit the uncontested fact that noncombatants often die in military attacks, however necessary and permissible this may be in the view of international law. Although the United States did not have a Qana I on its back, no nation is going to find it easy to defend any civilian casualties with the sophistry of proportionality. Finally, the Taliban campaign, like Hezbollah's, was militarily effective and inexpensive.

These cases exemplify the poles of information operations. At one end, insurgents in East Timor prosecuted a just war with little media manipulation. At the other end, guerrillas battled foreign occupation accompanied by varying degrees of media manipulation, fabrication, and censorship. In many cases, information operations are impressively effective, leading some commentators to attribute the tactical retreats of the Israelis in Jenin (2002) and Southern Lebanon (2006) and the United States in Fallujah (2004) to successful media campaigns by the Palestinians, Hezbollah, and Taliban respectively (Payne 2005; Peskowitz 2010). In each case, guerrillas could exploit collateral casualties with exaggerated claims of civilian deaths and subsequent charges of barbarism. That these charges were later refuted made no difference. Insurgents gained a significant strategic victory at little cost in men and materiel.

PUBLIC DIPLOMACY, ETHICS, AND *JUS IN BELLO*

The moral costs of public diplomacy abroad and public works at home are not as insignificant as one might think. Pulling enemies, compatriots, or bystanders to one's side usually entails pushing them away from the other. The push can be gentle, infused with material and ideological incentives, or it can be rough, attended by coercion, intimidation, and bald-faced lies. The following sections consider several morally problematic aspects of information operations: incitement, intimidation, exploitation, lying, and deception. Pushing beyond public works and information operations in the final section, I turn to overt interference with the media and the deliberate destruction of media facilities.

Incitement

What little international law has to say about propaganda and related information operations is buried in Article 20 of the *International Covenant on Civil and Political Rights* (1966). The article prohibits:

1. Any propaganda for war.
2. Any advocacy of national, racial or religious hatred that constitutes incitement to discrimination, hostility or violence.

The article remains obscure and unenforced because many nations lodged reservations fearing abridgement of free speech. Nevertheless, it underscores concern about the power of propaganda and wartime information operations.

In his study of this statute, Michael Kearney (2007) interprets war propaganda as any incitement of wars of aggression. Aggression, as defined
by the recently amended Rome Statute (2010:Article 8) comprises "the
use of armed force by a State against the sovereignty, territorial integrity or political independence of another State, or in any other manner
inconsistent with the Charter of the United Nations." Practically, this prohibits all uses of armed force with the exception of national self-defense
or Security Council sanctioned humanitarian intervention.

Clearly there are grounds to prohibit war propaganda because wars
of aggression are unlawful. Nevertheless, there are ample grounds for
states and non-states to propagandize during armed conflict. This would
include states defending themselves against aggression, states intervening on behalf of a persecuted people, and non-states pursuing their right
of national self-determination. As such, the prohibition on war propaganda hardly affects information operations in contemporary armed
conflict. Still, one has to consider the limits that Article 20 (2) imposes
on "any advocacy of national, racial or religious hatred that constitutes
incitement to discrimination, hostility or violence." Clearly, no one will
countenance propaganda that exhorts civilians or military personnel to
violate the rights of noncombatants by committing genocide, murder,
sexual assault, or wanton destruction of property. Propaganda of this sort
is condemnable because the acts they advocate are unlawful and unjust.
Yet, acts of self-defense are the proper subject of information operations.
However distasteful it may seem, there are no grounds to prohibit clarion
calls to slaughter enemy combatants, inflict collateral harm on civilians,
or threaten nuclear annihilation to deter an adversary from future aggression. Credible threats demand convincing communication strategies, otherwise deterrence fails. Because deterrent postures, as well as the other
acts of war just described, are lawful and ethical in the pursuit of just war,
skillful information campaigns on their behalf are likewise permissible.
The challenge for just war theorists and jurists alike is to draw a very firm
line between war propaganda and unlawful incitement, on the one hand,
and permissible agitation on behalf of a just war, on the other.

This distinction is important. Both incitement and agitation can precipitate an impassioned response, but agitation only stimulates a general desire to act, while incitement engenders criminal wrong doing. In
2000, for example, French TV aired a video of a young boy, Mohammed

al-Dura, as he crouched beside a wall with his father, pinned down by fierce cross fire as Israelis battled Palestinian militants in Gaza during the Second Intifada. For many, the enduring image of al-Dura dying in his father's arms provoked and sustained what became a brutal Palestinian terror campaign. While the veracity of the video clip remains the subject of considerable contention, I will assume for the sake of argument that the video was fabricated and the boy did not die. And, I will assume that its purpose was to enflame the Palestinians and elicit deep disdain toward Israel within the world community. Whether the video constitutes an appropriate communication strategy depends upon how it was manipulated by Palestinian leadership and their supporters. And here it does not matter whether the video was fabricated or authentic. If used to spur terrorism, vengeance, or, as some claim, "genocidal attacks" (Poller 2011), then it, indeed, violates the principle of incitement. If, on the other hand, the video was successfully exploited to initiate and sustain an armed struggle (Harel and Issacharoff 2004:27–28) that could be waged by means respectful of the rights of the participants, then it falls within the purview of appropriate information operations.

While Palestinians ruthlessly violated the rights of noncombatants during the second Intifada, other guerrillas do not. Putting things into a different context, imagine how one might react to news that the Santa Cruz massacre was a staged media event. While we would rightfully condemn any attempt to exploit the video to slaughter Indonesian civilians, I doubt that many would object if the video was used to mobilize the East Timorese in their struggle for national self-determination and demonstrate (by artful reenactment) the brutality of Indonesian occupation to the world community. The line here is thin, and it is impossible to know the intentions of everyone who airs a video. Nevertheless, it remains the responsibility of the political leadership, whether of states or non-states, to condemn and restrain incitement. Otherwise, they cannot claim recourse to permissible public diplomacy.

Once belligerents turn from incitement, the range of deception is only limited by one's imagination. Scot MacDonald (2006) devotes an entire book to real life and speculative illustrations of "altered images" that include staged beheadings, doctored atrocities, photographs of liaisons with enemy agents that never occurred, fake terrorist attacks, and bogus weapons of mass destruction (particularly 139–154). The purpose in all cases is to gain some strategic advantage and throw one's enemy off balance. The rules governing such manipulations are the same as those governing the al-Dura incident and demand that deception reject incitement and respect the rights of combatants and noncombatants. These are neither impossible nor unreasonable conditions to demand. But as propaganda serves just war, other questions remain. The first returns to the question of exploitation and intimidation, and the second addresses the permissibility of lying or otherwise manipulating or distorting the truth in the cause of self-defense, the very question the al-Dura video raises most provocatively.

Exploitation and Intimidation

Exploitation and intimidation are staples of behavior modification when guerrillas lack coercive state institutions. In previous discussions of how guerrillas may permissibly enlist civilian support, military recruits, and human shields; deter defection; and impose taxes and raise funds, I tried to draw a sharp line between violent coercion (threats of injury and loss of life) and the coercive sanctions that track those permitted to states. It is not hard to see that physical intimidation is beyond permissible bounds because no person loses the right to life or bodily integrity simply by supporting one public works project or shunning another. On the other hand, military organizations cannot be expected to stand by while the local population tries to make up its mind. This is a particularly acute problem for guerrillas who anchor their support and legitimacy in the local population. State armies have formidable tools at their disposal to coerce compliance when citizens refuse to report for duty or pay taxes. While insurgents may not resort to bodily harm any more than states may, there is place to consider other means of coercion including social sanctions, fines, incarceration, or restrictions on access to the local economy, roads, or other infrastructures. While occupying armies are prevented by the laws of belligerent occupation from employing such coercive measures to prevent the local population from supporting insurgents (or inducing them to aid an occupying force) (Dinstein 2009:53–55, 133, 174), guerrillas may indeed utilize them to the extent that they are applied fairly, without discrimination, and with some measure of consent from constituents. These conditions preclude night letters and assassination but permit the more subtle and less harmful forms of coercion as adversaries pursue public works and conduct information operations.

Lying and Deception

When state armies complain about the legal, political, and ethical principles that constrain their ability to get their message across, they echo a refrain common in Western nations: while non-state organizations can lie at will, state armies must tell the truth. Obviously, this is not always the case. Knowing of Saddam Hussein's reliance on CNN, for example, the U.S. government fed false information to the station to convince Saddam of U.S. preparations to launch an amphibious landing to oust him from Kuwait (Moorcraft and Taylor 2007). At the same time, the CIA disseminated disinformation in the Arab press to impugn Saddam's character and downplay the number of civilian casualties caused by U.S. led bombings (Macdonald 2007:161). These are, however, two different forms of lying. The first is a ruse, permitted for strategic reasons so long as it is not perfidious and abusive of immunities due others (Chapter 4). The second is propaganda and, more specifically, a false accusation. Its purpose is to improve America's image among Iraqis, its Middle Eastern neighbors, and the world community. It, too, is permissible insofar as it does not incite murder or wanton violence. What seems to bother state armies is that guerrillas, unfettered by aggressive journalists, nosy citizens, and tenacious bloggers, have more latitude to shade the truth. For the purposes of argument, I will concede this point. The real question is, so what? What is wrong with information operations infused with half lies and shaded truths? Deception and lying fuel the rising indignation many bear toward the media tactics of guerrilla groups. But lying is not usually unlawful, only unethical. So, why all the fuss? Why should anyone care about lying during war?

Information operations characterized by gray propaganda carry two dangers. First, there is a strong aversion to government lying in democratic nations. Compatriots expect their government to conduct itself transparently, respect its popular mandate, and shun abuse of power. Second, lying denies citizens the epistemic resources necessary to judge the workings of state. The truth, as it comes from the media and the organs of government, is a necessary condition of solid citizenship, critical understanding, and constructive dialogue (Beitz 1989; Black 2001;

Cunningham 1992; Ellul 1981). Lacking the truth, citizens are denied the means to the make autonomous and judicious decisions necessary for honest moral choices. Nevertheless, there is little doubt that governments become less transparent during national emergencies as they hide information that may impair the war effort. A delicate balancing act ensues because governments bear a double burden: protecting national security and providing the information necessary so citizens can evaluate the justice, progress, and cost of the war they must fight and die in. Truth telling is the norm, while lying is the exception that governments must carefully justify by sustained appeals to necessity and national security. Often this accounting can only come *post bellum.*

Truth telling, therefore, is an important norm that only national security can override when shading the truth is effective and necessary in the pursuit of just war and when disclosure follows war's end. However, the importance of truth telling and disclosure is strongest among compatriots. Lying to the enemy is a different issue. After all, deception and ruses are the way of war (Caddell 2004). What ethical restrictions bind the U.S. or Western forces as they fight in Iraq or Afghanistan? When lying to the enemy, few restrictions matter. One is perfidy. Fearing a spiral of abuse, Additional Protocol I (1977:Article 37[1]) forbids, for example, "the feigning of an incapacitation by wounds or sickness" with the intent to cause an adversary to drop his guard and to then kill him. Sisella Bok (1999:142–143) cites fears that deceit may backfire and lead to a cascade of lies that draws in friends and enemies alike or bring "severe injuries to trust" that may impair future peace negotiations. Some of these fears are overstated. It is true that lies to an enemy may rebound among those compatriots who understand and believe what they hear (despite the inevitable language barrier). On the other hand, it is unlikely that the local supporters of the Taliban, Hamas, or Hezbollah (or citizens of Israel or the United States) will think any less of their leaders for their deceptive media campaigns. In fact, they would probably argue that their enemies simply *deserve* to be lied to or, in the very least, that lying is necessary for national security. The extent to which lying may affect peace negotiations is, to my knowledge, unproven. Given the brutal indignities that adversaries inflict upon one another during war, lying seems to be the least of all impediments to a conflict's negotiated end. This leads us to consider that reservations about lying are not concerned with enemies but with third parties.

The fact remains that deceptive information operations are largely aimed at third parties whose right to truth telling should not be questioned. And indeed, other nations, the United Nations, the world press, and international NGOs have a prima facie right to know the truth. However, their right seems considerably weaker than compatriots' rights. Citizens of a nation at war require the truth to fulfill their roles as dutiful citizens bound together by a social contract laden with fiduciary responsibilities. What, then, of foreigners? While one might agree that rules of honest and sincere discourse "provide the basis for predictability and comprehensibility that make orderly social life possible (Mor 2012:410)," the international system is anarchic and none too orderly. Why should the United States tell the truth to Great Britain or to the United Nations? One reason is simple pragmatism. Truth telling garners support for U.S. foreign policy, while lying to allies is easily counterproductive if it undermines the trust that serves mutual interests. Reduced to expediency, however, nations may sometimes shade the truth they owe other nations when lying first and apologizing later brings greater benefits.

Moreover, it is not entirely clear what "lying first" means. In the aftermath of a drone strike, for example, there is sufficient ambiguity so that the first impressions are frequently the weightiest. Engineering these impressions is the Taliban's and Hezbollah's true talent. They effectively spin initial events so by the time the truth emerges (if it ever does), no one particularly cares. Military necessity may override the right to know the truth when deception, manipulation, and misinformation are effective, necessary, incur little cost, and violate no superior rights of combatants or noncombatants. Rights violations occur when lying endangers combatants through perfidy or conceals grave violations of humanitarian law. These are important conditions. If the injunction against abusing the rights of combatants and noncombatants is to have any teeth at all, violators cannot be shielded from publicity. Public diplomacy cannot hide war crimes. In many instances, however, propaganda does not cultivate perfidy or cover up crimes. Rather it plays out in the context of routine warfare. Following an airstrike that kills noncombatants, for example, a guerrilla group finds it can skillfully manipulate information to bring enough public pressure to curtail enemy operations. Guerrillas stymie their enemy and win a tactical victory at little cost. Only the truth is sacrificed. Contrast this with a concerted counteroffensive that incurs heavy costs in blood and treasure to achieve the same result. Under these circumstances, necessity can be a very potent and decisive argument for an information campaign infused with lies and half-truths. Nevertheless, there should be considerable concern when the sole purpose of propaganda is to falsely incriminate the innocent.

When the consequences of an accusation, smear campaign, or night letter abridge a person's right to due process or right to life, then the military benefits of lying do not trump its ills. The challenge, as always, is to weigh the costs and benefits of information operations and understand that the benefits, however great, cannot supersede superior rights of either combatants or noncombatants. At the same time, any attempt to understand the prohibition on war propaganda to mean a sweeping ban on any form of propaganda during armed conflict is too broad and infringes upon an aggrieved people's right to fight. Under these circumstances, considerations of justice not only permit the artful dissemination of information but demand it. The bounds of permissible dissemination are, of course, only half the story. Belligerents often do their utmost to censor information, restrict reporters' access, or disable communications facilities to prevent any dissemination whatsoever.

Censoring Information and Restricting Battlefield Access

Accurate information is the lifeblood of constructive discourse and democratic deliberation. How else might citizens judge the merits of public policy? War, however, poses significant challenges to the flow of information. Insufficient information puts citizens at a decided disadvantage when they seek to judge the war they or their children must fight. On the other hand, too much information may undermine the ability of a nation to wage war successfully. Citizens may then find themselves recipients of partial information with the understanding that full disclosure will, hopefully, come at an appropriate and not too distant time when political leaders must give an account and justify the restrictions they imposed.

The restrictions that states, particularly democratic states, impose on journalists nearly always grate. During the Second Lebanon War, for example, the Israeli government expressed considerable concern that reporters disclosed sensitive information about troop placements. At the same time, they worried about overly restricting the free press that a democracy should enjoy. No solution was reached. In later fighting, however, Israel restricted access to the war zone. Journalists denied access have nothing interesting to report thereby making censorship unnecessary. Those journalists allowed access were carefully monitored and unable to wander the battlefield at will. This practice became known as "embedding."

The policy of embedding journalists with front line troops for weeks at a time generated much controversy during the Second Gulf War (Moorcraft and Taylor 2011:196–203). Embedding reporters with Coalition soldiers gave them a front row seat to the fighting and unparalleled access to warfighters. However, reporters were not free to roam. Restricted to their embedded units, journalists could only gain a very narrow and tactical view of the fighting. "Watching through straws," as one embed put it (Moorcraft and Taylor 2011:199), deprived journalists of the big, strategic picture. More importantly, charged critics, embeds identified so closely with their newfound comrades-in-arms as to obviate any criticism of the U.S. military (Banville 2010; Limor and Nossek 2006; Moorcraft and Taylor 2007). Department of Defense guidelines further restricted coverage of wounded compatriot soldiers. As a result, censorship was very much self-imposed by a friendly press anxious to serve the war effort, protect their friends, and preserve their place at the front lines. This amiable relationship where the media and military accommodated one another in post–9/11 fighting was a far cry from their hostile relationship during the Vietnam War (Limor and Nossek 2006; Pew 2007).

For guerrillas, options to embed reporters are less feasible. Nevertheless, this has not stopped some from demanding equal time for reports from journalists embedded with the other side (Banville 2010). But these cases are rare, and when militants successfully embedded an Al Jazeera reporter in a hospital in Fallujah in 2004, U.S. military authorities were decidedly unhappy. Whether realistic or not, however, the call for equal embedding may miss the point entirely. Rather than try to promote some balanced version of the truth, why not just allow each side access to the same communication techniques? This excludes incitement but leaves considerable room for censorship, deception, and media manipulation. Hezbollah, for example, excelled in such efforts during their 2006 war with Israel. Carefully controlling media access to the battlefield, Hezbollah militants supervised interviews with local militia, "tidied up" bomb sites, staged media events, doctored photographs, and, in general, led visiting journalists on what CNN's Charlie Moore (2006) called a "dog and pony" show.

There is little doubt that media manipulation brings concrete strategic benefits, while the costs of deception accrue largely to third parties and not compatriots. As a result, it is reasonable to suppose that compatriots may not complain much at being deceived. Nevertheless, an information campaign freighted with staged events, censorship, and falsehoods may impair the trust and authority individuals impart to guerrillas or the state. To mitigate these outcomes, political leaders, whether guerrilla or government, would do well to focus on what they should disseminate rather than on what they might restrict. Clearly there is a range of information compatriots deserve that once provided goes a long way to preserving personal autonomy and maintaining trust in leaders. This information includes:

1. The avowed goals of the war.
2. The projected and ongoing costs of war.
3. The availability of alternative means to war.
4. An assessment of the war's effectiveness and necessity.
5. An assessment of the tactics used in war, and their conformity to the principles of just guerrilla warfare.

These guidelines force a sharp dichotomy between the kind of information necessary to maintain legitimate authority and the vast field of public diplomacy. While the former demands the truth (although not necessarily the whole truth), the latter permits anything but the truth and is open to a wide range of strategically useful machinations. In this way, the emphasis shifts from the kind of information guerrillas or state armies censor or manipulate (for there will always be times when this is justifiable) to the kind of information they must provide compatriots. And here, I think, one will see a correlation between an abundance of verifiable, relevant information that states and guerrillas must deliver to compatriots and the practice of just war. Otherwise, it would be impossible to make the case for legitimate authority as detailed in Chapter 2.

THE STATUS OF JOURNALISTS AND MEDIA FACILITIES

Referring to 1999 Department of Defense *Assessment of International Legal Issues in Information Operations* that hints that force may be used to shut down a "civilian radio for the sole purpose of undermining the morale of the civilian population," Payne (2005:90) offers a stark choice: "If the media are behaving impartially, then they are entitled to treatment as civilians. Where they are not, the assessment of the general counsel suggests that they can be targeted militarily." This is a striking claim because I have argued that if the media eschew incitement, they may behave as biased as strategically necessary without becoming legitimate targets of deadly force. Nevertheless, there is no doubt that as information operations gain traction and form an increasingly significant component of contemporary warfare, weighty questions arise surrounding its practitioners. Are journalists, the purveyors of information, liable to direct harm?

The short answer is yes. Journalists are liable for the harm they cause and the threat they pose in the service of information operations. In sharp contrast to armed combatants who provide war-fighting capabilities, however, journalists only supply war-*sustaining* aid. The former presents a direct lethal threat; the latter only provides the means to maintain it. With the exception of those guilty of incitement, most journalists disseminate war-sustaining propaganda and are only liable to destruction of their infrastructures by precision attacks or electronic (e.g., cybernetic) means and/or to disabling by arrest, detention, or expulsion. Just media warfare prohibits media manipulation that incites others to murder or other egregious human rights violations, a stand consistent with the ban on war propaganda. This leaves, however, a vast area for journalists to till. As they do, most media facilities are insufficiently threatening so that anything more than minimal collateral injury to noncombatants as infrastructures are destroyed is unwarranted. Attacks on media facilities are permissible but widespread collateral harm is not. But this argument is premature. The first question still remains: Are attacks on the media effective?

Experience shows that they are not. In recent years the United States, Russia, Israel, NATO, and Syrian rebels have bombed television stations in Belgrade, Baghdad, Chechnya, Beirut, Libya, and Syria with loss of life. In none of these cases did transmission halt for any appreciable period of time (Cordone and Gidron 2000; Gebauer 2006; Mackey 2012; Norton-Taylor 2003; Ryan and Brunnstrom 2011; Thomas 2000). The attacks were entirely futile and, therefore, killed noncombatants without cause. Nevertheless, strikes against the media are not necessarily misguided if, as argued, media targets are not immune from attack. The question, as always, it to appropriately adjust the use of force to the threat the media poses. Conflicting interpretations of some events make this difficult. To illustrate this point, consider NATO's account of its aerial attack on Libyan telecommunications facilities that killed three employees in 2011:

Our intervention was necessary as TV was being used as an integral component of the regime apparatus designed to systematically oppress and threaten civilians and to incite attacks against them. Qaddafi's increasing practice of inflammatory broadcasts illustrates his regime's policy to instill hatred amongst Libyans, to mobilize its supporters against civilians and to trigger bloodshed. (NATO 2011)

In contrast, David D. Kirkpatrick of the New York Times (2011) described how the goal of Libyan TV was to, "urge Libyans to resist NATO and march against the rebels … and remind [them] that [Qaddafi] is alive and in charge."

These remarks are instructive in several regards. In the first description, journalists stand accused of inciting unlawful attacks against non-combatants, thereby justifying deadly strikes against media facilities and their employees. In the second description, journalists are just doing their job. While this allows NATO to disable media facilities, attacks must spare employees and bystanders from death and injury. As in all instances of war, the abiding challenge is to thread the needle and determine whose liability lies where. It is doubtful that everyone in the facility incited genocidal violence. Nevertheless, one should also consider whether UNESCO Director-General Irina Bokova is on solid moral grounds when she declared categorically:

Media outlets should not be targeted in military actions. The NATO strike is also contrary to the principles of the Geneva Conventions that establish the civilian status of journalists in times of war even when they engage in propaganda. Silencing the media is never a solution. Fostering independent and pluralistic media is the only way to enable people to form their own opinion. (UN News Centre 2011)

Bokova is right but for the wrong reasons. The attacks were not permissible but not because they violated the right to propagandize, but because they were not effective and inflicted unnecessary civilian casualties. Nevertheless and notwithstanding Additional Protocol I, Article 59 that affirms the civilian immunity status of journalists, belligerents should have some recourse against information operations beyond simply redoubling their own propaganda efforts. Hard power and military force are one answer but its exercise must be effective and respectful of noncombatant rights. Satisfying both NATO and United Nations concerns would require NATO to make a strenuous effort to effectively disable telecommunications facilities with little or no loss of life, incarcerate journalists for the duration of the war, and try those suspected of unlawful incitement for war crimes. Subject to these caveats, there is no reason to think that guerrillas and insurgents do not enjoy similar recourse against media facilities.

THE ALLURE OF PUBLIC DIPLOMACY

Some observers have argued that public diplomacy must be conducted ethically, that is objectively and accurately, or it will fail. The United States, therefore, cannot hope to woo the Muslim world unless it describes the virtues of democracy honestly and endeavors to "undermine myths and stereotypes" (Seib 2009:782). Such general assessments fail to capture the complexity of public diplomacy in wartime. The obligation of governments toward compatriots is different from what they owe an enemy. There is no doubt that legitimate governments owe their compatriots the truth (or at least most of it), without which citizens are unable to exercise the responsibility expected of them. Here, indeed, the ethics of public diplomacy turn on objectivity, transparency, and balance insofar as these do not have immediate and severe security repercussions. Objectivity requires honesty and accuracy; transparency demands the disclosure of relevant information, while balance makes room for conflicting and competing ideas as long as they meet the conditions of objectivity and transparency. Enemies, and to a lesser extent, third-party nations, however, deserve no such temperance. The flow of information is not only subject to national security interests but is put to its use and manipulated for its benefit. Information or disinformation becomes a tool of war when, for example, it may *not* be prudent to undermine myths and stereotypes but perpetuate them. The calculation is largely utilitarian: belligerents will manipulate the truth insofar as the benefits outweigh the costs, a calculation that may change as conflicts persist, improve, or deteriorate. On one hand, objectivity and balance may sometimes stimulate peace initiatives or temper armed conflict (Galtung 2007). On the other, there are certainly costs to dishonesty and disinformation. The jarring loss of British support for a U.S. attack on Syria in 2013, which some attributed to Bush's misrepresentations about Iraqi WMDs a decade earlier, highlights the pitfalls of black propaganda. With the exception of murderous incitement, however, these costs are rarely absolutely or self-evidently unethical.

While some truths remain a necessary casualty of legitimate public diplomacy, it is important to remember that successful media campaigns incur significantly lower costs than the exercise of military force. The results can be impressive, particularly in a media environment that makes room for numerous players but no longer offers the large, institutional news organizations the time to check stories thoroughly lest they are "scooped" by fast moving blogs and internet based media outlets. These same multiple media entry points make it easy to quickly inject propaganda into the public arena that the public and many journalists are ill suited to verify, particularly in wartime (Macdonald 2007:128–134). In the short run, guerrillas can manage to obfuscate and manipulate facts and use the media to stymie a state's military operations. For this reason alone, an effective and well-oiled public diplomacy campaign is an ethical imperative of just war. If the last resort condition governing the use of armed force is to have any teeth at all, nations must seek out alternative means to reach their objectives. Public diplomacy, propaganda, and public works often fit the bill. In some cases they do the job better than military force. This is especially true for guerrillas who often lack the means to destroy or seriously cripple state media facilities.

Nevertheless several caveats are in order. First, public diplomacy alone wins few wars. The task is always to find the right balance between military force and soft power, turning to the former when no less destructive means are available to achieve crucial military or political goals and choosing the latter when they are not. Second, hard and soft tactics sometimes reinforce one another. Militants in Israel, Lebanon, and Afghanistan have shown over again that well-timed information campaigns can act as a force multiplier, amplifying the effects of a local attack against better armed enemies well beyond the immediate casualties or damage they caused. Third, manipulation, exploitation, and fabrication are permissible components of public diplomacy when effective, necessary, and respectful of fundamental human rights as a belligerent pursues a just and lawful war of self-defense. These provisos repudiate violent coercion, incitement to take innocent lives, devastate livelihoods or destroy property, or attempts to conceal egregious violations of international law. If the injunction against abusing the rights of combatants and noncombatants is to have any bite, then violators cannot be shielded.

While third-party journalists should comply strictly with principles of truth and objectivity (McGoldrick 2006; Streckfuss 1990) to support their claims to neutrality and the protection it infers, media workers affiliated with a party to the conflict have no absolute obligation to objective or truthful reporting during wartime. As long as they abjure vicious incitement, disclose war crimes, and support a just war these journalists may permissibly propagandize despite codes of journalistic ethics that enjoin reporters to seek the truth and avoid "misleading re-enactments or staged news events" (Society of Professional Journalists 1996; also Christians and Nordenstreng 2004; Perkins 2002). Journalists are, after all, the foot soldiers of media warfare and, as I demonstrate in the following chapter, a crucial component of nonviolent resistance when guerrillas choose to set aside armed conflict entirely as they pursue national self-determination.

10

Civil Disobedience and Nonviolent Resistance

While this chapter comes at the end of the book, many might expect to see it at the beginning. Civil disobedience, demonstrations, and nonviolence are, after all, the first resort that we demand of disaffected groups before they turn to armed force. It is part of our preoccupation with terrorism, however, to condemn guerrilla movements for turning to violence before giving other avenues of redress a real chance. While this is only sometimes true, civil disobedience and nonviolent resistance remain an important, but overshadowed, tactic of struggles for national self-determination.

While nonviolent tactics often accompany armed force, they are usually the purview of an organization's political wing. The IRA's Sinn Fein, for example, organized demonstrations and hunger strikes but also campaigned for political office to promote the republican cause (English 2003:202–211, 224, 245). The electoral option is, of course, rare. Many insurgents do not fight democratic regimes, much less participate in their political processes during an armed struggle. Instead, the political wings of movements in East Timor, Kosovo, and the Western Sahara adopted nonviolent tactics, peaceful demonstrations, and diplomacy to augment their struggle for independence. In the late 1980s, the Palestinians waged a relatively nonviolent campaign against Israel during the first "Intifada" and, like the Kosovars, built an impressive shadow government of parallel social and educational institutions to replace those of Israel and Serbia respectively. Later, and in spite of recourse to terrorism, Palestinians could effectively initiate calls to boycott Israel and call on supporters to launch a 2010 blockade-breaking flotilla that was one of the most successful, although morally flawed, examples of civil disobedience in recent years.

As with economic warfare, cyberterrorism, and public diplomacy, just war theory has had little to say about the ethical conduct of civil disobedience and other nonviolent tactics. Given the moral preeminence nonviolence assumes in the struggle against injustice, this might not be surprising. What, after all, could be morally complex about nonviolent resistance? Answering this question requires a closer look at the dynamics of nonviolent resistance and its place in the struggle for national self-determination. Assessing these oddly neglected issues, one must confront the maxim that war must be a last resort and carefully consider its correlate that nonviolence be the penultimate resort. These are enormously complicated demands, but if the last-resort condition is to have any bite at all, alternative avenues of redress must be feasible. Among these avenues are vigorous campaigns of nonviolent resistance finely timed for politically opportune situations.

If the broad moral question is whether or not insurgents must (and can) resort to nonviolent resistance before turning to armed force, particular moral questions arise at the tactical level. Especially vexing is the question of provocation. Many forms of nonviolent resistance depend on "backfire," that is, a harsh, disproportionate, and violent state response to nonviolent protest that may put unarmed demonstrators at considerable risk. This risk, often foreseen by organizers, is analogous to the danger guerrillas can expect when employing human shields. Successful shielding, however, brings no casualties, while successful backfire may bring many. Backfire – a tactic of nonviolent protest – therefore poses a greater danger than shielding, a tactic of armed struggle. Herein lies the paradox and moral challenge of a campaign that professes nonviolence. To lay the groundwork for this discussion, the following section investigates the role nonviolent resistance plays in pursuit of national self-determination and the fight against occupation.

NONVIOLENT RESISTANCE

Chenoweth and Cunningham (2013:271) define nonviolent resistance as "the application of unarmed civilian power using nonviolent methods such as protests, strikes, boycotts, and demonstrations, without using or threatening physical harm against the opponent." Novel adjuncts to nonviolent resistance include "hacktivism" – disruptive cyber-attacks that may potentially turn into cyberterrorism (Chapter 7) and "lawfare" – vigorous prosecution of state officials for war crimes through the international justice system (Austin and Kolenc 2006; Dunlap 2009). Nonviolent resistance is not pacifism – that is, a moral repugnance of violence – but a strategic choice when the costs of waging an armed struggle are high and its chances of success against a militarily superior adversary are low. Nonviolent resistance can be no less coercive than armed violence is. Its goal is to undermine the prevailing asymmetry of power and force a state to yield to an aggrieved group's demands by threatening the state's material interests and international image. The commitment to nonviolence is pragmatic rather than principled, adopted when useful and abandoned when it is not. At best, nonviolent resistance is a precursor of violence that allows insurgents to assure the world community that they have only come to armed struggle reluctantly and without feasible alternatives to settle their just claims.

Tactics and Aims of Nonviolent Resistance

Researchers distinguish among three general tactics of nonviolent resistance: "(1) symbolic forms of nonviolent protest (vigils, marches, and flying flags); (2) noncooperation (social and economic boycotts, labor and hunger strikes, political noncooperation and civil disobedience); and (3) nonviolent intervention and physical obstruction (nonviolent occupations, blockades and the establishment of self-reliant institutions and rival parallel government infrastructures)" (Dudouet 2013:403; Sharp 1989:4). These tactics highlight political, social, and psychological power rather than military or material assets, and aim for a variety of goals as activists weigh the sensibilities of compatriots, enemies, and the international community.

Turning to the aggrieved community, solidarity and self-help are overriding aims of nonviolent resistance. Nonviolent resistance can solidify national identity, instill a sense of pride, and mitigate the daily humiliation of life under occupation. More encompassing tactics see nascent national communities building a state-like infrastructure through rudimentary welfare, educational, economic, and health institutions. Focusing next on their adversary, nonviolent protesters pursue conversion, accommodation, coercion, and disintegration. These outcomes represent the degree to which a state makes room for protesters' demands (Dajani 1995:108–110; King 2009:247–250). In the best case, a state will recognize the injustice of the status quo and accept insurgents' demands for self-determination (conversion). In other cases, insurgents hope that the state will seek accommodation on specific issues, often in an effort to contain the costs of continued military or police actions. Greater success comes when nonviolent protests coerce the state to relinquish or dramatically curtail its rule as may occur when well-functioning local institutions usurp those of an occupying power. Finally and most dramatically, nonviolent resistance will bring its state opponent to disintegration and collapse.

Success varies. Agents and citizens of an occupying power may accept some or all of the protesters' demands solely on the basis of law and morality. In Indonesia and Israel, for example, local human rights organizations and some military personnel supported insurgents' demands and initiated mutual dialogue (Hill 2002; Sharp 1989). Moral factors notwithstanding, rising costs may compel military and police officials to yield to protesters and accommodate some of their demands. The first Palestinian Intifada, for example, probably pushed up the timetable for self-rule and local elections (Schiff and Ya'ari 1990). Coercion and disintegration are rarer outcomes of nonviolent resistance. While Kosovo's shadow government and the nonviolence it embraced helped establish Kosovo's demands for independence, only armed force eventually forced Serbia's hand. Elsewhere, governments in Indonesia, Israel, and Morocco successfully resisted nonviolent demands for independence, eventually yielding to international force in Indonesia but remaining intransigent in Israel and Morocco.

Finally, consider the place of the international community whose support is crucial for any successful insurgency. Nonviolence helps internationalize a conflict, enlist worldwide support, and foster an underdog image. International backing assumes many forms including media support, diplomatic recognition, funding, and military assistance. How successfully insurgents can utilize nonviolent resistance depends on the tactics they choose and the intensity of the response they can generate when states armies respond to nonviolent resistance with disproportionate force.

The Outcomes of Nonviolent Resistance

Recent research suggests that nonviolent campaigns are twice as likely to bring significant political change as violent resistance (Chenoweth and Stephan 2011:7–11). Successful nonviolent resistance mobilizes large numbers of supporters and encourages defections and loyalty shifts in the military. In contrast to armed struggle, nonviolent campaigns encounter fewer moral qualms, lower participation costs, and less risk when they recruit followers. As mobilization gains traction, participation costs drop further and support can snowball. Defections are crucial and are most likely to occur when the armed forces are multiethnic, their soldiers poorly compensated, and the regime weak or outcast (Nepstad 2013). Apart from mass mobilization and defection, additional factors improve the chances of successful nonviolence. These include support from state and international organizations, access to the world media (often via social media), and, paradoxically, violent repression that serves to inflame an oppressed population (Dudouet 2013; Schock 1999, 2013).

Key strategies to mobilize support and encourage defection are those creating moral discomfort among the enemy or the world community. Chief among these are dilemma actions and backfire. *Dilemma actions* arise when armed troops, for example, face off against protesting women or members of their own ethic community and "place authorities in a position where they must either concede political space to opponents or order an action that undermines regime claims to power" (Gould and Moe 2012:141). Dilemma actions stop short of violence, but when violence erupts and the authorities respond with heavy-handed and disproportionate force, the result is backfire. *Backfire* underscores the moral outrage and international condemnation directed against a state that responds brutally to nonviolent protest. Backfire "increases the resistance, sows problems in the opponents' own camp, and mobilizes third parties in favor of the nonviolent resisters" (Sharp 1989:5; also Stephan and Chenoweth 2008:11–12). Backfire may strengthen solidarity among insurgents, sully a state's image at home and abroad, undercut international support for an occupying or repressive nation, and, ideally, force concessions.

Despite these trends, nonviolent resistance performs less successfully when aggrieved groups pursue "territorial objectives" including "anti-occupation, self-determination ... and secession" (Chenoweth and Stephan 2011:7). The factors explaining this outcome are the mirror images of those that characterize successful nonviolent resistance. Because proponents of secession are often ethnic minorities, strong states can identify, isolate, and suppress dissenters; thwart mass mobilization; stem loyalty shifts among members of the armed forces; and suppress media reports of repression – that is, halt the very factors that allow nonviolence to successfully break a despotic regime (Cunningham 2013; Svensson and Lindgren 2011). Nevertheless, nonviolence offers a useful adjunct to armed force. Past and present conflicts in the Palestinian territories, East Timor, Kosovo, and the Western Sahara exemplify how nonviolent resistance can augment conventional warfare.

NONVIOLENT RESISTANCE IN GUERRILLA WARFARE

In the quest for national self-determinations, violent and nonviolent resistance often work together. This is particularly true among guerrilla organizations that maintain a military and political wing; the former pursues armed struggle and the latter nonviolent resistance and political organization. But this is not a neat dichotomy: sometimes the wings work in tandem while at other times one entirely overshadows the other. Contrary to the normative preference for commencing a struggle with nonviolence, some guerrilla organizations pursue armed struggle first and then give it up when it proves ineffective. In the Western Sahara, Polisario guerrillas fought Morocco for decades to conduct a UN-approved referendum on independence until, so suppressed by 1987, insurgents turned to a nonviolent campaign of demonstrations, boycotts, hunger strikes, and educational programs. At present, the Polisario can only monitor Moroccan military activity and continue to lobby for their long-promised referendum (Stephan and Mundy 2006; UN Secretary General 2012:§77–78, 94, 100–101).

In East Timor, too, guerrillas fought a fierce but unsuccessful war against Indonesian occupation. After years of fighting, East Timorese guerrillas reorganized in 1989 to establish a vibrant Diplomatic Front to wage a vigorous campaign of public diplomacy and air East Timorese grievances to a worldwide audience. Locally, a Clandestine Front drew on students and young people to conduct a nonviolent struggle through mass demonstrations, educational activities and public diplomacy (Weldemichael 2013:195–217). By offering workshops, educational programs, and social events, the Clandestine Front bolstered East Timorese identity and helped train East Timorese professionals to manage a future independent state. In short order, "The Clandestine Front became the driving force behind the pro-independence resistance" (Stephan 2006:60–61).

The Palestinians, too, pursued a nonviolent campaign, or Intifada, alongside an armed struggle. By most accounts, the Intifada represented a well of widespread, popular anger fomented by crushing poverty, daily humiliation, and political frustration. Local leaders utilized and then expanded existing community infrastructures to mobilize nonviolent resistance against Israel (Dajani 1999). A Unified National Command embracing diverse political factions coordinated demonstrations, boycotted Israeli products and industries, encouraged tax evasion, organized strikes by Palestinian employees of the civil administration, and shut down local businesses for days at a stretch (Arens and Kaufman 2012; King 2007; Sharp 1989; Zunes 1999). Such tactics allowed Palestinians to withdraw support from the apparatus of occupation consistent with the demands of theorists that an occupying regime cannot survive without the consent and obedience of an occupied or repressed group (Sharp 1973:25–31). Frequent rallies against military occupation disrupted the public order, and, although protesters foreswore firearms, they showed no reluctance about showering troops with stones and firebombs. Demonstrators suffered many casualties and deaths.

Apart from confrontational tactics, the Palestinians developed alternative institutions to oversee education, health care, agricultural assistance, public safety, and commerce. These ad hoc facilities allowed Palestinians to shake off (the literal meaning of "Intifada") the Israeli civilian administration and reassert a limited sphere of self-control. Such endeavors corresponded to Kosovo's efforts to establish a shadow government. Like the Palestinians, Kosovar activists drew on existing infrastructures, professional associations, and neighborhood organizations to lay the groundwork for a fledging administration (Clark 2000:95–121). Kosovar parallel institutions, however, differed from those of the Palestinians in at least two ways. First, Kosovo conscientiously pursued nonviolent resistance for many years *before* resorting to armed struggle (Rogel 2003).

Lacking weapons and fearing ruthless Serbian reprisals, Kosovars had little hope of a successful armed campaign (Pula 2004). Second, Kosovo's shadow government was far more entrenched and institutionalized than the Palestinians' was. Following a national referendum, a declaration of independence and elections in 1991–1992, Kosovo's president, Ibrahim Rugova, and his LDK party built a government that repudiated armed struggle in favor of slowly strengthening local institutions and international support (Stephan 2006). If the Palestinian effort was largely grassroots driven and welfare oriented, Kosovo would forge a well-functioning, nationally governing apparatus that would hold elections, impose taxes, provide health, education, and welfare, manage media services, and pursue international diplomacy. In each case, the parallel institutions in the Palestinian territories and in Kosovo offered insurgents a nascent sense of what Rugova called "internal, psychological freedom" and "an avenue to assert their sovereignty while repudiating a repressive regime" (Rugova 1994:175–176 cited in Clark 2000:115). Ultimately, however, both the Palestinians and Kosovars would turn to armed force to press their claims. In what way, then, was nonviolent resistance effective in the pursuit of self-determination?

The Effectiveness of Nonviolent Resistance

Aggrieved groups turn to nonviolent resistance to achieve some measure of national self-determination be it independence or far-reaching autonomy. Internally, they hope to solidify support for a national struggle of liberation, maintain morale, nurture national identity, and lay the foundation for state-like institutions. Externally, they strive to debilitate and eject the occupying power, demoralize or enlist their adversary's armed forces, recruit support among their enemy's civilian population, publicize their grievances to the world community, and procure humanitarian and military aid for their cause. This is a very tall order.

In Kosovo, nonviolent resistance neither prevented Serbian aggression against Kosovo's people nor precipitated Serbia's ruthless retaliation against the civilian community that eventually brought NATO bombers to Belgrade. Nevertheless, the near-decade-long nonviolent campaign and robust shadow government laid an important foundation for the future state. Politically and socially it separated Kosovars from Serbs, prevented Serbianization of Albanian youth, and strengthened Kosovo's national identity (Clark 2000:187). While Kosovo's grievances and nonviolent tactics resonated, albeit weakly, among Serbians, the international community was far more receptive. Nevertheless, "international sympathy," concludes Reitan (2000:78), "stopped well short of recognizing and rewarding their nonviolent struggle for statehood." More than a decade later, this criticism is muted. Today, Kosovo enjoys de facto independence, an outcome that owes itself, in part, to nonviolent resistance but no less to the Kosovo Liberation Army's armed struggle, Serbian brutality and intransigence, and NATO's willingness to intervene.

Nonviolence, on the other hand, was probably the key to East Timor's drive for independence. While Kosovars could make few inroads among the Serbians, the East Timorese could effectively forge ties with Indonesian human rights organizations and successfully stage joint demonstrations in Indonesia (Weldemichael 2013:230–235). Against the backdrop of Indonesian brutality, these cooperative efforts led to "loyalty shifts" among Indonesian students, business elites, and younger officers who grew increasingly hostile to the occupation of East Timor (Stephan and Chenoweth 2008:31–32). Internationalization proved equally successful and East Timorese leaders won great praise, and a Noble Peace Prize, for their nonviolent efforts to achieve self-determination. The East Timorese struggle was not entirely nonviolent; FALANTIL guerrillas continued to pursue hit-and-run strikes against Indonesian military units (Weldemichael 2013:260). The bulk of their efforts were, however, geared toward public diplomacy. Nor did guerrillas turn to armed resistance following Indonesia's relentless campaign of devastation and displacement that came after the East Timorese voted for independence in 1999. Such restraint in the face of overwhelming brutality eventually led the United Nations to authorize military intervention and ensure the transition to statehood.

In the Palestinian territories outcomes were less successful. In Souad Dajani's view, the immediate goals of the Intifada were "to render the Occupied Palestinian Territories nongovernable ... and, to establish indigenous social structures that would serve as the foundations of the future Palestinian state" (Dajani 1999:55–56). The Palestinian territories were never rendered ungovernable. Israel could thwart tax protests by confiscating property (King 2007:232–233). Boycotts and strikes failed for lack of resources to sustain strikers through long periods of unemployment (Schiff and Ya'ari 1991:263). Nevertheless, many self-help committees flourished and the Intifada fostered a short-lived sense of grassroots solidarity that enervated and empowered many Palestinians while proving that local (rather than Diaspora) leaders could mobilize large segments of the population. Although the Palestinians were unable to halt arrests, torture, home demolitions, and deportations as they demanded in 1987, their pleas resonated in Israel where peace protests erupted and some Israelis refused military service in the occupied territories (Schiff and Ya'ari 1991:145–149, 169, 291–292). Nevertheless, the effects of the Intifada upon the general Israeli public were muted because the Palestinians could never entirely foreswear terrorism or overcome their factionalism.

In spite of these shortcomings, the Intifada dominated the Israeli political agenda for the coming years and produced some astonishing results. In July 1988, Jordan's King Hussein renounced claims to the West Bank and recognized the Palestine Liberation Organization (PLO) as the sole, legitimate representative of the Palestinian people. By December 1988, the PLO declared independence, renounced violence, and recognized Israel. In response, the United States tacitly recognized the PLO and urged Israel to open negotiations that lead to the 1991 Madrid talks and the 1993 Oslo Accords. The latter established the self-ruling Palestinian Authority and provided the first blueprint for Palestinian self-determination. "The Intifada smashed the status quo beyond repair," conclude Schiff and Ya'ari (1991:328), "brought partial gains in [the Palestinian] struggle against foreign domination ... sparked a deep crises of confidence" in Israel and showed "that there was no way to bypass the PLO." Despite these achievements, a settlement ensuring Palestinian self-determination remains elusive.

In many cases, the success of nonviolent resistance depends critically on how regimes respond. "Repression against nonviolent campaigns in the Philippines and East Timor," write Stephan and Chenoweth (2008:42), for example, "resulted in well-timed international sanctions against the opponent regime, which proved instrumental in the success of these nonviolent campaigns." Factors affecting success highlight the nonviolent nature of a campaign, the disproportionate nature of the state's response, and an attuned international community seasoned by ongoing and well-publicized acts of violence that arouse fierce moral indignation. Ultimately, violence and nonviolence work hand in glove to precipitate a harsh and disproportionally brutal response from state adversaries. In East Timor, the Palestinian territories, Kosovo, and elsewhere, this response often backfired and brought condemnation on state leaders while catapulting guerrilla organizations to unprecedented success. Under these circumstances provoking a violent response may prove exceptionally useful to insurgents as they toy with dilemma actions and backfire.

Provoking Violence: The Ethics of Dilemma Actions and Backfire

Successful dilemma actions stop just short of violence. John Gould and Edward Moe (2012:141) describe how in Serbia,

organizers often placed women in the front ranks of the protests – exploiting patriarchal Balkan constructions of gender that place wives, daughters, and mothers in a protected role. Organizers thus confronted security forces with a no-win choice: either do nothing – allowing the demonstrators to seize political space previously controlled by the regime – or attack a protected group and risk causing a popular backlash.

This dilemma action is little different from using human shields to protect a military asset from attack. In each case, insurgents aim to deter a state from exercising armed force by putting noncombatants at risk while exploiting their adversary's sense of justice to refrain from harming protestors. In contrast to shields, however, demonstrators protect no military target and are free from the threat of direct or severe collateral harm. Nonetheless, a repressive state may target protesters in the ostensible interests of public safety. As such, the same conditions guiding shielding apply to nonviolent resistance. Organizers must secure some measure of consent from demonstrators and field sufficiently large numbers of protesters to deter their opponent. Consent is sufficiently elastic to include socially acceptable forms of nonphysical coercion that might include social sanctions, religious injunctions, or fines. Circumstances change considerably, however, when dilemma actions fail to bring strategic gains. At that point, nonviolent resisters may provoke backfire to shock the conscience of the world.

Unlike dilemma actions, successful backfire requires violence. One way to precipitate overwhelming violence is with nonviolent provocation: political noncooperation, demonstrations, strikes, and so on. In Kosovo, the suppression of national aspirations invigorated the movement and its military wing, attracted considerable international sympathy, and eventually helped bring international intervention. In East Timor, the Clandestine Front "provoked the Indonesian military" (Kilcullen 2009:208) without resorting to significant violence. The resulting 1991 Santa Cruz massacre in Dili transformed their struggle for national self-determination (Chapter 9). In its wake, major supporters of Indonesia – the Netherlands, Canada, Denmark, and the United States – curtailed or suspended aid to Indonesia and later supported military intervention (Fukuda 2000). In its early years, the Free Aceh Movement (GAM) too faced an increasingly harsh response from Indonesian authorities that would attract considerable popular support, invigorate the movement, and rejuvenate its military wing (Aguswandi 2008).

In the Palestinian territories, on the other hand, backfire failed. There is elaborate theorizing among some Palestinians, for example, about the symbolic value of slinging rocks at the Israel's Goliath; how stones "manifest the absence of weaponry" and "symbolize nakedness against the occupier" (King 2007:259–260). Rocks are sometimes deadly, however; foreswearing firearms is not the same as nonviolence. Whether disproportionate or not, the Israeli response and ensuing melee did not evoke much sympathy for Palestinians. Thus, insurgents who utilize even moderate violence face a tough route: they want to precipitate a repressive response without looking like they deserve it. Resorting to terrorism, the Palestinians did not manage this, and, despite their modest political successes, their national aspirations were stymied. By mixing nonviolence with violence and, eventually, horrific terrorist attacks, the Palestinians left the world community sufficiently enraged to decline aid and support. During the first Intifada, ostensibly a period of relative nonviolence from 1987 to 1993, 1,087 Palestinians and 160 Israelis lost their lives in pursuit of goals that remain unfulfilled 25 years later (B'Tselem nd.).

As these cases suggest, backfire succeeds best when peacefully provocative. To get a strategically useful state response, protestors must incite violence without embracing violence themselves. How far, then, may provocation go and what obligations do organizers owe protestors who may ultimately provide the grist for an effective public relations campaign?

Provoking Violence Deliberately

Whether a nonviolent campaign may deliberately provoke violence sufficiently extreme to fundamentally transform an adversary's policies is
a question that makes theorists uncomfortable. Gene Sharp, one of the
field's leading thinkers, offers no answer. "Nonviolent actionists," writes
Sharp, "aware that brutal repression may produce unease, dissent and
opposition within the opponent group have on occasion provoked the
opponent to violence deliberately ... This type of provocation, however, has limited utility and contains its own dangers" (Sharp 1973:677).
Sharp, however, does not elaborate and leaves us to wonder what these
dangers are. Provocation may be of limited utility but only if the state's
response is proportionate. If protestors turn to some measure of violence
to provoke their adversary, then the army or police have that much more
leeway to quell demonstrators. This was the fate of the Palestinians. If, on
the other hand, protestors remain entirely nonviolent, as Sharp advises,
then they must find other ways to provoke or expose their adversary's
brutality. Sometimes a resigned willingness to take abuse is sufficient.
Gandhi's 1930 march on the Dharasana Salt Works is a dramatic example of the moral hazards of provocation:

> In complete silence the Gandhi men drew up and halted a hundred yards from
> the stockade ... The column silently ignored the warning [to disperse] and slowly
> walked forward ... Suddenly, at a word of command, scores of native police
> rushed upon the advancing marchers and rained blows on their heads with their
> steel-shod lathis [batons]. Not one of the marchers even raised an arm to fend
> off the blows. They went down like ten-pins ... Those struck down fell sprawling,
> unconscious or writhing in pain with fractured skulls or broken shoulders. In
> two or three minutes the ground was quilted with bodies. Great patches of blood
> widened on their white clothes. The survivors, without breaking ranks, silently
> and doggedly marched on until struck down. (Miller 1930, cited in Hess and
> Martin 2006)

In less dramatic instances, activists may incite opponents with verbal
abuse or insulting symbols, block roads, or stage embarrassing demonstrations for foreign dignitaries (Hess and Martin 2006).

In contrast to nonviolent protesters, armed insurgents face severe condemnation when they deliberately goad their opponent into harming
their compatriots disproportionately. In Chapter 6, I censured attempts
by the LTTE to draw fire upon fleeing civilians at the end of their war

with Sri Lanka. There, several factors were salient: the LTTE had lost legitimacy, their provocation served no military purpose, and the LTTE could no longer secure consent or prevent disproportionate harm. But if using violence to provoke violence against the innocent undermines noncombatant rights in the most fundamental way, it is odd that nonviolent resistors enjoy such tolerance. This seems out of place, but to my knowledge, renowned figures like Mohandas Gandhi and Martin Luther King, Jr. draw no moral censure when they "accept [and], sometimes solicit, violent retaliation because, as victims, they will be better able to gain sympathetic support from bystanders" (Gurr 2000:156).

Accepting violent retaliation is one thing, soliciting it is quite another. While neither King nor Gandhi lost their legitimacy or effectiveness, their actions remain bound by the moral constraints of consent and proportionality and what we make of the term "provocation."[1] The previous discussion suggests at least three kinds of provocation. Stone throwing in the first Intifada brought a response that demonstrators both intended and reasonably foresaw, while the Dili massacre was neither intended nor, quite possibly, foreseen. Gandhi's march on the salt works falls in between: the response was unintended but quickly foreseen. Nevertheless, the demonstrators did not desist, so it soon became clear that their intent was to precipitate ongoing violence. In each case, one may ask whether there were grounds for positioning demonstrators to provoke violence. The first Intifada raises hard but instructive questions about exposing civilians to certain harm. Here, for example, one may assume that demonstrators were aware of the danger and tendered consent. Organizers published pamphlets and calls to action but did not physically coerce anyone to join the demonstrations. In due time, however, it became clear that the tactic was failing and the community was slowly withdrawing its support. Organizers rightly proved sensitive to these developments and eventually abandoned their confrontational tactics. Some turned to alternative nonviolent tactics; others turned to terrorism.

The Santa Cruz massacre holds a different lesson. Here, activists organized a demonstration following the shooting of several activists who unfurled East Timorese flags and taunted troops. They might have

---

[1] My thanks to my students in my graduate seminar "Opposition to War: Civil Disobedience and Nonviolent Resistance" for prompting this discussion. I am especially indebted to Ameer Fakhourey, Nora Kopping, and David Reis for their incisive comments.

anticipated some level of violence. Nevertheless, East Timorese activists claim that the Indonesian response was entirely unexpected (Pinto and Jardine 1997:194). Under these circumstances, there was no apparent strategy to provoke a *deadly* reaction, although it is hard to rule out any provocative aim. The question remains whether organizers made this known to the participants and whether they dispersed the demonstration once the situation deteriorated and people's lives were unwillingly at risk. Apart from some assurance of success, informed consent and concern for safety are the minimal obligations that organizers have toward demonstrators. This returns us to the hardest case, Gandhi's march on the salt works. Assuming there was some reasonable chance of success (it was, in fact, a rousing success) and demonstrators had some idea what was in store (which they soon did), one might reasonably ask whether organizers should have, nonetheless, retreated and avoided further provocation.

Answering this question calls to mind the calculations of political leaders as they risk their compatriots' lives during war. Protestors and demonstrators are foot soldiers, and the criteria for a legitimate nonviolent struggle are the same as for armed struggles: just cause, legitimate authority, a reasonable chance of success, equitable conscription, consent, and proportionate human costs in view of the goals insurgents seek. Proportionate human costs demand that organizers of nonviolent protests distinguish between a *violent* response and a *deadly* response and convey this information to participants. Only extreme conditions might justify deadly confrontations and, in the prior examples, they were largely unintended. The prospect of a violent but nonlethal response offers protest organizers more latitude and greater ability to avert harm by redirecting protesters to less dangerous venues. Nevertheless, consent remains in the forefront as leaders enlist or conscript protestors.

Recruiting demonstrators is no different than recruiting soldiers, civilian supporters, or human shields. In previous chapters, I have noted how guerrilla organizations lack the coercive institutions of states to conscript soldiers or levy taxes. In their place, guerrillas may legitimately sway their compatriots with reasoned arguments; positive incentives, which include financial compensation, social solidarity, and fellow feeling; or negative incentives, which may include subtle social pressures or the prospect of fines or prison. Organizers must be careful to protect minors (and others who cannot consent) and provide adults with the information about risks and hazards while ensuring they make decisions free from overt coercion and physical intimidation. Just as in war, a people must be able to depend on their leaders for sound decision making and assurances that their sacrifice is necessary, proportionate, and worthwhile. In India it apparently was, but this will not be true in every case. It is for this reason precisely that legitimate guerrilla organizations, like states, require mechanisms to gauge public opinion. These caveats also apply to other forms of nonviolent resistance that hope for backfire including boycotts, civil disobedience, and hunger striking.

Boycott, Divestiture, and Sanctions (BDS)

There is much debate and hand-wringing surrounding the boycott, divesture, and sanction efforts of the Palestinians. Modeling the boycott campaign of South African guerrillas, Palestinian activists seek to maintain "non-violent punitive measures until Israel meets its obligation to recognize the Palestinian people's inalienable right to self-determination and fully complies with the precepts of international law" (BDS Movement 2005). Despite lofty goals of ending the occupation of the West Bank and Gaza and returning Palestinian refugees to Israel, BDS organizers emphasize its symbolic and promotional value as high- profile entertainers and academics refuse to appear in Israeli venues and international organizations divest themselves of Israeli assets (Awwad 2012; Erakat 2012). Critics, on the other hand, condemn BDS as the odious legacy of discriminatory boycotts of marginalized minorities in repressive regimes.

In the hands of insurgents confronting an occupying state regime, BDS-type tactics have little potential to do serious economic harm. Instead, calls for boycotts, divestiture, and sanctions exploit nonviolent means to discomfit states and draw international attention to insurgents' cause. Nevertheless, some individuals may suffer financial or other losses as guerrilla organizations boycott local products or interfere with academic and cultural exchanges, two campaigns the Palestinians successfully employ. To assess these measures, as well as the claims and counterclaims of BDS advocates and critics, it is first important to establish participatory liability and, second to assess the proportionality of the harm BDS causes.

In the Palestinian case, as in others, three groups suffer harm: participating civilians, enemy noncombatants, and compatriot noncombatants.

West Bank settlers, no less than Indonesian settlers in East Timor or Moroccan settlers in the Western Sahara, stymie self-determination in the most fundamental way by occupying territory unlawfully, preventing the territorial contiguity necessary for an emerging state or autonomous region, and diminishing the supply of scarce resources. When occupation is unjust, settlers bear the liability of participating civilians and are subject to the kind of proportionate nonlethal harm that comes from *discriminate* economic measures (boycotts of West Bank products or Israeli companies that provide services to the West Bank, for example). On the other hand, noncombatants, including academics and music enthusiasts, are also affected by cultural and academic boycotts. While the effects of the former are usually nothing more than inconvenience, some academics suffer adversely when research ties founder or funding dries up. And, while Palestinians may achieve impressive symbolic gains (Lim 2012:219–231), it is hard to see how noncombatants are liable to anything but collateral harm. But that is not the case here. Instead noncombatants, particularly academics, suffer directly. When they do, academic boycotts violate their rights.

Finally, one must consider the economic losses that the local population incurs when they boycott what are very often the sources of their livelihood (by refusing to work in West Bank settlements, for example). The prospect of unemployment and lack of alternative means of sustenance were partially responsible for the failure of strikes and boycotts during the first Intifada. In this regard, the BDS movement is more cautious and largely directs its appeal to the international community while calling on Palestinians to boycott settlement products "ranging from foodstuffs to construction materials" (Rigby 2010:72). While this boycott might impinge upon the Palestinians economic well-being, the BDS movement enjoys widespread support (BDS National Committee 2009). Continued evidence of support among Palestinians is essential to justify the financial losses they may suffer. Civil disobedience raises similar demands.

International Civil Disobedience, Backfire, and the Marmara Incident

Ordinarily, civil disobedience is the purview of citizens who violate their nation's law hoping to change public policy for the better. Its goal is restorative, to reset the majority's moral compass and reestablish an errant sense of justice when a nation fails to uphold its prevailing principles of democracy. As Rawls (1971:363) suggests, civil disobedience is most suited to the "special case of a nearly just society." In the context of a guerrilla campaign for national self-determination, protesters may violate international law in an attempt to wring concessions from their state adversary and gain international sympathy. Here, too, they flirt with violence, and the Marmara flotilla, undertaken in 2010 to break Israel's blockade of Gaza, offers an exceptional case study of the moral and physical hazards of backfire.

Following the election of the Hamas government in Gaza (2006), Israel imposed a land and naval blockade to prevent shipments of missiles and other war materiel (UN 2012). By the United Nation's account, the Israeli naval blockade met the minimum conditions of legality: it was effective, proclaimed publically, and enforced impartially (UN Secretary General 2011). In 2010, a flotilla set sail to break the blockade, deliver foodstuffs, toys, and medical supplies to the Palestinians, and call the world's attention to the growing humanitarian crisis in Gaza. While the smaller ships heeded Israeli warnings and turned away, the largest of the flotilla, the Mava Marmara, announced its intent to run the blockade. Intercepting the ship, Israeli commandos boarded and took control by force. About forty passengers resisted the takeover with improvised weapons, knives, and axes. When the smoke cleared, nine Turkish passengers lay dead and many wounded.

The flotilla was wildly successful. The deaths of Turkish activists ruptured Israel's relations with Turkey and brought Israel a flood of negative media attention, international condemnation, and high-level commissions of inquiry in the UN, Israel, and Turkey (Turkel 2010; UN Secretary General 2011). Following the incident, Israel eased the passage of goods significantly, and the Egyptians opened their border with Gaza allowing trade and smuggling to flourish and the economy to improve steadily (UNOCHA 2011). In short, the Marmara was a transformative event, breaking Israel's hold on Gaza, trumpeting the Palestinian cause around the world, enhancing the stature of Hamas, and laying grounds for a vibrant black market economy that would soon fill Gaza with advanced military hardware. In light of such success, was the blockade run a permissible tactic of war?

The Right to Civil Disobedience

While many commentators took a close look at the legality of the blockade and the proportionality of the Israeli response, few considered the merits of civil disobedience. Did the passengers have the right to run the blockade and did they have the right to respond with armed force when commandos boarded the ship? Because civil disobedience incorporates unlawful acts, it does not matter whether the blockade was legal or illegal (and in fact, the commissions of inquiry were split on this question). What matters is whether the naval blockade and, more importantly, the status quo it protected were unjust, arbitrary, and in violation of basic human rights. Established to interdict weapons, the naval blockade was a legitimate security measure that alone caused little hardship. There was considerable concern, however, about overland restrictions that prevented the passage of goods into Gaza and brought widespread unemployment, poverty, meager medical care, high rates of business failures, and poorly functioning water and sanitation infrastructures (UNOCHA 2011; UN Secretary General 2011:§153). Conjoining the effects of the naval blockade and the overland restrictions does not render the former unlawful, but it may offer a moral justification if there was a reasonable expectation that a blockade run might alter policy at the border crossings by generating sufficient media furor. And, as subsequent events showed, this expectation was, indeed, reasonable.

Confronting the protestors, one UN commission of inquiry declared, "There is no right ... to breach a lawful blockade as a right of protest" (UN Secretary General 2011:71). But this legal argument carries no ethical weight in the context of civil disobedience. By definition, civil disobedience is unlawful and may easily require that activists violate "laws of trespass" to press their grievances (Rawls 1971:364–365). But if civil disobedience theorists might be unconcerned about unlawfulness of the flotilla's action, some would certainly object to its violence. Most flotilla activists harbored no violent intentions (UN Secretary General 2011:4) and, presumably, expected that nonviolence would serve as a bond of their sincerity. Many flotilla activists (most of whom were not Palestinian) also hoped that nonviolent protest would open the world's eyes to the grievances of a persecuted people. As the smaller ships were towed ashore with barely a peep of public indignation, it would soon become clear that if left to its own devices, nonviolence would fail miserably. Violence became the key in this case. Activists would hardly have achieved the astonishing results they did without the deaths of nine passengers any more than the Dili protests or Salt Marches would have resonated without the attendant injuries. While guerrillas may certainly claim the right to civil disobedience, the harm befalling protesters is as problematic on the Marmara as it was in East Timor or India.

Violence and the Marmara

While organizers publically professed a nonviolent and peaceful blockade run, they privately prepared to resist the ship's takeover with substantial force. While this led to impressive results, the price in terms of the deaths, injuries, incarceration, and loss of property that befell unprepared peace activists is unjustifiable. They did not consent to this. Most activists committed themselves to nonviolence and, like proper civil disobedients, were prepared "to suffer the consequences of arrest" (UN Secretary General 2011:46–48, §88) but not physical injuries. Many of the Marmara activists were, it seems, intentionally misled about the organizers' commitment to nonviolence. They did not sign on for a gunfight, and organizers did not take the means necessary to protect nonviolent protesters from harm.

It is no wonder then that Israel concluded that many of those on board were not civilian activists but "direct participants in hostilities" whose violent actions permitted Israel to use armed force to subdue them (Turkel 2010:19–21). From Israel's perspective, the Marmara blockade run was not an act of civil disobedience but an act of war, no different from an insurgent attack on a military facility. Fighting among the nonviolent activists, militants attacked troops with bars and knives. In the ensuing melee, commandos could only make a minimal effort to protect noncombatants and avoid disproportionate casualties as they fought for their lives. Nevertheless, one cannot divorce the localized fracas on the Marmara from its larger context of legitimate civil disobedience in the context of a struggle for national self-determination. The blockade run was not a war-fighting operation. It was not even a war-sustaining operation except in the sense of mobilizing international public opinion to the cause. As it was, the Marmara carried little of anything. Although most of those killed were probably militants, there was tremendous fuss when they died. The answer as to why is clear: because the ship itself posed no major threat. There was a point that Israel exceeded early on where the right action in the face of resistance would have been to back off and let the ship go on its way.

Had Israel permitted the Mave Marmara to pierce its blockade, the Palestinians would have scored an important symbolic victory. To do so, however, insurgents need to raise the costs of boarding the ship so that either Israel let the ship pass or reacted with the disproportionate force necessary to create backfire. Contemplating their strategies, insurgents might have pursued three routes. First, they may have opted for a nonviolent blockade run. In spite of the inherent dangers of running a lawful military blockade in wartime, nonviolent civil disobedience would have probably ended peacefully and without backfire. Witness the ships towed quietly to port. Alternatively, they might have thought to defend themselves by the aggressive but nonlethal tactics that some ocean-going environmental activists adopt: maloderants (waste water, urine, etc.), flash grenades, disabling acoustical devices, or unarmed physical confrontation (Khatchadourian 2010). Should this fail to achieve the desired response, insurgents may up the ante. But turning to hand weapons, even while eschewing firearms, is no longer nonviolent resistance but a well-orchestrated violent confrontation that requires insurgents to protect noncombatants. Marmara organizers did not do this.

Backfire is essentially a game of brinksmanship that leaves organizers considerable flexibility to pursue various degrees of nonviolence *and* violence to achieve their goals. There are obvious benefits to precipitating backfire by nonviolent or minimally violent means: insurgents preserve their underdog image and enjoy significant international sympathy. But these measures may not work. As a result, nonviolent resisters are justifiably pragmatic, hewing to nonviolence when it works, resorting to violence when it does not. In the context of a legitimate armed struggle, there is nothing morally suspect about this. Insurgents are free to contemplate the costs and benefits of armed force when nonviolence fails. Choosing either, the usual caveats apply: necessity, effectiveness, and proportionality.

Resort to violence, then, was not the major issue in the Marmara case. While neutral nations are banned from violating blockades, parties to an armed conflict may certainly do so. When nonviolence proves ineffective, armed force is a reasonable tactical decision. Lack of consent among the nonviolent protesters, however, presents a more formidable obstacle when organizers recruit what amounts to human shields and put them at risk for life or limb without their consent. And while guerrillas hope human shields will not suffer harm, protest organizers often hope they will. Had shielding worked successfully, Israel would never have boarded the Marmara. But without boarding the Marmara, there would be no backfire. This would have left the Palestinians with a successful blockade run but deprived of the many benefits of backfire. In the final analysis, Marmara militants could only generate backfire by misleading the nonviolent activists under their protection. The Marmara is an extreme example but amply demonstrates the moral hazards of backfire. Hunger striking raises similar concerns.

HUNGER STRIKING AND THE PROBLEM OF CONSENT

Hunger strikes are brinksmanship no different than dilemma actions or backfire as strikers put their own health on the line to force concessions from the state. As strikes progress, states have only three options: accommodate strikers, let them die, or feed them by force. Following frequent attempts to successfully force-feed hunger strikers at Guantanamo Bay by strapping them to a chair and snaking a feeding tube into their gut, the International Committee of the Red Cross (2013) and the World Medical Association (2006) have condemned force-feeding detainees. Each organization, however, leaves room to feed a hunger striker medically using intravenous lines or artificial food supplements when one's life is in danger, the individual is no longer competent, and on the condition that the striker has not left binding instructions to the contrary. Because the striker is unconscious, medical feeding is less violent than the force-feeding at Guantanamo. Nevertheless, each practice is coercive and taken against a striker's wishes. Opposite this rancorous debate about the ethics of force-feeding hunger strikers is the near-inaudible concern about the ethics of hunger striking. Although professing nonviolence, strikers may trigger considerable violence when they are force-fed, treated against their will, or left to die.

As many observers point out, hunger strikers profess no wish to die (Annas et al. 2013; Gross 2013; Howe et al. 2009). Rather, they hope that the prospects of their deaths, like those of shields, will force states to cede political concessions. This hope is sometimes dashed. In 1981, the British government allowed ten members of the IRA to die of hunger after denying prisoners' demands for recognition as political detainees (English 2003:263–274, 280–283). British policy backfired thunderously, however, bringing Britain worldwide condemnation, sweeping support for the IRA, a vicious, decade-long campaign of terror, and the successful entry of the Sinn Fein into British politics. Wary of similar mistakes, states facing hunger-striking insurgents tread more warily today. Recent hunger strikes by Palestinian and PKK insurgents in Israel and Turkey have ended with modest concessions that hunger strikers were happy to accept (Krajeski 2012; Khoury 2013).

Severe problems arise when strikers make excessive demands and/ or states prove intransigent. The United States, for example, refuses to negotiate with striking inmates at Guantanamo Bay and force-feeds detainees within days of their refusal to eat. Elsewhere in the United States, prison authorities in California put off force-feeding until inmates were incapacitated. Considering the state's request to feed strikers against their will, a California court permitted artificial feeding provided the inmate was deathly ill, incapacitated, left no valid order to refrain from resuscitation or was coerced to strike (Plata v. Brown 2013). The last point is telling. Authorities feared gang leaders (or, by extension, guerrilla leaders) ordered detainees to maintain their strike. In which case, ruled the courts, there are no grounds to respect the detainees' decision to refuse food. Underlying this debate is the idea that hunger strikers must agree to participate in a hunger strike and to risk their lives. If strikers do not consent but instead act under duress, then, reasoned the court, the hunger strike is impermissible and physicians may force-feed detainees. Is the court correct or may guerrilla leaders nevertheless order a hunger strike?

Answering this question depends on how we understand legitimate coercion, a question that bedevils not just hunger striking but the conscription and recruitment of combatants, civilians, and human shields as well. In medicine, the notion of consent is very strong and, as the California court notes, requires a patient to receive all the information necessary to understand the ramifications of his or her actions and to assent freely, untainted by external influences when deciding to accept or refuse food and fluids. Medically valid consent is far more solipsistic than consent tendered in response to positive or negative incentives and suggests that consent driven by payment, social pressure, obedience to religious, military, or political authorities, or by the threat of fines or imprisonment is somehow defective. While perhaps this is true in some rarefied sense of extreme individualism, consent remains very much a social construct, the product of one's shared political, social, and moral environment. Decision making is always responsive to norms of fidelity, social cooperation and peer pressure, religious or political duties, mutual responsibility, and personal well-being. While some might be less comfortable attributing consent to those who accept payment for services (e.g., organ donors) or those aware of possible sanctions or punishments (e.g., taxpayers), this criticism is too severe. A person's motives are often mixed and a taxpayer or hunger striker who agrees to comply may be acting freely as he or she considers collective responsibility, political or religious duties, and sanctions.

In many cases individuals also exercise higher, "second-order autonomy" when they freely entrust their decisions to others. This, too, is a form of consent and particularly salient in wartime as hunger strikers, no less than many other compatriots, entrust their leaders with the task of pursuing their collective good and protecting the welfare of those who risk their lives. It is, for this reason, that legitimate authority is so crucial. Just as nonviolent resisters expect protest organizers to walk a fine line between constructive provocation and violent backfire, rank-and-file guerrillas must enjoin their leaders to exploit the bodies of hunger strikers sparingly while keeping their political demands sufficiently feasible to merit serious attention by the state. In this way, hunger striking joins the repertoire of nonviolent resistance that may, nonetheless, wreak violence upon protesters. There is no reason to think that the decision to participate in nonviolent resistance is not given freely unless accompanied by violent coercion and physical intimidation. In these circumstances, agents lose their power to opt out at a reasonable cost and can no longer act freely. Overt coercion exceeds the bounds of legitimate authority.

LAST RESORT, NONVIOLENCE, AND GUERRILLA WAR

There is much to recommend nonviolent resistance. By stressing nonviolence, theorists hope to minimize the human and material costs of political confrontation, mitigate animus, and, hopefully, pave the way for a more lasting peace. Reaching for avenues of understanding, nonviolent activists seek cooperation, conversion, and reconciliation, not competition and compromise. These hopes are not without foundation: peace is often more durable and politics more democratic following successful campaigns of nonviolent resistance (Schock 2013).

Among the conflicts described in this chapter, success along these lines is difficult to evaluate. Conflicts that saw concerted efforts at nonviolence in Kosovo and East Timor degenerated into armed violence that left the earth scorched and the population violently displaced. While these conflicts ended with a firm measure of national self-determination, the efforts of the Palestinian Intifada have yet to bear any significant fruit. Elsewhere – in Chechnya, Southern Sudan, or Eritrea, for example – guerrillas rarely pursued nonviolent resistance. Although violence seems endemic, if not preponderant, in many conflicts, one may still ask whether nonviolent resistance should be the avenue of first resort when a people battles for the right of national self-determination and against grave human rights abuse.

The relationship between states and non-states is intensely asymmetric. Unlike rival sovereign states that enjoy the vigor of an institutionalized civilian and military administration, non-state actors are bereft of the advantages of sovereignty. Living under occupation or suffering maltreatment and exploitation, non-state actors have no facility to undertake armed violence as a very first resort. Guerrilla movements, whether political or military, do not emerge rapidly in the wake of abuse and violence, but emerge gradually after taking time to organize compatriots, mobilize local and international support, and establish legitimacy. In the interim, members of aggrieved groups find themselves wholly at the mercy of local authorities and the whims of the international community whose policies and priorities vary over time. Calls for referenda in East Timor and the Western Sahara went unheeded for many years. Attempts by the United States to pressure Israel and broker a peace deal with the Palestinians have been entirely unsuccessful. Chechnya did not go to war for independence until 1994, fully three years after the dissolution of the Soviet Union. And, in Kosovo, Albanian separatists could defer armed resistance for nearly a decade after Serbia unilaterally abrogated Kosovar autonomy. In these and other cases, diplomatic efforts and sanctions preceded armed conflict. Do these efforts satisfy the measures that must come before war? Does it matter that the diplomatic and other efforts just described came at the behest of third parties or do insurgents have some obligation, consistent with their chances of success, to defer armed struggle until they (and not others) have exhausted other means?

Answering this question depends on the direness of the situation, the material means available to insurgents, and the chances of success. While these same conditions apply to states, just war theory has never asked how the principle of last resort applies to guerrillas. Direness, in the case of guerrilla organizations, is not a function of what may occur in the wake of an immediate armed attack but in the wake of ongoing and long-term oppression. Direness varies from place to place. Ethnic cleansing, indiscriminate bombing, and destruction of infrastructures are considerably more urgent than affronts to self-determination whether abrogated (as in Kosovo) or never fully realized (as in the Palestinian territories or Chechnya). Less pressing, the latter may afford opportunities for alternative responses; the prospect of genocide does not. All things being equal, nonviolent resistance should be the first response when non-state actors face obstacles to fuller autonomy or independence that do not put their populations at risk for massive physical harm, depredation, or impoverishment.

Two factors impinge on this moral imperative. First, large *n* statistical studies suggest that nonviolent resistance alone is rarely effective when a people fights occupation or pursues self-determination, succession, or expansive autonomy (Stephen and Chenoweth 2008). In response, insurgents may find it advantageous to mix violent and nonviolent tactics. Second, and given that a nonviolent campaign can be a long process, when is enough enough? When can a guerrilla organization's military wing finally usurp the political wing and carry on the struggle by violent means? Was a decade in Kosovo or half decade in the Palestinian territories enough to establish that nonviolent resistance does not work? There is no single answer because there is an obvious difference between these two cases. The KLA, a military organization, appeared on the scene as the LDK, a political organization, exhausted itself and failed to curb Serbian terror, move the international community closer to recognizing demands for independence (witness the failure to address Kosovar demands at Dayton in 1995), or engender any significant loyalty shifts among Serbian soldiers or civilians. The KLA's timing was opportune; nonviolent resistance had indeed run its course. Palestinians, on the other hand, never embraced nonviolent tactics or repudiated terrorism entirely. Perhaps if they had, their efforts might have borne fruit or, as in Kosovo, paved the way for legitimate – that is, non-terrorist – armed force and international interventions. As such, the obvious answer to when is enough enough, is when nonviolence irrefutably fails. This demands, however, that insurgents first give nonviolence a chance.

Clouding these moral issues is one of simple expediency. Most guerrilla organizations choose nonviolent resistance for pragmatic, not principled, reasons. Simply put, they have no other choice. Guerrillas are neither able to battle a state army effectively nor willing to risk the catastrophic consequences of war. This renders the question of last resort moot in some cases. To demand that a state look to diplomacy before deploying gunboats means that both are viable options. War may be costly for states, but it is usually feasible. It is precisely for this reason that just war theory demands prior attention to other forms of conflict resolution. Among guerrilla movements, this is not always the case. The Kosovars never attempted armed struggle before pursuing their parallel government and East Timor guerrillas were thoroughly routed by Indonesian troops. Nonviolent resistance was a viable option and, for a while, perhaps the only option as insurgents in East Timor and Kosovo faced growing existential threats and saw their right to national self-determination fall by the wayside.

This was not true in other cases. With time, both Eritrea and Chechnya managed to field relatively formidable armies. Yet, neither faced existential threats that might justify immediate recourse to armed force. Here, one may indeed ask the same question one asks of states: Were effective and less destructive means than warfare available to guerrillas? And, if so, should they avail themselves of nonviolence before going to war? Many of the nonviolent tactics described – symbolic protests, boycotts, strikes, civil disobedience, and shadow governments – are usually practicable to some degree. But this is not usually what we think of when we demand that states try to settle their differences peacefully. Here, the repertoire we expect includes threats of armed attack, economic and financial sanctions, embargoes, hostile alliances, legal proceedings, and other coercive measures. Non-states have none of these tools. When armed force is a feasible strategic choice for guerrillas, nonviolent tactics are often paltry in comparison and justifiably rejected.

What then does nonviolent resistance and armed guerrilla warfare tell us about just war theory's "last resort" imperative? Must non-state parties exhaust nonviolent means before turning to war? In principle, the answer is yes, but the necessary conditions for meeting these requirements differ considerably from those constraining states. First, both war and its alternative must be feasible options. Often they are not. In some instances, non-states do not have recourse to war and so choose nonviolence. But in other cases, armed struggle is feasible but nonviolent resistance is not. Second, one must weigh the question of just cause and only then the consequences of war. Consider Michael Walzer's (2004:88) discussion of the First Gulf War, "Assuming that war was justified in the first instance, at the moment of the invasion," he writes, "then it is justifiable at *any* subsequent point when its costs and benefits seem on balance better than those of the available alternatives." This is an important point. Last resort is a consequentialist principle, and war is justifiably pursued at any time after just cause is established.

Despite legitimate grievances, guerrillas do not always have cause to wage war. When they do, the method they choose should be the best: the one that promises to rectify legitimate grievances at the least cost. Sometimes this requires nonviolence, but not always. To demand nonviolence, it must be more effective than armed struggle and not stem merely from a moral aversion to war. Successful nonviolent resistance requires mass mobilization, an adversary that is both brutal but sensitive to local and worldwide public opinion and a supportive international community. It also requires breathing room. States cannot usually defer war when they face, much less endure, immediate aggression and rampant destruction. Under the same circumstances, non-states will not defer war either unless an armed struggle promises to be manifestly suicidal, in which case nonviolent resistance may be their only and best alternative. Nonviolence is neither the first nor the last resort; it is the only resort. Here, as in other cases, insurgents' choice of means is pragmatic but constrained by moral norms. Just guerrilla warfare strives for national self-determination and, like state warfare, will pursue those means, violent or nonviolent, that are cost-effective and respectful of the fundamental rights of combatants and noncombatants. Nonviolent resistance remains an important adjunct to just guerrilla warfare, but it is neither decisive nor morally imperative.

PART IV

CONCLUDING REMARKS

11

Just War and Liberal Guerrilla Theorizing

Authors of books about just war theory have two motives: one is theoretical; the other is practical. Theoretically, we are vitally interested in answering the question: What is just war? What aims and practices of war are morally permissible? In doing so, we hope to be eminently practical as well and swell with pride when we hear statesmen proclaim, as did President Obama in 2009, that "the problems of war ... require us to think in new ways about the notions of just war and the imperatives of a just peace." His remarks link the successful prosecution of war with its just conduct. Most just war theorists hope this is true.

Scholars of civil wars, less prone to wax moralistic, are also engaged by the theoretical and practical questions of war. Theoretically, they seek to understand the causes of civil war and the violent enmity it releases while grappling with those factors that perpetuate war or bring it to an end. Practically, they, too, hope to offer statesmen and policy makers the tools to prosecute wars successfully, conduct effective counterinsurgencies, and realize an enduring peace.

Amid these pressing concerns, a question hovers over this book: Who is it for? Who will listen to or care about what might be called a liberal theory of guerrilla warfare? A generation ago, guerrilla theoreticians called on Marx and Lenin, while today some turn to radical religious teachings. But none might be called liberal, the product of an American education infused by such luminaries as John Rawls, Brian Barry, or Alan Gewirth. Contemporary just war theorists writing for their liberal leaders and compatriots, on the other hand, freely appropriate liberal political principles. But guerrillas usually do not. Civil war is exceedingly violent, and there is no doubt that many insurgents, rebels, guerrillas, and states reject the normative constraints of war as they pursue a litany of atrocities – terrorism, kidnapping, torture, sexual assault, and starvation – whose aim is to "sow fear and discord, to instill unbearable memories of what was once home [and] to desecrate whatever has social meaning" (Kaldor 1997:16). This only makes the quest for principles of just guerrilla warfare all the more pressing.

The just conduct of guerrilla warfare should serve the interests of liberal statesmen, just war theorists, *and* guerrillas aspiring to join the international community. The ability to recognize just guerrilla war enables policy makers and political leaders to know whom among insurgents to honor with their military, economic, and political support and whom to repudiate, sanction, or crush. An appeal to international law (rather than ethics) is not immediately helpful because insurgents can only comply erratically. Some insurgents plea for international assistance while pursuing only nonviolent resistance and before undertaking an armed struggle. They, seemingly, violate no law. Other insurgents violate law and custom egregiously, while still others face compelling constraints that make compliance difficult. Among other things, inaccurate weapons hinder compliance with the principle of discrimination, material asymmetry forces renewed interest in human shields, and war-sustaining civilian facilities present advantageous targets for direct attack. Nevertheless, and in spite of deviating from the law, many guerrillas, particularly those who enjoy just cause, seek their place among the nations. Waging just war serves insurgents' interests, while the flagrant violations of noncombatant rights that come with terrorism rightfully condemns guerrillas to outlaw status and unworthy of any consideration whatsoever. Guerrillas need to know how the international community expects them to conduct war justly. But guerrillas also contend, and rightfully so, that "ought" implies "can." The requirements of just war must be feasible. As they evolve, the laws and norms of armed conflict cannot make it impossible for guerrillas to pursue wars of national self-determination or prevail against oppressive states. The onus then falls on just war theorists to elaborate a set of principles and practices to guide the world community's response to insurgency cognizant of the constraints guerrillas encounter as they face state armies.

THE PRACTICE OF JUST GUERRILLA WARFARE

At first glance, the permissible practices that just guerrilla warfare endorses are alarming. Consider the brief recap provided in Table 11.1. Column 1 indeed summarizes what are often unlawful practices of war. Column 2, on the other hand, introduces provisos that render the practice morally permissible.

This table summarizes some of the main arguments presented in the previous chapters, and several points are worth noting. First, the repertoire of war is astonishingly variegated. Not all guerrilla groups sink to atrocity, torture, or murder. Terrorism, in fact, is neither the sole nor the chief tactic guerrillas employ. It is a feature of contemporary discourse surrounding asymmetric war to tar every insurgency with the crime of terrorism. To be sure, this is true of pan-national Islamic groups such as Al Qaeda, who are excluded from this study. It is for this reason that the criteria of just cause and legitimate authority are crucial to any investigation of just guerrilla warfare. While many armed groups populate the battlefield, just guerrilla warfare and, hence, the practices outlined in the preceding chapters remain permissible means of warfare only for groups pursuing national self-determination and/or freedom from rights-repressing regimes. Among many of these groups, terrorism is not the tactic of choice. Some insurgents disavow terrorism entirely (Eritrean People's Liberation Front, FALANTIL of East Timor, or the Polisario of the Western Sahara, for example), while others use terrorism as a modest adjunct to more comprehensive military and political campaigns. This was true in Kosovo, Chechnya, and Sri Lanka where insurgents largely preferred to battle military troops. The Palestinians, on the other hand, have, until recently, shown a particular affinity for terrorism. Enjoying the highest profile of contemporary guerrilla movements, the Palestinians have indelibly imprinted modern insurgency with its terrorist label. This, however, is misleading. Most insurgencies reject widespread terrorism, and the Palestinians are taking notice.

Second, the table provides a comprehensive view of the hard war and soft war tactics guerrillas employ and the principles that constrain each. Making room for the tactics described here does not demand legal changes but points instead to a more accommodating moral evaluation of existing practices. Among hard war tactics, IEDs, rockets, and

TABLE 11.1. *Just Guerrilla Warfare: Practices and Provisos*

Practice	Proviso
HARD WAR	
Improvised explosive devices	Discrimination: Supervise detonation to prevent indiscriminate harm
Rockets and missiles	Discrimination: Target military targets only
	Proportionality: Desist if collateral harm is disproportionate
Human shields	Consent: Secure explicit/implicit consent of shields
	Credible deterrence: Recruit sufficient shields to deter attack
Assassination: Military/political targets	Liability: Target liable persons only
	Proportionality: Desist if collateral harm is disproportionate
Assassination: Informers and collaborators	Due process: Conduct hearings or trials for accused
Prisoners of war, hostage taking	Humane treatment: Provide minimally humane conditions for captives
SOFT WAR	
Terrorism	Appropriate force: Utilize nonlethal/cyber force only
	Liability: Target war-sustaining facilities only
Economic warfare	Appropriate Force: Utilize cyber, nonlethal, or limited kinetic force
	Proportionality: Desist if collateral harm is disproportionate
Propaganda	Non-incitement: Ban incitement of genocide/war crimes
Nonviolent resistance	Protection: Keep participants from excessive harm.
THE ECONOMY OF WAR	
Taxation, tolls	Moderate enforcement means: No physical intimidation
Drug trafficking, smuggling	Utility: Ensure that collective benefits outweigh costs
	Moderate enforcement means: No physical intimidation

targeted killings are not unlawful. Nevertheless, constraining provisos call attention to collateral harm, the liability of targets, and the fundamental rights of combatants. The principle of proportionate collateral harm hopes to eliminate excessive injury among noncombatants who have done nothing to put themselves in harm's way. Considering the vagaries and dangers of war, the best one might do is strive to hold noncombatant harm to some level of what ordinary people consider reasonable. Taking into account the danger surrounding improvised explosive devices, rockets, and human shields, this is not always easy. Sometimes guerrillas abandon roadside bombs recklessly, fire their rockets wildly, or risk the lives of human shields unnecessarily. While states are right to condemn these practices, such defects can be remedied. Permissible human shielding demands consent from shields and obliges guerrillas to bring sufficient numbers of shields to deter state armies. IEDs and rockets require measures to ensure they hit military targets and avoid disproportionate harm. These demands are neither unreasonable nor unfeasible.

Liability is another lynchpin of just war. The notion remains vexing, contentious, and ambiguous. Liability means "deserving of attack or direct harm." In one sense, liability is material. Combatants, therefore, are permissible targets because they endanger enemy soldiers and threaten military installations. Noncombatants pose no threat and, therefore, incur no liability. Noncombatants stand protected from direct harm, while those unfortunate enough to find themselves in the way of a necessary military operation may suffer collaterally. Liability is also moral and turns on responsibility for an unjust threat. Regimes that make a dignified life impossible for some or all of their inhabitants, therefore, are the proper target of an armed struggle.

This book employs both forms of liability. On one hand, just cause is vitally important for guerrillas if they wish to avoid the stigma of criminality. *Jus ad bellum*, the right to make war, is an intensely moral notion that places a significant burden on guerrilla organizations. Insurgents may only go to war if they enjoy just cause. The object of their wrath is not any saber-rattling neighbor but the repressive state that denies them a dignified life. On the other hand, material liability is usually sufficient to drive *jus in bello* and the permissible tactics of war. Consider assassination, a double-edged sword aimed at enemy soldiers and civilians as well as compatriot informers and collaborators. Assassination is an endemic feature of civil war that many condemn as extrajudicial execution – and sometimes it is. But this need not be the case. If states increasingly condone the targeted killing of insurgents and their leaders, there is no reason to think that insurgents should not enjoy the same privilege, subject to the same constraints: proper identification and proportionate collateral harm. Assassinating informers and collaborators is a far more troubling business because some guerrillas conduct nothing more that summary executions. This, too, might be remedied, and there are sufficient examples of rudimentary due process to establish liability and warrant punishment of informers and spies.

Hard war also raises concerns for humane treatment. Guerrillas do not usually have access to such banned weapons as blinding lasers or poison gas that cause inhuman and unnecessary suffering. Instead, questions of humane conditions arise when they find themselves unable to provide Geneva Convention–like conditions for their prisoners. Taking prisoners is a feature of conventional war that takes odd turns in asymmetric war. Many guerrilla groups have no facilities for taking prisoners and instead kidnap soldiers to exchange them for large numbers of compatriots that their enemy holds captive. Some insurgents – Eritrean or Palestinian, for example – hold their prisoners in relatively benign conditions, while others – and the Chechens are among the most notorious – deny their prisoners the rudiments of decent care. Although state armies fulminate against guerrillas who take captives and hold them incommunicado, only those insurgents who fail to provide minimally humane conditions warrant moral condemnation.

Terrorism, too, merits unequivocal condemnation. Terrorists perpetrate lethal attacks against noncombatants to coerce political concessions from a state or non-state by instilling widespread fear. Terrorism has no place in the repertoire of just guerrilla warfare. In this study, however, terrorism bridges the tactics of hard and soft war by making room for nonlethal attacks against participating civilians, that is, those who provide war-sustaining financial, legal, or logistic support to states or insurgents. Increasingly, state armies seek out such civilian targets, and there is no reason to deny guerrillas the same latitude as they fight. Because war-sustaining targets provide crucial support but pose no direct lethal threat, participating civilians are only liable to nonlethal harm. Belligerents may use nonlethal force to restrain, arrest or expel participating civilians or turn to chemical agents and electromagnetic technologies to incapacitate civilians who sustain a war effort. Alternatively, cyber technologies may now offer guerrillas the means to disable war-sustaining facilities without harming the civilians who work there. Utilizing nonlethal measures, guerrillas, like state armies, hope to force concessions without bringing widespread devastation or casualties.

Similar goals motivate all soft war and non-kinetic tactics. Like states, guerrillas may adopt economic sanctions, public diplomacy, and nonviolent resistance to press their claims. And, no less than states, insurgents require the resources to sustain an armed struggle. Unlike sovereign nations, however, insurgents lack coercive institutions and therefore turn to force to procure funds through taxation, tolls, drug trafficking, and smuggling. However unlawful such practices are, they are not condemnable until terrorism takes innocent lives, economic warfare brings excessive pain and suffering, propaganda incites genocide or ethnic cleansing, nonviolent resistance puts lives at risk, and war economies turn to extortion and physical intimidation. Hence, there is the need for provisos. They delimit a zone of permissible maneuverability between what is unlawful and what is morally proscribed. These provisos circumscribe just guerrilla warfare.

At this point, one will reasonably ask: Are such provisos practicable? Navigating between practices and provisos is a fine line to walk and may require that guerrillas assume considerable risk as they closely supervise the deployment of IEDs or assure that kinetic attacks on economic targets avoid noncombatant casualties. The demand to protect noncombatants is equally, if not more, compelling when guerrillas employ non-kinetic force and soft war tactics. Harm is not always obvious, and it is therefore easy for guerrillas to ignore the downstream costs of cyber strikes, forget that predatory war economies victimize many compatriots, and overlook that provocative nonviolent resistance may cause considerable casualties among demonstrators.

Will guerrillas respect such provisos? Or have I given away too much and gifted guerrillas a concession that insurgents will only exploit without ever accepting the law of war? Interestingly enough, states ask themselves the same questions as they consider adjustments to the current norms regulating collateral harm, targeted killing, interrogation, economic sanctions, and media warfare. Among states, however, particularly democratic states, vibrant public discourse engages executive, legislative, and judicial institutions and affects policy outcomes. There is no similar process among most guerrilla organizations. Ad hoc debates among insurgents, however, reflect some concern for these issues. Some Islamic scholars, for example, condemn terror attacks on enemy and compatriot civilians as a gross violation of Islamic "norms for honorable combat" (Kelsay 2007:140–142; 202). Some armed groups do forswear terrorism, conduct rudimentary trials, search out military targets, protect prisoners, or entertain nonviolent resistance. The International Committee of the Red Cross, moreover, reports moderate success as they work to convince armed groups to sustain these practices (Sivakumaran 2011). The motivations are complex but not so different from those that drive states to comply with the law: self-interest, the desire for international support, norms of honor and courage, the prospect of international law enforcement, and the fear of reciprocal harm.

ENFORCEMENT AND COMPLIANCE

Any reference to provisos or constraints immediately raises the question of enforcement, particularly as some observers see many armed groups as nothing but free-riding brigands. Enforcement at the international level is notoriously inefficient, but non-states can be motivated to comply just as states can. While often weak and arduously lengthy, war crime prosecutions by an empowered international criminal court will certainly carry some weight when insurgents and state officials find themselves in the dock. More importantly, perhaps, are the strengthening norms of international intervention following the United Nations' 2005 commitment to its responsibility to protect. While the international community certainly intervened in prior years, there is growing willingness to violate another state's sovereignty when a nation not only threatens world peace and security but also violates the human rights of its citizens. National liberation takes many forms and, indeed, armed struggles have lurched from anti-colonialism to ethnic or national secession and, most recently, to democratization and liberation from autocratic regimes. Guerrillas fighting for independence or rebels now fighting for regime change in North Africa or the Middle East will take increasing notice of the UN commitment to its responsibility to protect and demand military and financial aid, peacekeeping troops, and reconstructive assistance. To do so, an aggrieved people must convince the world community that its cause and means of warfare are just. This is particularly true when a people strive for independence. They, by definition, require international recognition if they wish to join the family of nations.

Recently minted states in Eritrea, East Timor, and Southern Sudan and de facto states or near-states in Kosovo and Palestine exemplify the need for constrained warfare, although not all, obviously, fully comply.

Eritrea and East Timor forswore terrorism, the Kosovars pursued nonviolence before turning to armed resistance, while Palestinians cycled from violence to nonviolence and to vicious terrorism before recently settling down to pursue public diplomacy and the periodic shooting match. Other insurgencies fared worse when they could not renounce violence. Insurgents in neither Chechnya nor Sri Lanka showed much concern for many of the provisos of hard war or any inclination to pursue the less violent means of soft war. These lapses alone, of course, are not the only reason for their failures. For one, the legal basis for Chechen and Tamil claims were far weaker than those the Palestinians, Kosovars, Eritreans, and East Timorese enjoyed. As a result, the international community was inclined to steer clear of what were regarded as internal conflicts. Nor does fighting well and/or legal recognition promise success. Witness Western Sahara's fruitless attempt to get Morocco to hold a long-promised and UN-sanctioned referendum on independence. Nevertheless, there are good reasons to think that a guerrilla group that repudiates terrorism and gratuitous violence will find it easier to secure the support of the international community than those who do not. Such support will sometimes, but not always, facilitate success.

THE PROSPECTS FOR JUST GUERRILLA WARFARE

This study is exploratory, not confirmatory. It is a normative and not an empirical investigation. Perusing the tactics that guerrillas adopt, a normative exploration can only suggest that moral behavior matters, particularly when peoples pursue a goal defined in terms of justice, dignity, and self-determination. To achieve their goals, insurgents must convince the international community that their war is just and allow observers to readily distinguish among adversaries so they will lend their support to the side most deserving. Unfortunately, the means legally available to guerrilla groups are not always sufficient to allow them to persevere. To no little extent, international law denies just guerrillas the right to a fighting chance. This right is not a blank check, but embraces the very practices just described, together with the provisos that protect the rights of participants. Thus the other audience of this book: state leaders and policy makers whose support just guerrillas require. They, no less than insurgents, need to comprehend what practices guerrillas may legitimately pursue, and to understand that when these practices violate the law, the international community should consider altering the law accordingly.

This is not a new enterprise. Writing in the *International Review of the Red Cross* in 1976 at the height of the debate surrounding Protocol I and Protocol II, Michel Veuthey (1976:292–293) pointed to the still unresolved difficulty that humanitarian law poses for guerrilla warfare:

> To wish to retain the law in its present discriminatory form as regards guerrilla warfare, in defiance of the facts, would be virtually to put an end to humanitarian law, in the tradition of Moliere's doctors, who preferred people to die under their treatment rather than to recover without it ... The essential objective in formulating humanitarian law, as in actual humanitarian activities, seems to be to establish the fundamental similarity of humanitarian interests of the parties and the humanity they share, beyond the divergences and differences that make them adversaries.

This challenge abides. Jurists, like generals, are always ready for the last war. Lawmakers woke up and recognized partisans as lawful combatants only after partisan fighters disappeared from modern warfare. Scrambling to confront the exigencies of post–World War II colonial warfare, the international legal community bestowed its grace on peoples fighting against colonial, alien, and racist regimes just as these regimes were breathing their last. As the codified norms of war lag behind the reality on the field, state armies facing terrorism and insurgency complain that current laws and practices make it impossible to wage war effectively. In response, state armies have, over the years, looked for the latitude to conduct aggressive interrogation, rendition, targeted killing, and military intervention. Guerrilla organizations face similar constraints as they turn to human shields, targeted killing, improvised missiles and explosive devices, and cyber-warfare to confront a much stronger state adversary. The law has yet to rise to all these challenges. Doing so first requires a firm moral foundation for just guerrilla warfare.

The moral principles governing armed conflict should allow guerrillas the right to purse a fighting chance while protecting civilians and other participants from the worst ravages of civil war. When any group goes to war, leaders and their compatriots must carefully weigh its costs and benefits. It is important to remember that the costs and benefits of just guerrilla warfare turn not on lives lost or saved, but on *dignified* lives lost or saved. When fighters, hunger strikers, human shields, or protesters put their lives on the line, they are threatening to end an undignified life, one they no longer cherish in the absence of political freedom. Above all, guerrillas fighting for self-determination are fighting for their right to lead dignified lives, lives that offer the possibility of fulfilling their capabilities as human beings. This translates into a wide-reaching right to fight, but only as long as they respect the fundamental rights of those they confront.

With this in mind, the next step will be to move beyond the lessons of wars of national liberation and secession to think how the now growing disenchantment among peoples with their governments may translate into just guerrilla war. Like a disenfranchised or persecuted minority, these populations face a regime that denies their right to a dignified life. An armed struggle may be a reasonable alternative assuming that insurgents can garner legitimate authority and regulate their practices by the provisos necessary to pursue a just war. This is not an easy task. Many civil wars degenerate into ethnic rivalries, pitting one corrupt and debased regime against another and whose leaders take control through intimidation, murder, and extortion. Regardless of who prevails, these governments will remain the objects of assault by an embittered people. At one point, one can only hope that they will either wear their governments down or take up the challenge of waging a just war. But doing this requires the forbearance of the international community and a wider tolerance of who may justifiably fight and how.

References

Abdel-Jawad, S. (2001). The classification and recruitment of collaborators. In *The phenomenon of collaborators in Palestine, proceedings of a Passia workshop* (pp. 17–28). Jerusalem: Palestinian Academic Society for the Study of International Affairs.

Abi-Saab, G. (1979). "The legal status of wars of national liberation." Wars of national liberation in the Geneva conventions and protocols. *Collected Courses of the Hague Academy of International Law*, 165, 353–448. Martinus Nijhoff Publishers.

Abrahms, M. (2011). Does terrorism really work? Evolution in the conventional wisdom since 9/11. *Defence and Peace Economics*, 22(6), 583–594.

Abresch, W. (2005). A human rights law of internal armed conflict: The European Court of Human Rights in Chechnya. *European Journal of International Law*, 16(4), 741–767.

Additional Protocol I (API). (1977). Protocol Additional to the Geneva Conventions of August 12, 1949, and relating to the Protection of Victims of International Armed Conflicts (Protocol I), June 8, 1977.

Additional Protocol I (API) Commentary. (1977). Protocol Additional to the Geneva Conventions of August 12, 1949, and relating to the Protection of Victims of International Armed Conflicts (Protocol I), June 8, 1977. Commentary.

Additional Protocol II (APII). (1977). Protocol Additional to the Geneva Conventions of August 12, 1949, and relating to the Protection of Victims of Non-International Armed Conflicts (Protocol II), June 8, 1977.

Aguswandi, W. Z. (2008). *From politics to arms to politics again: The transition of the Gerakan Aceh Merdeka (Free Aceh Movement – GAM)*. Berlin: Berghof Research Center for Constructive Conflict Management.

Aharonovitz, E. (2010). Hezi Shai, who fell prisoner in the battle of Sultan Yaakov, returns to those dark days. *Haaretz*, March 19. http://www.haaretz.co.il/misc/1.1193851.

Ahmed, A. and Sahak, S. (2013). Drone and Taliban attacks hit civilians, Afghans say. *New York Times*, August 8. http://www.nytimes.com/2013/09/09/world/asia/two-deadly-attacks-in-afghanistan.html?pagewanted=print.

Akande, D. (2010). Clearing the fog of war? The ICRC's interpretive guidance on direct participation in hostilities. *International and Comparative Law Quarterly,* 59(01), 180–192.

al-Libi, Abu-Yahya. (2008). Website posts Abu-Yahya al-Libi's research on human shields in jihad." *World News Connection, National Technical Information Service,* April 10. http://hdl.handle.net/10066/4607.

Amble, J. C. (2012). Combating terrorism in the new media environment. *Studies in Conflict & Terrorism,* 35(5), 339–353.

Amnesty International. (2011a). Israel-Hamas prisoner swap casts harsh light on detention practices of all sides. http://www.amnesty.org/en/news-and -updates/israel-hamas-prisoner-swap-casts-harsh-light-detention-practices -all-sides-2011-10-.

(2011b). Questions around operation against Osama bin Laden. http://www .amnesty.org/en/news-and-updates/questions-around-operation-against -osama-bin-laden-2011-05-04.

(2011c). Support campaign to end the suffering of Gilad Shalit and his family. http://www.amnesty.org/en/appeals-for-action/end-suffering-gilad-shalit -family.

Annas, G. J., Crosby, S. S., and Glantz, L. H. (2013). Guantanamo Bay: A medical ethics–free zone? *New England Journal of Medicine,* 369, 101–103.

Arens, O. and Kaufman, E. (2012). The potential Impact of Palestinian Nonviolent Struggle on Israel: Preliminary Lessons and projections for the Future. *The Middle East Journal,* 66(2), 231–252.

Ariely, G. (2008). Knowledge management, terrorism, and cyber terrorism. In L. Janczewski and A. Colarik (Eds.), *Cyber warfare and cyber terrorism* (pp. 7–16). Hershey, PA: IGI Global.

Arkin, W. (2007). *Divining victory: Airpower in the Israel-Hezbollah war.* Maxwell Airforce Base: Air University Press.

Army Recognition. (nd). Fadjr-5 333mm multiple rocket launcher system. http:// www.armyrecognition.com/iran_iranian_army_artillery_vehicles_systems _uk/fadjr-5_fajr5_333mm_multiple_rocket_launcher_system_technical _data_sheet_specifications.htm.

Arquilla, J. and Karasik, T. (1999). Chechnya: A glimpse of future conflict? *Studies in Conflict and Terrorism,* 22(3), 207–229.

Aspinall, E. (2009). *Islam and nation: Separatist rebellion in Aceh, Indonesia.* Stanford, CA: Stanford University Press.

Aspinall, E. and Crouch, H. A. (2003). *The Aceh peace process: Why it failed.* Washington, DC: East-West Center.

Asprey, R. B. (1994). *War in the shadows: The guerrilla in history.* New York: William Morrow and Company.

Asser, M. (2006). Qana makes grim history again. July 31. http://news.bbc. co.uk/2/hi/5228554.stm.

Austin, W. C. and Kolenc, A. B. (2006). Who's afraid of the big bad wolf? The International Criminal Court as a weapon of asymmetric warfare. *Vanderbilt Journal of Transnational Law,* 39, 291–346.

Awwad, H. (2012). Six years of BDS: Success. In A. Lim (Ed.), *The case for sanctions against Israel* (pp. 77–84). London: Verso.

Bakker, P. (2001). New nationalism: The internet crusade (paper prepared for International Studies Association Annual Convention, Chicago, Illinois, February 20–24). http://www. tamilnation. org/selfdetermination/nation/bakker.pdf.

Ballentine K. and Nitzschke, H. (2005). *The political economy of civil war and conflict transformation.* Berlin: Berghof Research Center for Constructive Conflict Management. http://www.berghof-handbook.net/documents/publications/dialogue3_ballentine_nitzschke.pdf. 2005.

Bangerter O. (2011a). A collection of codes of conduct issued by armed groups. *International Review of the Red Cross,* 93(882), 483–501.

Bangerter, O. (2011b). Reasons why armed groups choose to respect international humanitarian law or not. *International Review of the Red Cross,* 93(882), 353–384.

Banville, A. (2010). Embedded war reporting cannot escape its own bias. *The Guardian,* April 18. http://www.guardian.co.uk/commentisfree/2010/apr/18/embedded-war-reporting-iraq-afghanistan.

Bar Joseph, U. (2007). Israel's military intelligence performance in the Second Lebanon War. *International Journal of Intelligence and Counterintelligence,* 20, 583–601.

Barilan, Y. M. (2012). *Human dignity, human rights, and responsibility: The new language of global ethics and biolaw.* Cambridge: The MIT Press.

Barilan, Y. M. and Zuckerman, S. (2013). Revisiting medical neutrality as a moral value and as a doctrine of international law. In M. L. Gross and D. Carrick (Eds.), *Military medical ethics for the 21st century* (pp. 97–110). Farnham, UK: Ashgate.

Barnard, A. and Rudoren, J. (2014). Israel says that Hamas uses civilian shields, reviving debate. *New York Times,* July 23. http://www.nytimes.com/2014/07/24/world/middleeast/israel-says-hamas-is-using-civilians-as-shields-in-gaza.html.

Barzashka, I. (2013). Are cyber-weapons effective? *The RUSI Journal,* 158(2), 48–56.

BDS Movement. (2005). Palestinian civil society call for BDS. July 9. http://www.bdsmovement.net/call.

BDS National Committee. (2009). Palestinian trade unions unanimously support boycott movement. *BDS National Committee, Press Release,* November 26. http://www.leedspsc.org.uk/palestinian-trade-unions-unanimously-support-boycott-movement/.

Be'er, Y. and Abdel-Jawad, S. (1994). *Collaborators in the Occupied Territories: Human rights abuses and violations.* Jerusalem: B'Tselem.

Beaton, A., Cook, M., Kavanagh, M., and Herrington, C. (2000). The psychological impact of burglary. *Psychology, Crime and Law,* 6(1), 33–43.

Becker, J. (2011). Beirut bank seen as a hub of Hezbollah's financing. *New York Times,* December 12. http://www.nytimes.com/2011/12/14/world/middleeast/beirut-bank-seen-as-a-hub-of-hezbollahs-financing.html?pagewanted=all.

Bedau, H. (1968). The right to life. *The Monist,* 52(4), 550–572.

(1971). Military service and moral obligation. *Inquiry,* 14(1–4), 244–266.

Beggs, C. (2008). Cyber-terrorism in Australia. *IGI Global*, 108–113.

Beitz, C. R. (1989). Covert intervention as a moral problem. *Ethics & International Affairs*, 3(1), 45–60.

Bekaj, A. R. (2010). *The KLA and the Kosovo war: From intra-state conflict to independent country*. Berlin: Berghof Research Center for Constructive Conflict Management.

Bekkers, R. and Wiepking, P. (2007). Generosity and philanthropy: A literature review. Available at SSRN. http://dx.doi.org/10.2139/ssrn.1015507.

Bellamy, A. J. (2012). *Massacres and morality: Mass atrocities in an age of civilian immunity*. Oxford: Oxford University Press.

Benbaji, Y. (2013). Justice in asymmetric wars: A contractarian analysis. *The Law & Ethics of Human Rights*, 6(2), 172–200.

Bennet, H. (2010). Detention and interrogation in Northern Ireland, 1969–75. In S. Scheipers (Ed.), *Prisoners in war* (pp. 187–204). Oxford: Oxford University Press.

Beres, L. R. (2002–2003). Assassinating Saddam Hussein: The view from international law. *Indiana International & Comparative Law Review*, 13, 847–870.

Betz, D. J. (2006). The more you know, the less you understand: The problem with information warfare. *The Journal of Strategic Studies*, 29(3), 505–533.

Black, J. (2001). Semantics and ethics of propaganda. *Journal of Mass Media Ethics*, 16(2–3), 121–137.

Blake, N. and Pole, K. (Eds.). (1984). *Objections to nuclear defence: Philosophers on deterrence*. London: Routledge and Kegan Paul.

Blau, P. M. (1963). Critical remarks on Weber's theory of authority. *The American Political Science Review*, 57(2), 305–316.

Bleich, A., Gelkopf, M., Melamed, Y., and Solomon, Z. (2006). Mental health and resiliency following 44 months of terrorism: a survey of an Israeli national representative sample. *BMC Medicine*, 4(1), 21. http://www.biomedcentral.com/1741-7015/4/21.

Bleich, A., Gelkopf, M., and Solomon, Z. (2003). Exposure to terrorism, stress-related mental health symptoms, and coping behaviors among a nationally representative sample in Israel. *The Journal of the American Medical Association*, 290, 612–620.

Bok, S. (1999). *Lying: Moral choice in public and private life*. New York: Vintage.

Bozarslan, H. (2001). Human rights and the Kurdish issue in Turkey: 1984–1999. *Human Rights Review*, 3(1), 45–54.

Bracken, P. (2007). Financial warfare. *Orbis*, 51(4), 685–696.

Branche, R. (2010). The French in Algeria: Can there be prisoners of war in a 'domestic' operation? In S. Scheipers (Ed.), *Prisoners in war* (pp. 173–186). Oxford: Oxford University Press.

Bray, J., Lunde, L., and Murshed, S. M. (2003). Nepal: Economic drivers of the Maoist insurgency. In K. Ballentine and J. Sherman (Eds.), *The political economy of armed conflict: Beyond greed and grievance* (pp. 107–132). London: Rienner.

Brown, B. B. and Harris, P. B. (1989). Residential burglary victimization: Reactions to the invasion of a primary territory. *Journal of Environmental Psychology*, 9(2), 119–132.

Brown, C. (2003). Selective humanitarianism: In defense of inconsistency. In D. K. Chatterjee and D. Scheid (Eds.), *Ethics and foreign intervention* (pp. 31–52). Cambridge: Cambridge University Press.

Brown, J. (2008). Public diplomacy and propaganda: Their difference. *American Diplomacy*, September 16. http://go.galegroup.com/ps/i.do?id=GALE%7CA187797702&v=2.1&u=nysl_me_newyorku&it=r&p=AONE&sw=w&asid=16eed8550b74aad7f678208574c9e225.

B'Tselem. (The Israeli Information Center for Human Rights in the Occupied Territories). (2009). B'Tselem's investigation of fatalities in Operation Cast Lead. September 9. http://www.btselem.org/download/20090909_cast_lead_fatalities_eng.pdf.

(2012a). Statistics on Palestinians in the custody of the Israeli security forces. http://www.btselem.org/hebrew/statistics/detainees_and_prisoners.

(2012b). Five years without visits – Will family visits be renewed for prisoners from Gaza? http://www.btselem.org/gaza_strip/20120515_prison_visits.

(2013). Fatalities before, during and since Operation Cast Lead and in the First Intifada. http://www.btselem.org/statistics.

(nd.). Fatalities in the first Intifada. Accessed November 20, 2013. http://www.btselem.org/statistics/first_Intifada_tables.

Buchanan, A. (1997). Theories of secession. *Philosophy & Public Affairs*, 26(1), 31–61.

Buchanan, J. M. (1984). The ethical limits of taxation. *The Scandinavian Journal of Economics*, 86(2), 102–114.

Buchholz, M. B. (2013). The human shield in Islamic jurisprudence. *Military Review*, May-June, 48–52.

Bush, G. W. (2010). *Decision points.* New York: Random House.

Byrd, W. and Mansfield. D. (2012) Drugs in Afghanistan – A forgotten issue? Implications and risks for transition. United States Institute of Peace *PEACEBRIEF*, 126, May 18, p. 4. http://www.usip.org/files/resources/PB-126.pdf.

Byrd, W. and Ward, C. (2004). Drugs and development in Afghanistan. *World Bank, Social Development Papers, Conflict Prevention and Reconstruction*, Paper No. 18, December. http://www.giz.de/Themen/de/dokumente/en-wb-drugs-development-afg.pdf.

Caddell, J. W. (2004). *Deception 101: Primer on deception.* Carlisle, PA: Strategic Studies Institute, US Army War College.

Canetti, D., Gross, M. L., and Waismel-Manor, I. (in press). Immune from cyber fire? The psychological & physiological effects of cyber war. In F. Allhoff, A. Henschke, and B. J. Strawser (Eds.). *Binary Bullets: The Ethics of Cyberwarfare.* Oxford: Oxford University Press.

Canetti-Nisim, D., Halperin, E., Sharvit, K., and Hobfoll, S. E. (2009). A new stress-based model of political extremism personal exposure to terrorism, psychological distress, and exclusionist political attitudes. *Journal of Conflict Resolution*, 53(3), 363–389.

Caney, S. (1998). National self-determination and national secession: Individualist and communitarian approaches. In P. B. Lehning (Ed.), *Theories of secession* (pp. 151–181). New York: Routledge.

Carley, S. D. and Mackway-Jones, K. (1997). The casualty profile from the Manchester bombing 1996: a proposal for the construction and dissemination of casualty profiles from major incidents. *Journal of Accident & Emergency Medicine*, 14(2), 76–80.

Carter, A. (1998). Liberalism and the obligation to military service. *Political Studies*, 46(1), 68–81.

Casalin, D. (2011). Taking prisoners: Reviewing the international humanitarian law grounds for deprivation of liberty by armed opposition groups. *International Review of the Red Cross*, 93(883), 743–757.

Cass, D. Z. (1992). Re-thinking self-determination: A critical analysis of current international law theories. *Syracuse Journal of International Law and Commerce*, 18, 21–40.

Cassese, A. (1995). *Self-determination of peoples: A legal reappraisal*. Cambridge: Cambridge University Press.

Cavelty, M. D. (2012). Cyber-threats. In M. D. Cavelty and V. Mauer (Eds.), *The Routledge handbook of security studies* (pp. 180–189). London: Routledge.

Central Intelligence Agency (CIA). (2007, 2012). The world factbook. https://www.cia.gov/library/publications/the-world-factbook/rankorder/2091rank.html.

Chadwick, A. (2012). Lebanon: The 2006 Lebanon War: A Short History Part II. *Small Wars Journal*, September 12. http://smallwarsjournal.com/jrnl/art/the-2006-lebanon-war-a-short-history-part-ii.

Chaumont, C. (1977). *Additional Protocol I, Commentary*, Article 44, page 519, note 40.

Checchia, M. (2012a). Improvised explosive devices: A Global Review. Civil-military fusion centre counter-Improvised explosive devices. January & February. https://www.cimicweb.org/cmo/afg/Documents/Security/CFC_IED-Trends-and-Issues_March2012.pdf.

(2012b). Improvised explosive devices: A Global Review. Civil-military fusion centre counter-Improvised explosive devices.March, April & May. http://reliefweb.int/sites/reliefweb.int/files/resources/CFC_IED-Trends-and-Issues_June2012.pdf.

Chenoweth, E. and Cunningham, K. G. (2013). Understanding nonviolent resistance: An introduction. *Journal of Peace Research*, 50(3), 271–276.

Chenoweth, E. and Stephan, M. J. (2011). *Why civil resistance works: The strategic logic of nonviolent conflict*. New York: Columbia University Press.

Chesser, S. G. (2012). Afghanistan casualties: Military forces and civilians. Congressional Research Service. Washington, DC: Library of Congress. http://www.fas.org/sgp/crs/natsec/R41084.pdf.

Christians, C. and Nordenstreng, K. (2004). Social responsibility worldwide. *Journal of Mass Media Ethics*, 19(1), 3–28.

Christiansen, D. and Powers G. F. (1995). Economic sanctions and the just-war doctrine. In D. Cortright and G. A. Lopez (Eds.), *Economic sanctions:*

Panacea or peacebuilding in a post-Cold War world? (pp. 97–117). Boulder, CO: Westview.

Cirino, J. A., Elizondo, S. L., and Wawro, G. (2004). Latin America's lawless areas and failed states. In P. D. Taylor (Ed.), *Latin American security challenges: A collaborative inquiry from North and South, Newport Papers, 21* (pp. 7–48). Newport, RI: Naval War College. http://www.stormingmedia.us/52/5240 /A524034.html.

Clark, H. (2000). *Civil resistance in Kosovo*. London: Pluto Press.

Cliffe, L. (1989). Forging a nation: The Eritrean experience. *Third World Quarterly*, 11(4), 131–147.

Coady, C. A. J. (1988). Deterrent intentions revisited. *Ethics*, 99(1), 98–108.

——— (1989). Escaping from the bomb: Immoral deterrence and the problem of extrication. In H. Shue (Ed.), *Nuclear deterrence and moral restraint: Critical choices for American strategy* (pp. 163–225). Cambridge: Cambridge University Press.

——— (2004). Terrorism, morality, and supreme emergency. *Ethics*, 114(4), 772–789.

Cohen, A. (2009). Economic sanctions in IHL: Suggested principles. *Israel Law Review*, 42, 117–149.

Connell, D. (2001). Inside the EPLF: The origins of the 'people's party' & its role in the liberation of Eritrea. *Review of African Political Economy*, 28(89), 345–364.

Coogan, T. P. (2002). *The IRA*. London: HarperCollins UK.

Coovadia, H. M. (1999). Sanctions and the struggle for health in South Africa. *American Journal of Public Health*, 89(10), 1505–1508.

Cordesman, A. H. (2003). *The military balance in North Africa*. Washington, DC: Center for Strategic and International Studies. http://csis.org/files/media/ csis/pubs/mbnafrica.pdf.

——— (2007). *Lessons of the 2006 Israeli-Hezbollah war*. Washington, DC: Center for Strategic and International Studies. http://csis.org/files/publica-tion/120720_Cordesman_LessonsIsraeliHezbollah.pdf.

Cordone, C. and Gidron A. (2000). Was the Serbian TV station really a legitimate target. *Le Monde Diplomatique*. http://mondediplo. com/2000/07/03kosovo.

Corlett, J. A. (2003). *Terrorism: A philosophical analysis* (vol. 101). Dordrecht: Kluwer Academic Publishers.

Council on Foreign Relations (CFR). (2011). Hamas. http://www.cfr.org/israel /hamas/p8968#p6.

Cunningham, K. G. (2013). Understanding strategic choice: The determinants of civil war and nonviolent campaign in self-determination disputes. *Journal of Peace Research*, 50(3), 291–304.

Cunningham, S. B. (1992). Sorting out the ethics of propaganda. *Communication Studies*, 43(4), 233–245.

Daadaoui, M. (2008). The Western Sahara conflict: Towards a constructivist approach to self-determination. *The Journal of North African Studies*, 13(2), 143–156.

Daboné, Z. (2012). International law: Armed groups in a state-centric system. *International Review of the Red Cross*, 93(882), 395.

Dadpay, A. (2010). A review of Iranian aviation industry: Victim of sanctions or creation of mismanagement? (paper presented at the Third Conference on Iran's Economy, University of Chicago, Chicago, October). http://iraneconomy.csames.illinois.edu/full%20papers/Dadpay%20-%20IranAviation.pdf.

Dajani, S. R. (1995). *Eyes without country: Searching for a Palestinian strategy of liberation*. Philadelphia, PA: Temple University Press.

(1999). Nonviolent resistance in the Occupied Territories: A critical reevaluation. In S. Zunes, S. B. Asher, and L. Kurtz (Eds.), *Nonviolent social movements: A geographical perspective* (pp. 52–74). Oxford: Blackwell.

Darwish, R., Farajalla, N., and Masri, R. (2009). The 2006 war and its inter-temporal economic impact on agriculture in Lebanon. *Disasters*, 33(4), 629–644.

Davison, N. (2009). *Nonlethal weapons*. Hampshire: Palgrave Macmillan.

Demmers, J. (2002). Diaspora and conflict: Locality, long-distance nationalism, and delocalisation of conflict dynamics. *The Public-Javnost*, 9(1), 85–96.

Denselow, J. (2010). The Iran sanctions dilemma. *The Guardian*, January 2. http://www.guardian.co.uk/commentisfree/cifamerica/2010/feb/01/iran-sanctions-us-airline.

Department of the Army. (2006). *Field manual FM 3-24, MCWP 3-33.5, Counterinsurgency*. Washington, DC: Department of the Army.

Department of the Navy. (2007). *Naval Warfare Publication, The Commander's Handbook on The Law of Naval Operations, NWP 1-14M 2007*. http://www.lawofwar.org/naval_warfare_publication_N-114M.htm.

Diamant, A. (2009). *Day after night: A novel*. New York: Simon and Schuster.

Diamond, G. M., Lipsitz, J. D., Fajerman, Z.and Rozenblat, O. (2010). Ongoing traumatic stress response (OTSR) in Sderot, Israel. *Professional Psychology: Research and Practice*, 41, 19–25.

Dietz, A. S. (2011). Countering the effects of IED systems in Afghanistan: An integral approach. *Small Wars & Insurgencies*, 22(02), 385–401.

Dinniss, H. H. (2012). *Cyber warfare and the laws of war*. Cambridge: Cambridge University Press.

Dinstein, Y. (2005) *War, aggression and self-defense (4th ed)*. Cambridge: Cambridge University Press.

(2009). *The international law of belligerent occupation*. Cambridge: Cambridge University Press.

(2010). *The conduct of hostilities under the law of international armed conflict* (2nd ed.). Cambridge: Cambridge University Press.

Dishman, C. (2001). Terrorism, crime, and transformation. *Studies in Conflict and Terrorism*, 24(1), 43–58.

Dörmann, K. (2004). Applicability of the Additional Protocols to computer network attacks. *War Studies*, 76. http://www.961.ch/eng/assets/files/other/applicabilityofihltocna.pdf.

Dudouet, V. (2008). *Nonviolent resistance and conflict transformation in power asymmetries*. Berlin: Berghof Research Center for Constructive Conflict Management

(2013). Dynamics and factors of transition from armed struggle to nonviolent resistance. *Journal of Peace Research*, 50(3), 401–413.

Dunlap Jr., C. J. (2009). Lawfare: A decisive element of 21st-century conflicts? *Joint Force Quarterly*, 54, 34–39.

Dunn, J. (2001). Against Humanity in East Timor, January to October 1999: Their nature and causes. *East Timor Action Network*. http://www.etan.org /news/2001a/dunn1.htm.

Eck, K. (2009). From armed conflict to war: Ethnic mobilization and conflict intensification. *International Studies Quarterly*, 53(2), 369–388.

(2010). Recruiting rebels: Indoctrination and political education in Nepal. In M. Lawoti and A. Pahari (Eds.), *The Maoist insurgency in Nepal: Dynamics and growth in the 21st century* (pp. 33–51). London: Routledge.

Ellul, J. (1981). The ethics of propaganda: Propaganda, innocence, and amorality. *Communication*, 6, 159–175.

Elster, J. (1989). *The cement of society: A survey of social order*. Cambridge: Cambridge University Press.

English, R. (2003). *Armed struggle: The history of the IRA*. Oxford: Oxford University Press.

Erakat, N. (2012). BDS in the USA: 2001–2010. In A. Lim (Ed.), *The case for sanctions against Israel* (pp. 85–100). London: Verso.

Erdbrink, T. (2012). For Iran's sick, sanctions turn lethal as drugs vanish. *International Herald Tribune*, November 3–4, p. 1.

Erlich, R. (2006). *Hezbollah's use of Lebanese civilians as human shields*. Ramat Hasharon: Intelligence and Terrorism Information Center at the Center for Special Studies (C.S.S). http://www.terrorism-info.org.il/en/article/19120.

Estes, A. C. (2013). Anonymous hits Israel with a massive cyber attack, Israel attacks back. *Yahoo News*. http://news.yahoo.com/anonymous-hits-israel-massive-cyber-attack-israel-attacks-005659174.html.

Estrin, D. (2012). Gilad Shalit, Freed Israeli soldier, speaks of Hamas captivity. *Huffington Post*. http://www.huffingtonpost.com/2012/10/12/gilad -schalit_n_1961095.html.

Ethics. (1985). Symposium on ethics and nuclear deterrence. *Ethics*, 95 (3), 409–739.

Etzioni, A. (2010). The case for drones. Email posting from *Communitarian International Relations*, 3.31.2010.

Even, S. (2010). Israel's defense expenditure. *Strategic Assessment* (Tel Aviv), 12(4), 37–55.

Fabre, C. (2012). *Cosmopolitan war*. Oxford: Oxford University Press.

Fattouh, B. and Kolb, J. (2006). The outlook for economic reconstruction in Lebanon after the 2006 war. *The MIT Electronic Journal of Middle East Studies*, 6, 97–111.

Federation of American Scientists (FAS). (2000). Sudan People's Liberation Army (SPLA); Sudan People's Liberation Movement (SPLM). http://www .fas.org/irp/world/para/spla.htm.

Fidler, D. P. (1999). International legal implications of nonlethal weapons. *The Michigan Journal of International Law*, 21, 51–100.

(2005). The meaning of Moscow: "Nonlethal" weapons and international law in the early 21st century. *International Review of the Red Cross*, 87. 525–552.

Finlay, C. J. (2010). Legitimacy and non-state political violence. *Journal of Political Philosophy*, 18(3), 287–312.

(2012). Fairness and liability in the just war: Combatants, non-combatants and lawful irregulars. *Political Studies*, 61(1), 142–160.

Fletcher, G. P. (2003). Ambivalence about treason. *North Carolina Law Review*, 82, 1611–1628.

Freedman, L. (2006). Strategic communications. *The Adelphi Papers*, 45(379), 73–93. http://dx.doi.org/10.1080/05679320600661715.

Friedman, R. B. (1990). On the concept of authority in political philosophy. In J. Raz (Ed.), *Authority* (pp. 56–91). New York: NYU Press.

Friedman, T. (2013). Without water, revolution. *New York Times*, May 19. http://www.nytimes.com/2013/05/19/opinion/sunday/friedman-without-water-revolution.html.

Fukuda, C. M. (2000). Peace through nonviolent action: The East Timorese resistance movement's strategy for engagement. *Pacifica Review: Peace, Security & Global Change*, 12(1), 17–31.

The Fund for Peace (2010). The Failed State Index: Indicators. http://ffp.statesindex.org/rankings-2010-sortable.

Galeotti, M. (2002). 'Brotherhoods' and 'associates': Chechen networks of crime and resistance. *Low Intensity Conflict & Law Enforcement*, 11(2–3), 340–352.

Galtung, J. (2007). Peace journalism as ethical challenge. *asteriskos, 3/4*, 7–16.

Ganor, B. (2005). Terrorism as a strategy of psychological warfare. *Journal of Aggression, Maltreatment & Trauma*, 9(1–2), 33–43.

Garfield, R. (1999). The impact of economic sanctions on health and well-being. *Relief and Rehabilitation Network*, London. http://www.essex.ac.uk/armedcon/story_id/The%20Impact%20of%20Econmoic%20Sanctins%20on%20Health%20abd%20Well-Being.pdf.

Garfield, R., Devin, J., and Fausey, J. (1995). The health impact of economic sanctions. *Bulletin of the New York Academy of Medicine*, 72(2), 454–469.

Gebauer, M. (2006). Hezbollah's Al-Manar: Broadcasting from the bunker. *Spiegel Online*, August 10. http://www.spiegel.de/international/spiegel/hezbollah-s-al-manar-broadcasting-from-the-bunker-a-430905.html.

Gelkopf, M., Berger, R., Bleich, A., and Silver, R. C. (2012). Protective factors and predictors of vulnerability to chronic stress: A comparative study of 4 communities after 7 years of continuous rocket fire. *Social Science & Medicine*, 74(5), 757–766.

Geneva Convention (IV). (1949). *Convention relative to the protection of civilian persons in time of war.* Geneva, August 12, 1949.

(1949). Commentary, convention relative to the protection of civilian persons in time of war. Geneva, August 12, 1949.

Gewirth, A. (1982). Individual rights and political-military objectives. In A. Gewirth (Ed.), *Human rights: Essays on justification and applications* (pp. 234–255). Chicago: University of Chicago Press.

Gilbert, P. (1998). Good and bad cases for national secession. In P. B. Lehning (Ed.), *Theories of secession* (pp. 208–226). New York: Routledge.

Gilboa, D. (2008). To be an Israeli prisoner of war. *Association of Israeli POWs*, November 20. (Hebrew). http://www.erim-pow.co.il/article.php?id=296.

Giustozzi, A. (2014). The Taliban's 'military courts.' *Small Wars & Insurgencies*, 25(2), 284–296.

Global Security. (2003). Improvised explosive devices – Iraq. http://www.globalsecurity.org/military/intro/ied-iraq.htm.

(nd.a). Hamas funding. http://www.globalsecurity.org/military/world/para/hamas-funds.htm.

(nd. b) Iranian artillery rockets, Fadjr-5 333mm multiple rocket launcher system. http://www.globalsecurity.org/military/world/iran/mrl-iran.htm.

Goodhand, J. (2004). From war economy to peace economy? Reconstruction and state building in Afghanistan. *Journal of International Affairs*, 58(1), 155–174.

Gordon, J. (1999a). Economic sanctions, just war doctrine, and the fearful spectacle of the civilian dead. *Cross Currents*, 49(3), 387–400.

(1999b). A peaceful, silent, deadly remedy: The ethics of economic sanctions. *Ethics & International Affairs*, 13(1), 123–142.

(2011). Smart sanctions revisited. *Ethics & International Affairs*, 25, 315–335.

Gottstein, U. (1999). Peace through sanctions? Lessons from Cuba, former Yugoslavia and Iraq. *Medicine, Conflict and Survival*, 15(3), 271–285.

Gould, J. A. and Moe, E. (2012). Beyond rational choice: Ideational assault and the use of delegitimation frames in nonviolent revolutionary movements. *Research in Social Movements, Conflict, and Change*, 34, 123–151.

Graham, G. (2007). People's war? Self-interest, coercion and ideology in Nepal's Maoist insurgency. *Small Wars & Insurgencies*, 18(2), 231–248.

Greenberg, H. and Vaked, E. (2004). Six soldiers killed by IED explosion in Gaza. *Ynet* (Hebrew). http://www.ynet.co.il/articles/0,7340,L-2914770,00.html

Gross, M. L. (1997). *Ethics and activism: The theory and practice of political morality*. Cambridge: Cambridge University Press.

(2006). *Bioethics and armed conflict: Moral dilemmas of medicine and war*. Cambridge, MA: MIT Press.

(2008). The Second Lebanon War: The question of proportionality and the prospect of nonlethal warfare. *Journal of Military Ethics*, 7(1), 1–22.

(2010a). *Moral dilemmas of modern war: Torture, assassination, and blackmail in an age of asymmetric conflict*. Cambridge: Cambridge University Press.

(2010b). Medicalized weapons & modern war. *Hastings Center Report*, 40(1), 34–43.

(2012). Civilian vulnerability in asymmetric conflict: Lessons from the second Lebanon and Gaza Wars. In D. Rothbart (Ed.), *Civilians and modern war* (pp. 146–164). London: Routledge.

(2013). Force-feeding, autonomy, and the public interest. *New England Journal of Medicine*, 369,103–105.

Gruf, Y. (2003). I was a prisoner. *Association of Israeli POWs*, October 3. (Hebrew). http://www.erim-pow.co.il/article.php?id=24.

Grynkewich, A. G. (2008). Welfare as warfare: How violent non-state groups use social services to attack the state. *Studies in Conflict & Terrorism*, 31(4), 350–370.

Gunaratna, R. (2003). Sri Lanka: Feeding the Tamil Tigers. In K. Ballentine and J. Sherman (Eds.), *The political economy of armed conflict: Beyond greed and grievance* (pp. 197–224). London: Rienner.

Gunning, J. (2009). *Hamas in politics: Democracy, religion, violence*. Oxford: Columbia University Press.

Gurr, T. R. (2000). Nonviolence in ethnopolitics: Strategies for the attainment of group rights and autonomy. *PS: Political Science and Politics*, 33(2), 155–160.

Guth, D. W. (2009). Black, white, and shades of gray: The sixty-year debate over propaganda versus public diplomacy. *Journal of Promotion Management*, 14(3–4), 309–325.

Haensel, P. (1941). War taxation. *Taxes*, 19(2), 67–69; 122–125.

Hardin. R. (1982). *Collective action*. Baltimore: Hopkins University Press.

Harel, A. and Issacharoff, A. (2004). *The seventh war* (Hebrew). Tel Aviv: Yedioth Ahronoth.

Harel A. and Issacharoff, A. (2008). Analysis: A hard look at Hamas' capabilities. *Haaretz*, December 26. http://www.haaretz.com/print-edition/news/analysis-a-hard-look-at-hamas-capabilities-1.260281.

Harff, B. (2003). No lessons learned from the Holocaust? Assessing risks of genocide and political mass murder since 1955. *American Political Science Review*, 97(1), 57–73.

Haslam, E. (2000). Information warfare: Technological changes and international law. *Journal of Conflict & Security Law*, 5, 157–175.

Heller, J. (1995). *Catch 22*. New York: Everyman's Library Random House.

Henriksen, R. and Vinci, A. (2007). Combat motivation in non-state armed groups. *Terrorism and Political Violence*, 20(1), 87–109.

Hercheui, M. D. (2011). A literature review of virtual communities: The relevance of understanding the influence of institutions on online collectives. *Information, Communication & Society*, 14(1), 1–23.

Hess, D. and Martin, B. (2006). Repression, backfire, and the theory of transformative events. *Mobilization: An International Quarterly*, 11(2). http://tavaana.org/sites/default/files/Repression,%20Backfire%20and%20the%20Theory%20of%20Transformative%20Events%20-%20PDF%20-%20English.pdf.

Hicks, M. H. R., Lee, U. R., Sundberg, R., and Spagat, M. (2011). Global comparison of warring groups in 2002–2007: Fatalities from targeting civilians vs. fighting battles. *PloS one*, 6(9)http://www.plosone.org/article/info:doi/10.1371/journal.pone.0023976#pone-0023976-g001.

Hiebert, R. E. (2003). Public relations and propaganda in framing the Iraq war: A preliminary review. *Public Relations Review*, 29(3), 243–255.

Hill, D. T. (2002). East Timor and the Internet: Global political leverage in/on Indonesia. *Indonesia*, 73, 25–51.

Hoff, D. L. and Mitchell, S. N. (2009). Cyberbullying: Causes, effects, and remedies. *Journal of Educational Administration*, 47(5), 652–665.

Holder, E. (2012). Attorney General Eric Holder Speaks at Northwestern University School of Law. Chicago, Monday, March 5, 2012. http://www.justice .gov/iso/opa/ag/speeches/2012/ag-speech-1203051.html.

Hollis, R. (1997). Europe and the Middle East: Power by stealth? *International Affairs* (Royal Institute of International Affairs), 73(1), 15–29.

Horgan, J. and Taylor, M. (1999). Playing the 'Green Card'–Financing the provisional IRA: Part 1. *Terrorism and Political Violence*, 11(2), 1–38.

(2003). Playing the 'green card'–Financing the provisional IRA: part 2. *Terrorism and Political Violence*, 15(2), 1–60.

Howe, E. G., Kosaraju, A., Laraby, P. R., and Casscells, S. W. (2009). Guantanamo: Ethics, interrogation, and forced feeding. *Military Medicine*, 174(1), iv-xiii.

Howie, L. (2007). The terrorism threat and managing workplaces. *Disaster Prevention and Management*, 16, 70–78.

Hughes, J. (2007). *Chechnya: From nationalism to jihad.* Philadelphia, PA: University of Pennsylvania Press.

Hultman, L. (2007). Battle losses and rebel violence: Raising the costs for fighting. *Terrorism and Political Violence*, 19(2), 205–222.

Human Rights Watch (HRW). (1991). *Evil days: Thirty years of war and famine in Ethiopia.* http://www.hrw.org/sites/default/files/reports/Ethiopia919 .pdf.

(1995). *Russia three months of war in Chechnya.* Vol. 7, No. 6. http://www.hrw .org/reports/pdfs/r/russia/russia952.pdf.

(1997). *East Timor – Guerrilla attacks.* June 4. http://www.hrw.org/news/1997/ 06/04/east-timor-guerrilla-attacks.

(2001). *Under orders: War crimes in Kosovo.* http://www.hrw.org/reports/2001 /kosovo/.

(2003). *International humanitarian law issues in a potential war in Iraq* (Human Rights Watch Briefing Paper, February 20). http://www.hrw.org/sites /default/files/reports/Iraq%20IHL%20formatted.pdf

(2006). Israel/Lebanon: Israel responsible for Qana attack. July 30. http:// www.hrw.org/news/2006/07/29/israellebanon-israel-responsible-qana -attack.

(2007a). *Civilians under assault: Hezbollah's rocket attacks on Israel in the 2006 war.* http://www.hrw.org/sites/default/files/reports/iopt0807.pdf.

(2007b). *Why they died: Civilian casualties in Lebanon during the 2006 War.* http:// www.hrw.org/en/reports/2007/09/05/why-they-died.

(2008–2011). *World report, Russia 2008–11.* http://www.hrw.org/legacy /englishwr2k8/docs/2008/01/31/russia17710.htm; http://www.hrw.org /en/node/79324; http://www.hrw.org/en/node/87531; http://www.hrw .org/en/world-report-2011/russia.

(2010). *Gaza: Do not resume executions.* http://www.unhcr.org/refworld/ country,,HRW,,ISR,,4bc2cceec,0.html.

(2011–2013). *World report, Sri Lanka 2011–2013.* http://www.hrw.org/world -report/2013/country-chapters/sri-lanka; http://www.hrw.org/world -report-2012/world-report-2012-sri-lanka; http://www.hrw.org/en/world -report-2011/sri-lanka; http://www.hrw.org/en/node/87402.

(2012). *Abusive system: Criminal justice in Gaza*. http://www.hrw.org/sites /default/files/reports/iopt1012ForUpload_0.pdf.

Human Security Report Project. (2013). Human security report 2013, The decline in global violence: Evidence, explanation, and contestation. Vancouver: Human Security Press.

Humphreys, M. and Weinstein, J. M. (2008). Who fights? The determinants of participation in civil war. *American Journal of Political Science*, 52(2), 436–455.

iCasualties.org. (2014). Coalition military fatalities by year. http://icasualties .org/oef.

Intelligence and Terrorism Information Center (ITIC). (2008). Preventing the harming of uninvolved persons. December 27. http://www.terrorism-info .org.il/malam_multimedia/Hebrew/heb_n/video/v12b.wmv. Archived with author.

(2009). Hamas modus operandi – Terrorist shooting from a roof of a house and using children as a human shield. 6 January. http://www.terrorism-info .org.il/malam_multimedia/Hebrew/heb_n/video/v9.wmv. Archived with author.

International Criminal Court (ICC). (2011). *Rome Statute of the International Criminal Court*. http://www.icc-cpi.int/nr/rdonlyres/add16852-aee9-4757 -abe7-9cdc7cf02886/283503/romestatuteng1.pdf

International Criminal Tribunal for the former Yugoslavia (ICCT). (nd.). *Case Information Sheet, IT-03-66, LIMAJ et al., The Prosecutor v. Fatmir Limaj, Isak Musliu & Haradin Bala*. http://www.icty.org/x/cases/limaj/cis/en /cis_limaj_al_en.pdf.

International Committee of the Red Cross (ICRC). (1994). Geneva Conventions of 12 August 1949 and Additional Protocols of 8 June 1977: Ratifications, accessions and successions as at 31 December 1994. http://www.icrc.org /eng/resources/documents/misc/57jmc7.htm.

(2006). A guide to the legal review of new weapons, means and methods of warfare: Measures to implement Article 36 of Additional Protocol I of 1977. *International Review of the Red Cross*, 88(864), 931–956.

(2010). Gaza closure: Not another year! *News Release*, 10/103, 06-14-2010. http://www.icrc.org/eng/resources/documents/update/palestine -update-140610.htm.

(2011). Annex: The Islamic Emirate of Afghanistan. The Layha [Code of conduct For Mujahids]. *International Review of the Red Cross*, 93(881), 103–120.

(2013). Hunger strikes in prisons: The ICRC's position. http://www.icrc.org /eng/resources/documents/faq/hunger-strike-icrc-position.htm.

(nd. a). *Customary international humanitarian law, Rule 8, definition of military objectives*. http://www.icrc.org/customary-ihl/eng/docs/v1_rul_rule8).

(nd. b). Customary international humanitarian law, Rule 65, perfidy. http:// www.icrc.org/customary-ihl/eng/docs/v1_rul_rule65

(nd. c). Customary international humanitarian law, Rule 62, improper use of the flags or military emblems, insignia or uniforms of the adversary. http:// www.icrc.org/customary-ihl/eng/docs/v1_rul_rule62

(nd. d). Customary international humanitarian law, Rule 96, hostage taking. http://www.icrc.org/customary-ihl/eng/docs/v1_cha_chapter32_rule96

(nd. e). Customary international humanitarian law, Rule 97, human shields. http://www.icrc.org/customary-ihl/eng/docs/v1_rul_rule97.

International Court of Justice (ICJ). (1975). *Western Sahara, Advisory Opinion.* October 16. http://www.icj-cij.org/docket/files/61/6195.pdf.

International Institute of Strategic Studies (IISS). (1999). The Kosovo Liberation Army. Volume 5, Issue 4. http://www.iiss.org/publications/strategic-comments/past-issues/volume-5—1999/volume-5—issue-4/the-kosovo-liberation-army/ 1999.

(2006). The military balance 2006. http://archive.is/web.archive.org /web/20080313084832/http://www.iiss.org/whats-new/iiss-in-the-press/ press-coverage-2006/july-2006/strength-of-israel-lebanon-and-hezbollah/.

International Security Assistance Force (ISAF). (2011). Effective civil partnerships: Community project yields unlikely find. August 25. http://www.isaf .nato.int/article/coin/effective-civil-partnerships-community-project -yields-unlikely-find.html.

Iraq Body Count. (2012). Iraqi deaths from violence 2003-2011. http://www .iraqbodycount.org/analysis/numbers/2011/.

Israel High Court of Justice (HCJ). (2005). *The Public Committee against Torture in Israel and the Palestinian Society for the Protection of Human Rights and the Environment v. The Government of Israel and others.* Case 769/02.

Israel Ministry of Foreign Affairs (MFA). (2009). *The operation in Gaza: Factual and legal aspects.* Jerusalem; MFA.

Israel Security Agency. (nd.). Analysis of attacks in the last decade, 2000–2010. http://www.shabak.gov.il/English/EnTerrorData/decade/Pages/default. aspx.

Issacharoff, A. (2009). Rights group: Most Gazans killed in war were civilians. *Haaretz,* September 9. http://www.haaretz.com/hasen/spages/1113402 .html.

Iyob, R. (1995). *The Eritrean struggle for independence: Domination, resistance, nationalism, 1941–1993.* Cambridge: Cambridge University Press.

Janes. (2007). Sudan People's Liberation Army (SPLA) (Sudan). *Jane's World Insurgency and Terrorism,* January 30.

Jensen, B. (2012). The Taliban's wedge strategy. *New York Times,* August 21. http://www.nytimes.com/2012/08/22/opinion/the-talibans-wedge-strategy.html?_r=0.

Johnson, D. H. (1998). The Sudan People's Liberation Army and the Problem of Factionalism. In C. Clapham (Ed.), *African guerrillas* (pp. 53–72). Bloomington, IN: Indiana University Press.

Johnson, T. H. (2007). The Taliban insurgency and an analysis of Shabnamah (Night Letters). *Small Wars and Insurgencies,* 18(3): 317–344.

Johnson, N. (2010). Small projects reap large gains in Helmand. *ISAF.* http:// www.isaf.nato.int/article/isaf-releases/small-projects-reap-large-gains-in -helmand.html.

Johnson, M. and Johnson, T. (1981). Eritrea: The national question and the logic of protracted struggle. *African Affairs*, 80(319), 181–195.

Johnson, T. and Bruno, G. (2012). The lengthening list of Iran sanctions. *Council on Foreign Relations*, July 31. http://www.cfr.org/iran/lengthening-list-iran-sanctions/p20258.

Judah, T. (2000). The Kosovo Liberation Army. *Perceptions*, Sep-Nov, 61–77.

Jurist. (2011). Spain court turns over Guantanamo torture investigation to US. *The Juris*, April 13. http://jurist.org/paperchase/2011/04/spain-court -turns-over-guantanamo-torture-investigation-to-US.php#.

Kalb, M. and Saivetz, C. (2007). The Israeli – Hezbollah war of 2006: The media as a weapon in asymmetrical conflict. *The Harvard International Journal of Press/Politics*, 12(3), 43–66.

Kaldor, M. (1997). Introduction. In M. Kaldor and B. Vashee, *Restructuring the global military sector. Volume 1: New wars* (pp. 3–33). New York: Pinter.

Kalyvas, S. N. (2003). The ontology of "political violence": Action and identity in civil wars. *Perspectives on Politics*, 1, 475–494.

(2006). *The logic of violence in civil war.* Cambridge: Cambridge University Press.

Kalyvas, S. N. and Kocher, M. (2007). How "free" is free-riding in civil wars? *World Politics*, 59(2), 177–216.

Kasfir, N. (2005). Guerrillas and civilian participation: The National Resistance Army in Uganda, 1981–86. *Journal of Modern African Studies*,43(2), 271–296.

Kasymov, S. (2011). The right of communities to self-determination. *Peace Review: A Journal of Social Justice*, 23(2), 221–227.

Katz, Y. (2012). Security and defense: Protecting the weakest link. *The Jerusalem Post*, June 1. http://www.jpost.com/Features/FrontLines/Article .aspx?id=252479.

Kaurin, P. (2010). With fear and trembling: An ethical framework for nonlethal weapons. *Journal of Military Ethics*, 9(1), 100–114.

Kearney, M. (2007). *The prohibition of propaganda for war in international law.* Oxford: Oxford University Press.

Kelsay, J. (2007). *Arguing the just war in Islam.* Cambridge: Harvard University Press.

Kelsey, J. T. (2008). Hacking into International Humanitarian Law: The prin-ciples of distinction and neutrality in the age of cyber warfare. *Michigan Law Review*, 106(7), 1427–1451.

Keynes, J. M. (1940). *How to pay for the War: A radical plan for the Chancellor of the Exchequer.* New York: Harcourt Brace and Company.

Khatchadourian, R. (2010). Street fight on the high seas posted. *The New Yorker*, January 12. http://www.newyorker.com/online/blogs/newsdesk/2010/01 /sea-shepherd.html.

Khoury. J. (2013). Analysis: For Palestinian prisoners in Israel, hunger strikes have become a winning strategy. *Haaretz*, April 24. http://www.haaretz .com/news/diplomacy-defense/for-palestinian-prisoners-in-israel-hunger -strikes-have-become-a-winning-strategy-1.517263.

Khoury-Machool, M. (2007). Palestinian youth and political activism: The emerging internet culture and new modes of resistance. *Policy Futures in Education,* 5(1), 17–36.

Kilcullen, D. (2009). *The accidental guerrilla: Fighting small wars in the midst of a big one.* Oxford: Oxford University Press.

King, M. E. (2007). *A quiet revolution: The first Palestinian Intifada and nonviolent resistance.* New York: Nation Books.

Kingsbury, D. (2000). East Timor to 1999. In D. Kingsbury (Ed.), *Guns and ballot boxes East Timor's vote for independence* (pp. 17–28). Victoria, Australia: Monash Asia Institute.

Kirkpatrick, D. D. (2011). NATO strikes at Libyan state TV. *New York Times,* July 30. http://www.nytimes.com/2011/07/31/world/africa/31tripoli. html?_r=1&pagewanted=pr.

Klare, M. T. (1992/1993). The next great arms race. *Foreign Affairs,* 72, 136–152.

Knoke, D. (1988). Incentives in collective action organizations. *American Sociological Review,* 53(3), 311–329.

Kober, A. (2007). Targeted killing during the second intifada: The quest for effectiveness. *Journal of Conflict Studies,* 27(1). http://journals.hil.unb.ca /index.php/JCS/article/viewArticle/8292/9353.

Kocher, M. (2007). The decline of PKK and the viability of a one–state solution in Turkey. Democracy and human rights in multicultural societies. *International Journal on Multicultural Societies,* 4(1), Societies. http://www.unesco.org /most/vl4n1kocher.pdf.

Koplow, D. A. (2006). *Nonlethal weapons.* Cambridge: Cambridge University Press.

Kopp, C. (2007). Technology of improvised explosive devices. *Defence Today,* 6(3), 45–47.

Kraja, G. (2011). *Recruitment practices of Europe's last guerrilla: Ethnic mobilization, violence and networks in the recruitment strategy of the Kosovo Liberation Army.* Unpublished senior thesis, Yale University.

Krajeski, J. (2012). After the hunger strike. *The New Yorker,* November 29. http:// www.newyorker.com/online/blogs/newsdesk/2012/11/after-the-kurdish -hunger-strike-in-turkish-prisons.html.

Kruglanski, A. W., Chen, X., Dechesne, M., Fishman, S. and Orehek, E. (2009). Fully committed: Suicide bombers' motivation and the quest for personal significance. *Political Psychology,* 30(3), 331–357.

Kymlicka, W. (1998). Is federalism a viable alternative to secession? In P. B. Lehning (Ed.), *Theories of secession* (pp. 109–148). New York: Routledge.

Landau-Tassero, E. (2006). Non-combatants in Muslim legal thought. Center on Islam, Democracy, and the Future of the Muslim World, Research Monographs on the Muslim. World Series No 1, Paper No 3, December. Washington, DC: Hudson Institute. http://www.currenttrends.org /docLib/20061226_NoncombatantsFinal.pdf.

Lango, J. W. (2010). Nonlethal weapons, noncombatant immunity, and combatant nonimmunity: A study of just war theory. *Philosophia* 38(3), 475–497.

Lappin, Y. (2009). IDF releases Cast Lead casualty numbers. *Jerusalem Post*, March 26. www.jpost.com/Israel/Article.aspx?id=137286.

Laqueur, W. (1998). *Guerrilla warfare: A historical and critical study*. New Brunswick/London: Transaction Publishers.

Lasswell, H. D. (1927). The theory of political propaganda. *The American Political Science Review*, 21(3), 627–631.

Lee, J. (2014). South Koreans seethe, sue as credit card details swiped. *Reuters*, January 21. http://www.reuters.com/article/2014/01/21/us-korea-cards-idUSBREA0K05120140121.

Leigh, D. (2010). US forces hit target 'with no civilian deaths' – but Afghans tell different tale. *The Guardian*, July 26. http://www.guardian.co.uk/world/2010/jul/26/afghanistan-war-logs-helmand-bombing.

Levin, D. (2004). *Memorandum for James B. Comey, Deputy Attorney General, Re: legal standards applicable under 18 U.S.C. §§ 2340-2340.* December 30. http://www.usdoj.gov/olc/18usc23402340a2.htm.

Levitt, M. (2006). *Hamas: Politics, charity, and terrorism in the service of jihad.* New Haven, CT: Yale University Press.

Levy, P. I. (1999). Sanctions on South Africa: What did they do? *The American Economic Review*, 89(2), 415–420.

Lewis, J. A. (2002). *Assessing the risks of cyber terrorism, cyber war and other cyber threats.* Washington, DC: Center for Strategic & International Studies.

(2010). *Thresholds for cyberwar.* Washington, DC: Center for Strategic and International Studies. http://csis.org/files/publication/101001_ieee_insert.pdf.

(2011). Cyberwar thresholds and effects. *Security & Privacy, IEEE* 9(5), 23–29.

Lieber, F. (1863). *Instructions for the Government of Armies of the United States in the Field (Lieber Code).* http://www.icrc.org/applic/ihl/ihl.nsf/Treaty.xsp?documentId=A25AA5871A04919BC12563CD002D65C5&action=openDocument.

Lieberman, E. (2012). *Reconceptualising deterrence.* London: Routledge.

Lim, A. (2012). *The Case for Sanctions against Israel.* London: Verso.

Limor, Y. and Nossek, H. (2006). The Military and the media in the twenty-first century: Towards a new model of relations. *Israel Affairs*, 12(3), 484–510.

Looney, R. E. (2005). The business of insurgency: The expansion of Iraq's shadow economy. *The National Interest*, 81, 67–72.

Lopez, G. A. (2012). In defense of smart sanctions: A response to Joy Gordon. *Ethics & International Affairs*, 26 (1), 135–46.

Lopez, G. A. and Cortright, D. (1997). Economic sanctions and human rights: Part of the problem or part of the solution? *The International Journal of Human Rights*, 1(2), 1–25.

Love, J. B. (2010). *Hezbollah: Social services as a source of power* (No. JSOU-10-05). Hurlburt Field, FL: Joint Special Operations University.

Lowe, V. and Tzanakopoulos, A. (2012). Economic warfare. In R. Wolfrum (Ed.), *Max Planck encyclopedia of public international law*. Oxford: Oxford University Press. http://papers.ssrn.com/sol3/papers.cfm?abstract_id=1701590.

Lutz, J. M. and Lutz, B. J. (2006). Terrorism as economic warfare. *Global Economy Journal*, 6(2), 1–20.

Ly, P. E. (2007). The charitable activities of terrorist organizations. *Public Choice*, 131(1–2), 177–195.

Lyall, R. (2008). Voluntary human shields, direct participation in hostilities and the international humanitarian law obligations of states. *Melbourne Journal of International Law*, 9, 313–334.

Macdonald, S. (2007). *Propaganda and information warfare in the twenty-first century: Altered images and deception operations*. London: Routledge.

Mackey, R. (2012). Attack on pro-Assad television studio raises questions on rules of war. *New York Times*, June 27. http://thelede.blogs.nytimes.com/2012/06/27/attack-on-pro-assad-television-studio-raises-questions-on-rules-of-war/.

Maguire, M. (1980). Impact of burglary upon victims. *The British Journal of Criminology*, 20, 261–275.

Maital, S. (1972). Inflation, taxation and equity: How to pay for the war revisited. *The Economic Journal*, 82(325), 158–169., P. (2010). Ireland revisited week 15. *The Examiner*, December 21. http://www.crossexaminer.co.uk/archives/5602.

Manheim, J. B. (1990). *Strategic public diplomacy: The evolution of influence*. New York: Oxford University Press.

Mantel, B. (2009). Terrorism and the internet. Should websites that promote terrorism be shut down? *CQ Global Researcher*, 3(11), 129–152.

March A. and Sil R. (1999). The "Republic of Kosovo" (1989–1998) and the resolution of ethno-separatist conflict: rethinking "sovereignty" in the post-cold war era. Working Paper Series #99-01, Political Science Department, University of Pennsylvania, Philidelphia, PA. http://www.sas.upenn.edu/penncip/Reports/MarchSil.htm.

Marcus, A. (2007). *Blood and belief: The PKK and the Kurdish fight for independence*. New York: New York University Press.

Margalit, A. and Raz, J. (1990). National self-determination. *The Journal of Philosophy*, 87(9), 439–461.

Margalit, A. and Walzer, M. (2009). Israel: Civilians & combatants. *New York Review of Books*, 56(8), 21–22.

Marks, T. A. and Palmer, D. S. (2005). Radical Maoist insurgents and terrorist tactics: Comparing Peru and Nepal 1. *Low Intensity Conflict & Law Enforcement*, 13(2), 91–116.

Marten, K. (2006/2007) Warlordism in comparative perspective, *International Security*, 31(3), 41–73.

(2011). Warlords. In H. Strachan and S. Scheipers (Eds.), *The changing character of war* (pp. 302–314). New York: Oxford University Press.

(2012). Patronage versus professionalism in new security institutions. *Prism*, 2(4), 83–98.

Martin, J. K. (2009). Dragon's claws: The improvised explosive device (IED) as a weapon of strategic influence. Unpublished masters' thesis, Naval Postgraduate School, Monterey, CA.

Martin, L. J. (1982). Disinformation: An instrumentality in the propaganda arsenal. *Political Communication and Persuasion*, 2(1), 47–64.

McCormick, G. H. and Fritz, L. (2009). The logic of warlord politics. *Third World Quarterly*, 30(1), 81–112.

McCormick, G. H. and Giordano, F. (2007). Things come together: Symbolic violence and guerrilla mobilisation. *Third World Quarterly*, 28(2), 295–320.

McGoldrick, A. (2006). War journalism and 'objectivity.' *Conflict & Communication online*, 5(2), 1–7. http://www.cco.regener-online.de/2006_2/pdf/mcgoldrick.pdf.

McKay, S. and Mazurana, D. (2004). *Where are the Girls? Girls in fighting forces in Northern Uganda, Sierra Leone and Mozambique: Their lives during and after war.* Montreal, Canada: International Centre for Human Rights and Democratic Development.

McMahan, J. (1985). Deterrence and deontology. *Ethics*, 95(3), 517–536.

(2009). *Killing in war*. Oxford: Oxford University Press.

Meisels, T. (2008). *The trouble with terror*. Cambridge: Cambridge University Press.

Melzer, N. (2009). *Interpretive guidance on the notion of direct participation in hostilities under international humanitarian law*. Geneva: International Committee of the Red Cross.

Metelits, C. (2004). Reformed rebels? Democratization, global norms, and the Sudan People's Liberation Army. *Africa Today*, 51(1), 65–82.

Middle East Media Research Institute (Memri). (2006). Special dispatch No. 1323. Islamist Websites Monitor no. 7. October 13, clashes between Al-Qaeda in Iraq and Iraqi mujahideen who are not members of Al-Qaeda or of Its Shura Council. http://www.memri.org/report/en/print1909.htm.

Miller, D. (1998). Secession and the principle of nationality. In M. Moore (Ed.), *National self-determination and secession* (pp. 62–78). Oxford: Oxford University Press.

Moloney, E. (2003). *A secret history of the IRA*. New York: WW Norton & Company.

Monro, D. H. (1982). Civil rights and conscription. In M. Anderson and B. Honegger (Eds.), *The military draft: Selected readings on conscription* (pp. 133–152). Stanford. CA: The Hoover Institute.

Moorcraft, P.L. and Taylor, P. M. (2007). War watchdogs or lapdogs? *British Journalism Review*, 18(4), 39–50.

Moorcraft, P. L. and Taylor, P. M. (2008). *Shooting the messenger: The politics of war reporting*. Sterling, VA: Potomic Books.

Moore, C. (2006). Our very strange day with Hezbollah. *CNN*. http://edition.cnn.com/CNN/Programs/anderson.cooper.360/blog/archives/2006_07_23_ac360_archive.html.

Moore, M. (1998). Introduction: The self-determination principle and the ethics of secession. In M. Moore (Ed.), *National self-determination and secession* (pp. 1–14). Oxford: Oxford University Press.

Moore, S. (2001). The Indonesian military's last years in East Timor: An analysis of its secret documents. *Indonesia*, 72, 9–44.

(2011). Taliban sleeper agent kills 9 at Afghan base. *Associated Press*, April 17. http://www.highbeam.com/doc/1G1-254293670.html.

Mor, B. D. (2007). The rhetoric of public diplomacy and propaganda wars: A view from self-presentation theory. *European Journal of Political Research*, 46(5), 661–683.

(2012). Credibility talk in public diplomacy. *Review of International Studies*, 38(2), 393–422.

Moreno, J. (1975). Che Guevara on guerrilla warfare: Doctrine, practice and evaluation. In S. C. Sarkesian (Ed.), *Revolutionary guerrilla warfare* (pp. 395–420). New Brunswick, NJ: Transaction Publishers.

Morris, B. (2011). *Righteous victims: A history of the Zionist-Arab conflict, 1881–1998.* New York: Random House Digital, Inc.

Mshvidobadze, K. (2011). State-sponsored cyber terrorism: Georgia's experience. Presentation to the *Georgian Foundation for Strategic and International Studies*, 1–7.

Mukhopadhyay, D. (2009). Disguised warlordism and combatanthood in Balkh: The persistence of informal power in the formal Afghan state. *Conflict, Security & Development*, 9(4), 535–564.

Nakashima, E., Miller, G., and Tate, J. (2012). US, Israel developed Flame computer virus to slow Iranian nuclear efforts, officials say. *The Washington Post* online, 1–4, June 19. http://www.washingtonpost.com/world/national -security/us-israel-developed-computer-virus-to-slow-iranian-nuclear-efforts -officials-say/2012/06/19/gJQA6xBPoV_story.html.

Namazi, S. (2013). Sanctions and medical supply shortages in Iran. *Woodrow Wilson Center, Viewpoints*, 20. http://www.wilsoncenter.org/sites/default /files/sanctions_medical_supply_shortages_in_iran.pdf.

Napoleoni, L. (2004). The new economy of terror: How terrorism is financed. In A. P. Schmid (Ed.), *Forum on Crime and Society, United Nations Office on Drugs and Crime 4* (1, 2), (pp. 31–48). New York: United Nations Publications. http://www.unodc.org/documents/data-and-analysis/Forum/V05-81059_ EBOOK.pdf

National Research Council (NRC). (2003). *An assessment of nonlethal weapons science and technology.* Committee for an Assessment of Nonlethal Weapons Science and Technology Naval Studies Board Division on Engineering and Physical Sciences. Washington, DC: The National Academies Press.

NATO. (2011). NATO strikes Libyan state TV satellite facility. July 30. http:// www.nato.int/cps/en/natolive/news_76776.htm.

Navias, M. S. (2002). Finance warfare as a response to international terrorism. *The Political Quarterly*, 73(S1), 57–79.

Naylor, R. T. (1993). The insurgent economy: Black market operations of guerrilla organizations. *Crime, Law and Social Change*, 20(1), 13–51.

Neff, S. C. (1988). Boycott and the law of nations: Economic warfare and modern international law in historical perspective. *British Yearbook of International Law*, 59(1), 113–149.

Nepstad, S. E. (2013). Mutiny and nonviolence in the Arab Spring: Exploring military defections and loyalty in Egypt, Bahrain, and Syria. *Journal of Peace Research*, 50(3), 337–349.

Nevins, J. (2005). *A not-so-distant horror: Mass violence in East Timor*. Ithaca, NY: Cornell University Press.

Nielson, K. (1998). Liberal nationalism and secession. In M. Moore (Ed.), *National self-determination and secession* (pp. 103–133). Oxford: Oxford University Press.

Niner, S. (2007). Martyrs, heroes and warriors: The leadership of East Timor. In D. Kingsbury and M. Leach (Eds.), *East Timor: Beyond independence* (pp. 113–128). Victoria, Australia: Monash University Press.

Nissen, T. E. (2007). *The Taliban's information warfare: A comparative analysis of NATO information operations (info ops) and Taliban information activities*. Copenhagen: Royal Danish Defence College.

Nordland, R. (2013). Attacks on aid workers rise in Afghanistan, UN says. *New York Times*, December 4. http://www.nytimes.com/2013/12/03/world /asia/attacks-rise-on-aid-workers-in-afghanistan.html.

Norton-Taylor, R. (2003). TV station attack could be illegal. *The Guardian*, March 26. http://www.theguardian.com/media/2003/mar/26/Iraqandthemedia. iraq2

Nuwayhid, I., Zurayk, H., Yamout, R., and Cortas, C. S. (2011). Summer 2006 war on Lebanon: A lesson in community resilience. *Global Public Health*, 6(5), 505–519.

Nyce, J. M. and Dekker, S. W. (2010). IED casualties mask the real problem: It's us. *Small Wars & Insurgencies*, 21(2), 409–413.

Nye, J. S. (2008). Public diplomacy and soft power. *The Annals of the American Academy of Political and Social Science*, 616(1), 94–109.

OAU. (1981). African [Banjul] charter on human and peoples' rights, adopted June 27, 1981. OAU Doc. CAB/LEG/67/3 rev. 5, 21 I.L.M. 58 (1982). http://www1.umn.edu/humanrts/instree/z1afchar.htm.

Obama, B. (2009). Remarks by the President at the acceptance of the Nobel Peace Prize. The White House, Office of the Press Secretary. http://www. whitehouse.gov/the-press-office/remarks-president-acceptance-nobel -peace-prize.

O'Driscoll, C. (2012). A fighting chance or fighting dirty? Irregular warfare, Michael Gross and the Spartans. *European Journal of Political Theory*, 11(2): 112–130.

Olsthoorn, P. and Bollen, M. (2013). Civilian care in war: lessons from Afghanistan. In M. L. Gross and D. Carrick (Eds.), *Military Medical Ethics in the 21st Century* (pp. 59–70). London: Ashgate Publishing.

Özdemir, H. (2012). Where do terror organizations get their money? A case study: Financial resources of the PKK, *International Journal of Security and Terrorism*, 3(2), 85–102.

Özerdem, A. (2003). From a 'terrorist' group to a 'civil defence' corps: The 'transformation' of the Kosovo Liberation Army. *International Peacekeeping*, 10(3), 79–101.

Palestinian Centre for Human Rights (PCHR). (2009). The dead in the course of the Israeli recent military offensive on the Gaza strip between 27 December 2008 and 18, January 2009. March 19, 2009. http://www.scribd.com /doc/22883962/The-Dead-in-the-course-of-the-Israeli-Military-offensive -on-the-Gaza-Strip-between-27-Dec-2008-and-18-Jan-2009.

Pape, R. (1997). Why economic sanctions do not work. *International Security,* 22(2), 90–136.

(2005). *Dying to win: The strategic logic of suicide terrorism.* New York: Random House Digital, Inc.

Parker, A. M. S. (2009). Cyberterrorism: The emerging worldwide threat. In D. Canter (Ed.), *The faces of terrorism: Multidisciplinary perspectives* (pp. 245–66). Chichester: Wiley-Blackwell.

Pateman, R. (1990). The Eritrean war. *Armed Forces & Society,* 17(1), 81–98.

Payne, K. (2005). The media as an instrument of war. *Parameters,* 35(1), 81–93.

Pedahzur, A., Perliger, A. and Weinberg, L. (2003). Altruism and fatalism: The characteristics of Palestinian suicide terrorists. *Deviant Behavior,* 24(4), 405–423.

Peimani, H. (2004). *Armed violence and poverty in Chechnya: Mini case study for the armed violence and poverty initiative.* Bradford, UK: University of Bradford, Center for International Cooperation and Security.

Peksen, D. (2009). Better or worse? The effect of economic sanctions on human rights. *Journal of Peace Research,* 46(1), 59–77.

Perkins, M. (2002). International law and the search for universal principles in journalism ethics. *Journal of Mass Media Ethics,* 17(3), 193–208.

Perritt, H. H. (2008). *Kosovo Liberation Army: The inside story of an insurgency.* Champaign, IL: University of Illinois Press.

Peskowitz, A. (2010). IO on the counterinsurgency battlefield: Three case studies. *Global Security Studies,* 1(2), 100–114.

Peters, J. (2010). The Gaza disengagement: Five years later. *Israel Journal of Foreign Affairs* 4(3), 33–44.

Petit, B. S. (2003). *Chechen use of the internet in the Russo-Chechen conflict.* Fort Leavenworth KS: Army Command and General Staff College.

Pew. (2007). Journalists in Iraq – A survey of reporters on the front lines embedding. *The Pew Research Center's Project for Excellence in Journalism.* November 28. http://www.journalism.org/node/8645.

Phillips, E. (2009).The business of kidnap for ransom. In D. Canter (Ed.), *The faces of terrorism: Multidisciplinary perspectives* (pp. 189–206). Hoboken, NJ: John Wiley & Sons.

Philpott, D. (1998). Self-determination in practice. In M. Moore (Ed.), *National self-determination and secession* (pp. 79–102). Oxford: Oxford University Press.

Physicians for Human Rights, Israel (PHR). (2008). Holding health to ransom: GSS interrogation and extortion of Palestinian patients at Erez Crossing. http://www.phr.org.il/uploaded/HoldingHealthToRandsom_4.pdf.

(2010). Humanitarian minimum: Israel's role in creating food and water insecurity in the Gaza Strip. http://www.phr.org.il/uploaded/Humanitarian%20 Minimum_eng_webver_H.pdf.

Pictet, J. 1985. *Development and principles of international humanitarian law.* Leiden, Netherlands: Martinus Nijhoff.

Pinto, C. and Jardine, M. (1997). *East Timor's unfinished struggle: Inside the Timorese resistance.* Cambridge, MA: South End Press.

Plaisance, P. L. (2005). The propaganda war on terrorism: an analysis of the United States' "shared values" public-diplomacy campaign after September 11, 2001. *Journal of Mass Media Ethics,* 20(4), 250–268.

Plata v. Brown. (2013). United States District Court, Northern District of California, C01-1351 THE, August 19.

Politkovskaya, A. (2003). *A small corner of hell: Dispatches from Chechnya.* Chicago: University of Chicago Press.

Poller, N. (2011). The Muhammad al-Dura hoax and other myths revived. *Middle East Quarterly.* http://www.meforum.org/3076/muhammad-al-dura-hoax.

Pratkanis, A. (2009). Public diplomacy in international conflicts. In N. Snow and P. Taylor (Eds.), *Routledge handbook of public diplomacy* (pp. 111–154). London: Routledge.

Primoratz, I. (2013). *Terrorism: A philosophical investigation.* Cambridge: Polity Press.

Public Affairs Guidance (PAG). (2003). On embedding media during possible future operations/deployments in the U.S. Central Commands (CENTCOM) area of responsibility, February. http://www.defense.gov/news/feb2003 /d20030228pag.pdf.

Pugh, M. C., Cooper, N., and Goodhand, J. (2004). *War economies in a regional context: The challenges of transformation.* Boulder, CO: Lynne Rienner Publishers.

Pula, B. (2004). The emergence of the Kosovo "parallel state," 1988–1992. *Nationalities Papers: The Journal of Nationalism and Ethnicity,* 32(4), 797–826.

Qazi, S. H. (2011). Rebels of the frontier: Origins, organization, and recruitment of the Pakistani Taliban. *Small Wars & Insurgencies,* 22(4), 574–602.

Radu, M. (1999). Don't arm the KLA. *FPRI (Foreign Policy Research Institute),* April 6. http://www.fpri.org/enotes/19990406.radu.dontarmkla.html.

Rapin, A. J. (2009). Does terrorism create terror? *Critical Studies on Terrorism,* 2(2), 165–179.

Rawls, J. (1971). *A theory of justice.* Cambridge, MA: Belknap Press.

Rawls J. (2001). *The law of peoples: With "the idea of public reason revisited."* Cambridge, MA: Harvard University Press.

Reitan, R. (2000). Strategic nonviolent conflict in Kosovo. *Peace & Change,* 25(1), 70–102.

Renz, B. (2010). The status and treatment of detainees in Russian Chechen campaigns. In S. Scheipers (Ed.), *Prisoners in war* (pp. 205–218). Oxford: Oxford University Press.

Reuters. (2007). FACTBOX-War in Lebanon, one year ago. July 9. http://www .reuters.com/article/2007/07/09/idUSL0959275.

Rid, T. and Hecker, M. (2009). *War 2.0: Irregular warfare in the information age.* Santa Barbara, CA: Praeger.

Riechmann, D. (2011). Shooter in Afghan army uniform kills NATO troops. *Associated Press.* July 16. http://www.washingtontimes.com/news/2011 /jul/16/shooter-afghan-army-uniform-kills-nato-trooper/.

Rigby, A. (1997). *The legacy of the past: The problem of collaborators and the Palestinian case.* Jerusalem: Palestinian Academic Society for the Study of International Affairs.

(2010). *Palestinian resistance and nonviolence.* Jerusalem: Palestinian Academic Society for the Study of International Affairs.

Roberts, L. D., Indermaur, D. and Spiranovic, C. (2013). Fear of cyber-identity theft and related fraudulent activity. *Psychiatry, Psychology and Law,* 20(3), 315–328.

Rockoff, H. (2012). *America's economic way of war: War and the US economy from the Spanish-American War to the Persian Gulf War.* Cambridge: Cambridge University Press.

Rodin, D. (2005). *War and self-defense.* Oxford: Oxford University Press.

Rogel, C. (2003). Kosovo: Where it all began. *International Journal of Politics, Culture, and Society,* 17(1), 167–182.

Rogers, P. (2000). Political violence and economic targeting aspects of provisional IRA strategy, 1992–7. *Civil Wars,* 3(4), 1–28.

Roling, K. (nd.). Provincial development council discusses projects in Ghazni. *ISAF (International Security Assistance Force).* http://www.isaf.nato.int/article /news/provincial-development-council-discusses-projects-in-ghazni.html.

Rome Statute of the International Criminal Court. (2010). *Annex I, Amendments to the Rome Statute of the International Criminal Court on the crime of aggression, Article 8 bis Crime of aggression, paragraph 2, 11 June.*

Rome Statute of the International Criminal Court. (2002). http://www.icc-cpi .int/nr/rdonlyres/ea9aeff7-5752-4f84-be94-0a655eb30e16/0/rome _statute_english.pdf.

Rosenberg, M. and Nordland, R. (2012). Abandoning hopes for Taliban peace deal. *New York Times,* October 2. http://www.nytimes.com/2012/10/02/world /asia/us-scales-back-plans-for-afghan-peace.html?_r=1&pagewanted=all.

Ross, M. L. (2005). Resources and rebellion in Aceh, Indonesia. In P. Collier and N. Sambanis (Eds.), *Understanding civil war: Evidence and analysis – Europe, Central Asia, and other regions* (pp. 39–62). Washington, DC: World Bank Publications.

Rubin, A. J. and Bowley, G. (2012). Suicide attack kills 9 in Eastern Afghanistan. *New York Times,* February 28. http://www.nytimes.com/2012/02/28/world /asia/suicide-attack-kills-9-in-eastern-afghanistan.html?_r=0.

Rubin, E. (2006). In the land of the Taliban. *New York Times,* October 22. http://www.nytimes.com/2006/10/22/magazine/22afghanistan. html?pagewanted=all

Rudner, M. (2010). Hizbullah terrorism finance: Fund-raising and money-laundering. *Studies in Conflict & Terrorism,* 33(8), 700–715.

Rule, S. (1987). Ethiopian guerrillas raid food convoy, halting drought aid. *New York Times,* October 27. http://www.nytimes.com/1987/10/27/world /ethiopian-guerrillas-raid-food-convoy-halting-drought-aid.html.

Royal United Services Institute (RUSI). (2009). Reforming the Afghan National Police. A joint report of the Royal United Services Institute for Defence and Security Studies (London) and the Foreign Policy Research Institute (Philadelphia). http://www.rusi.org/downloads/assets/ANP_Nov09.pdf.

Russett, B. M. (1969a). The price of war. *Society*, 6, 28–35.

(1969b). Who pays for defense? *The American Political Science Review*, 63(2), 412–426.

Ryan, M. and Brunnstrom, D. (2011). Libyan TV still on air despite NATO bombing. *Reuters*, July 30. http://www.reuters.com/article/2011/07/30/us-libya -nato-idUSTRE76T0KE20110730.

Sabo, L. E. and Kibirige, J. S. (1989). Political violence and Eritrean health care. *Social Science & Medicine*, 28(7), 677–684.

Sarkesian, S. C. (Ed.). (1975). *Revolutionary guerrilla warfare*. New Brunswick, NJ: Transaction Publishers.

Sarma, K. (2007). Defensive propaganda and IRA political control in Republican communities. *Studies in Conflict & Terrorism*, 30(12), 1073–1094.

Sassòli, M. (2011). Introducing a sliding-scale of obligations to address the fundamental inequality between armed groups and states? *International Review of the Red Cross*, 93 (882), 426–431.

Sassòli, M., Bouvier A. A. and Quintin A. (2011). *How does law protect in war? Cases, documents and teaching materials on contemporary practice* (3rd ed., vol. 1). Geneva: International Committee of the Red Cross. http://www.icrc.org /eng/resources/documents/publication/p0739.htm

Schaap, A. J. (2009). Cyber warfare operations: Development and use under international law. *Air Force Law Review*, 64, 121–173.

Scheipers, S. (2010). Introduction: Prisoners in war. In S. Scheipers (Ed.), *Prisoners in war* (pp. 1–22). Oxford: Oxford University Press.

Schiff, Z. and Ya'ari, E. (1991). *Intifada: The Palestinian uprising-Israel's third front*. New York: Touchstone Books.

Schindler, D. (1979). Wars of national liberation. The different types of armed conflicts according to the Geneva conventions and protocols. In *Collected Courses of the Hague Academy of International Law*, 117–164. Leiden: Martinus Nijhoff Publishers.

Schmid, A. (2005). Terrorism as psychological warfare. *Democracy and Security*, 1(2), 137–146.

Schmitt, M. N. (2004). Direct Participation in hostilities and 21st century armed conflict. In H. Fischer (Ed.), *Krisensicherung und Humanitärer Schutz (Crisis management and humanitarian protection)* (pp. 505–509). Berlin: Berliner Wissenschaftsverlag.

(2005). Precision attack and international humanitarian law. *International Review of the Red Cross*, 87(859), 445–466.

(2008). Human shields in international humanitarian law. *Israel Yearbook on Human Rights*, 38, 17–59.

(2010a). Interpretive guidance on the notion of direct participation in hostilities: A critical analysis. *The Harvard National Security Journal*, 1, 5–44.

(2010b). Deconstructing direct participation in hostilities: The constitutive elements. *New York University Journal of International Law and Politics*, 42, 697–739.

(2011). Cyber operations and the jus in bello. *Israel Yearbook on Human Rights*, 41, 113–135.

(Ed.). (2013). *Tallinn manual on the international law applicable to cyber warfare.* Cambridge: Cambridge University Press.

Schock, K. (1999). People power and political opportunities: Social movement mobilization and outcomes in the Philippines and Burma. *Social Problems,* 46(3), 355–375.

(2013). The practice and study of civil resistance. *Journal of Peace Research,* 50, 277–290.

Schulze K. E. (2003). The struggle for an independent Aceh: The ideology, capacity, and strategy of GAM. *Studies in Conflict & Terrorism,* 26(4), 241–271.

Schulze, K. E. (2004). *The free Aceh movement (GAM): Anatomy of a separatist organization.* Washington, DC: Policy Studies, East-West Center.

Sderot Media Center. (2009). What you need to know about qassams. March 3. http://sderotmedia.org.il/bin/content.cgi?ID=388&q=6&s=16.

Seely, R. (2001). *Russo-Chechen conflict: 1800–2000; A deadly embrace.* New York: Psychology Press.

Seib, P. (2009). Public diplomacy and journalism parallels, ethical issues, and practical concerns. *American Behavioral Scientist,* 52(5), 772–786.

Sharp, G. (1973). *The Politics of nonviolent action.* Boston: Porter Sargent.

(1989). The intifadah and nonviolent struggle. *Journal of Palestine Studies,* 19(1), 3–13.

Shore, M. (2013) The Jewish hero history forgot. *New York Times,* April 18. http://mobile.nytimes.com/2013/04/19/opinion/the-jewish-hero-history-forgot.html.

Shulman, M. R. (1998). Discrimination in the laws of information warfare. *Columbia Journal of Transnational Law* 37, 939–968.

Shultz, R. H. and Dew, A. J. (2006). *Insurgents, terrorists, and militias: The warriors of contemporary combat.* New York: Columbia University Press.

Silver, R. C., Holman, E. A., McIntosh, D. N., Poulin, M. and Gil-Rivas, V. (2002). Nationwide longitudinal study of psychological responses to September 11. *JAMA: The Journal of the American Medical Association,* 288(10), 1235–1244.

Sinclair, J. S. (2010). Fears of terrorism and future threat: theoretical and empirical considerations. In D. Antonius (Eds.), *Interdisciplinary analyses of terrorism and political aggression* (pp. 101–115). Newcastle upon Tyne: Cambridge Scholars Publishing

Sinclair, S. J. and Antonius, F. (2012a). *The psychology of terrorism fears.* Oxford: Oxford University Press

Sinclair, S. J. and LoCicero, A. (2007). Fearing future terrorism: Development, validation, and psychometric testing of the Terrorism Catastrophizing Scale (TCS). *Traumatology,* 13, 75–90.

Sivakumaran, S. (2009). Courts of armed opposition groups fair trials or summary justice? *Journal of International Criminal Justice,* 7(3), 489–513.

(2011). Lessons for the law of armed conflict from commitments of armed groups: Identification of legitimate targets and prisoners of war. *International Review of the Red Cross,* 93(882), 463–482.

Skerker, M. (2004). Just war criteria and the new face of war: Human shields, manufactured martyrs, and little boys with stones. *Journal of Military Ethics,* 3(1), 27–39.

Slater, J. (2012). Just war moral philosophy and the 2008–09 Israeli campaign in Gaza. *International Security,* 37(2), 44–80.

Society of Professional Journalists. (1996). *Code of ethics.* http://www.spj.org /ethicscode.asp.

Solà-Martín, A. (2007). The Western Sahara cul-de-sac. *Mediterranean Politics,* 12(3), 399–405.

Solis, G. D. (2010). *The law of armed conflict: International humanitarian law in war.* Cambridge: Cambridge University Press.

Spencer, M. E. (1970). Weber on legitimate norms and authority. *The British Journal of Sociology,* 21(2), 123–134.

Steinberg, G. M. (2006). Soft powers play hardball: NGOs wage war against Israel. *Israel Affairs,* 12(4), 748–768.

    (2011). The politics of NGOs, human rights and the Arab-Israel conflict. *Israel Studies,* 16(2), 24–54.

Steinhoff, U. (2004). How can terrorism be justified? In I. Primoratz (Ed.), *Terrorism: The philosophical issues* (pp. 97–112). Hampshire: Palgrave MacMillan.

Stephan, M. J. (2006). Fighting for statehood: The role of civilian-based resistance in the East Timorese, Palestinian, and Kosovo Albanian self-determination movements. *Fletcher Forum for World Affairs,* 30(2), 57–80.

Stephan, M. J. and Chenoweth, E. (2008). Why civil resistance works: The strategic logic of nonviolent conflict. *International Security,* 33(1), 7–44.

Stephan, M. J. and Mundy, J. (2006). A battlefield transformed: from guerrilla resistance to mass nonviolent struggle in the Western Sahara. *Journal of Military and Strategic Studies,* 8(3).

Streckfuss, R. (1990). Objectivity in journalism: A search and a reassessment. *Journalism & Mass Communication Quarterly,* 67(4), 973–983.

Studenski, P. and Krooss, H. E. (1963). *Financial history of the United States.* New York: McGraw Hill.

Suter, K. (1984). *An international law of guerrilla warfare: The global politics of law-making.* New York: St. Martin's Press.

Svensson, I. and Lindgren, M. (2011). Community and consent: Unarmed insurrections in non-democracies. *European Journal of International Relations,* 17(1), 97–120.

Tafoya, W. L. (2011). *Cyber* terror. http://www.fbi.gov/stats-services/publications /law-enforcement-bulletin/november-2011/cyber-terror.

Tesón, F. R. (2003). The liberal case for humanitarian intervention. In J. Holzgrefe and R. O. Keohane (Eds.), *Humanitarian intervention: Ethical, legal and political dilemmas* (pp. 93–129). Cambridge: Cambridge University Press.

Theoharis, A. G. (2006). A new agency: The origins and expansion of CIA covert operations. In A. G. Theoharis (Ed.), *The Central Intelligence Agency: Security under scrutiny* (pp. 155–188). Westport, CT: Greenwood Publishing Group.

Thom, W. G. (1974) Trends in Soviet support for African liberation. *Air University Review*, July-August. http://www.airpower.maxwell.af.mil/airchronicles/aureview/1974/jul-aug/thom.html#thom.

Thomas, T. (2000). Manipulating the mass consciousness: Russian & Chechen "information war" tactics in the Second Chechen-Russian conflict. In A. Aldis (Ed.), *The Second Chechen War* (pp. 112–129). Shrivenham, Wilts: Strategic and Combat Studies Institute (Great Britain).

Thruelsen, P. D. (2010). The Taliban in southern Afghanistan: A localised insurgency with a local objective. *Small Wars & Insurgencies*, 21(2), 259–276.

Tishkov, V. A. (2004). *Chechnya: Life in a war-torn society.* Berkeley, CA: University of California Press.

Tuck, D. (2011). Detention by armed groups: Overcoming challenges to humanitarian action. *International Review of the Red Cross*, 93(883), 759–782.

Tugwell, M. (1981). Politics and propaganda of the Provisional IRA. *Studies in Conflict & Terrorism*, 5(1–2), 13–40.

Turkel. (2010). *Turkel Commission report, summary, report of the public commission to examine the maritime incident of 31 May 2010 – Part One.* Jerusalem: The Public Commission To Examine the Maritime Incident of 31 May 2010. http://www.turkel-committee.com/files/wordocs/7896summary-eng.PDF.

UNICEF. (1999). Update. September 23. http://www.unicef.org/easttimor/timor2309.htm.

United Nations (UN). (2004). *A more secure world: our shared responsibility, report of the high-level panel on threats, challenges and change.* New York: The United Nations.

(2012). Office for the Coordination of Humanitarian Affairs occupied Palestinian territory. *Five years of blockade: The humanitarian situation in the Gaza Strip.* June. http://www.ochaopt.org/documents/ocha_opt_gaza_blockade_factsheet_june_2012_english.pdf

United Nations Assistance Mission in Afghanistan (UNAMA). (2012). *Afghanistan: Annual report 2011, Protection of civilians in armed conflict,* February 2012. Accessed http://www.refworld.org/docid/4f2fa7572.html.

United Nations General Assembly. (1947). *Resolution 181,* November 29. (Partition of Palestine into Jewish and Arab states).

(1966). *Resolution 2229 [XXI],* December 20. (Western Sahara).

(1972). *Resolution 2983,* December 14. (Western Sahara).

(1975). *Resolution 3485 (XXX),* December 12. (Question of Timor).

(1976). *Resolution 389,* Adopted by the Security Council at its 1914th meeting, on April 22, 1976, S/RES/389.

(2005). *World summit outcome,* A/60/L.1 Paragraph 139. http://daccessdds.un.org/doc/UNDOC/GEN/N05/487/60/PDF/N0548760.pdf?OpenElement.

United Nations Human Rights Council (UNHRC). (2009). *Human rights in Palestine and other occupied Arab territories. Report of the UN fact-finding mission on the Gaza conflict (The "Goldstone Report").* A/HRC/12/48, September 25. http://www2.ohchr.org/english/bodies/hrcouncil/docs/12session/A-HRC-12-48.pdf.

United Nations News Centre. (2011). *UN official deplores NATO attack on Libyan television station,* 8. August. http://www.un.org/apps/news/story .asp?NewsID=39255&Cr=Libya&Cr1

United Nations Office for the Coordination of Humanitarian Affairs, occupied Palestinian territory (UNOCHA). (2011). *Occupied Palestinian territory, special focus.* http://www.ochaopt.org/documents/ocha_opt_special_easing _the_blockade_2011_03_english.pdf.

(2012). *Five years of blockade.* http://www.ochaopt.org/documents/ocha_opt _gaza_blockade_factsheet_june_2012_english.pdf.

United Nations Office on Drugs and Crime (UNODC). (2012). *World drug report.* New York: United Nations.

United Nations Secretary General. (2011). *Report of the Secretary-General's Panel of Inquiry on the 31 May 2010 Flotilla Incident.* New York: United Nations. http://www.un.org/News/dh/infocus/middle_east/Gaza_Flotilla_Panel _Report.pdf.

(2012). *Report of the Secretary-General on the situation concerning Western Sahara.* United Nations S/2012/197, April 5.

United Nations Security Council (UNSC). (1990). *Resolution 661: The Situation between Iraq and Kuwait.* U.N. Doc. S/RES/661.

United States Code. (2006). *Title 18, Part I, Chapter 115, § 2381 – Treason.* http:// www.gpo.gov/fdsys/granule/USCODE-2011-title18/USCODE-2011 -title18-partI-chap115-sec2381/content-detail.html.

(2009). *Title 10§ 948a(7), (Supp. III 2009). Definitions: Unprivileged enemy belligerent.*

USA Today. (2010). Taliban fights back with bombs, using civilians as human shields. *USA TODAY,* February 18. http://usatoday30.usatoday.com /printedition/news/20100218/marjah18_st.art.htm.

Utzinger, J. and Weiss, M. G. (2007). Editorial: Armed conflict, war and public health. *Tropical Medicine & International Health,* 12(8), 903–906.

Vall, A. (2000). Can terrorism be justified? In A. Vall (Ed.), *Ethics in international affairs* (pp. 65–80). Lanham, MD: Rowman and Littlefield.

Veuthey, M. (1976). Guerrilla warfare and humanitarian law. *International Review of the Red Cross,* 16(183), 277–294.

Vidanage, H. R. (2004). Cyber cafes in Sri Lanka: Tamil virtual communities. *Economic and Political Weekly,* 39(36), 3988–3991.

Wall Street Journal (WSJ). (2012). 'Lawfare' loses big. *The Wall Street Journal,* January 28. http://online.wsj.com/news/articles/SB10001424052970203 7185045777181191271527180

Walzer, M. (1977). *Just and unjust wars: A moral argument with historical illustrations.* New York: Basic Books.

(2004). After 9/11: Five questions about terrorism. In M. Walzer, *Arguing about war* (pp. 130–142). New Haven, CT: Yale University Press.

Weinstein, J. M. (2007). *Inside rebellion.* New York: Cambridge University Press.

Weiss, T. G. (1999). Sanctions as a foreign policy tool: Weighing humanitarian impulses. *Journal of Peace Research,* 36(5), 499–509.

Weldemichael, A. T. (2013). *Third world colonialism and strategies of liberation: Eritrea and East Timor compared.* Cambridge: Cambridge University Press.

Wellman, C. H. (1995). A defense of secession and political self-determination. *Philosophy & Public Affairs,* 24(2), 142–171.

Wennmann, A. (2009). Grasping the financing and mobilization cost of armed groups: A new perspective on conflict dynamics. *Contemporary Security Policy,* 30(2), 265–280.

World Health Organization(WHO). (2010). *Health conditions in the occupied Palestinian territory, including east Jerusalem, and in the occupied Syrian Golan* A63/INF.DOC./J 13. http://unispal.un.org/UNISPAL.NSF/0/885BD85F 892778F28525772700503A4B

Wijedasa, N. (2011). Sri Lanka's ghosts of war. *New York Times,* December 30. http://www.nytimes.com/2011/12/31/opinion/sri-lankas-ghosts-of-war. html?_r=0.

Williams, P. (2007). Terrorist financing and organized crime: Nexus, appropriation of transformation. In T. J. Biersteker and S. El Eckert (Eds.), *Countering the financing of terrorism* (pp. 126–149). London: Routledge.

Wilson, C. (2006). Improvised explosive devices (IEDs) in Iraq and Afghanistan: Effects and countermeasures. CRS Report for Congress, The Library of Congress. http://research.fit.edu/fip/documents/secnews1.pdf.

Wilson, H. A. (1988). *International law and the use of force by national liberation movements.* Oxford: Clarendon Press.

Woldemikael, T. M. (1991). Political mobilization and nationalist movements: The case of the Eritrean People's Liberation Front. *Africa Today,* 38(2), 31–42.

Wood, R. M. (2010). Rebel capability and strategic violence against civilians. *Journal of Peace Research,* 47(5), 601–614.

Wood, T. (2007). *Chechnya: The case for independence.* Brooklyn: Verso Books.

World Medical Association. (2006) *Declaration of Malta on Hunger Strikers.* http:// www.wma.net/en/30publications/10policies/h31/.

Wright, J. and Bryett, K. (1991). Propaganda and justice administration in Northern Ireland. *Terrorism and Political Violence,* 3(2), 25–42.

Yehuda, R. (2002). Post-traumatic stress disorder. *New England Journal of Medicine,* 346(2), 108–114.

Yehuda, R., Bryant, R., Marmar, C., and Zohar, J. (2005). Pathological responses to terrorism. *Neuropsychopharmacology,* 30(10), 1793–1805.

Yildiz, K. and Breau, S. (2010). Historical background. In K. Yildiz and S. Breau (Eds.), *The Kurdish conflict: International humanitarian law and post-conflict mechanisms* (pp. 4–28). London: Routledge.

Young, R. (2004). Political terrorism as a weapon of the politically powerless. In I. Primoratz (Ed.), *Terrorism: The philosophical issues* (pp. 55–64). Hampshire: Palgrave MacMillan.

Zehr, N. A. (2013). Legitimate authority and the war against Al-Qaida. In A. F. Lang Jr., C. O'Driscoll, and J. Williams (Eds.), *Just war: Authority, tradition and practice* (pp. 97–114). Georgetown: Georgetown University Press.

Zemishlany, Z. (2012). Resilience and vulnerability in coping with stress and terrorism. *Israel Medical Association Journal*, 14, 307–309.

Zunes, S. (1999). Unarmed resistance in the Middle East and North Africa. In S. Zunes, S. B. Asher, and L. Kurtz (Eds.), *Nonviolent social movements: A geographical perspective* (pp. 41–51). Oxford: Blackwell.

Index

abduction, 60
Aceh guerrillas, 15, 83, 220
    assassination and, 105
    economic warfare by, 193
    Free Aceh Movement (GAM) and, 40,
        104, 112, 113, 193, 251
    kidnapping by, 207
    looting by, 205
    politicians as targets of, 104
    prosecution of informers by, 74, 112
    recruitment and, 55
    right to fight of, 36, 77, 219
    taxation by, 201
Additional Protocol I and II. See Geneva
    Conventions, 1977, Protocols I
    and II
Afghanistan, 1–2, 13, 25, 28, 102, 213.
    See also Taliban
    Afghani judicial system and, 113
    aid worker killings in, 207
    assassination of Afghan collaborators in
        by Taliban, 105
    human shields and, 139, 146
    IEDs as cause of combat deaths in, 86
    IEDs as cause of noncombatant deaths
        in, 90
    information operations in, 223–224, 238
    ISAF troops and, 85
    local public works projects in, 217, 220
    opium economy of, 201, 209
    propaganda and, 217
    recruitment in, 55
    smuggling in, 205
    UN Assistance Mission and, 74
    uniforms of Afghan army and, 84, 85
aid workers, 85, 86, 191, 207
Al Jazeera, 221
Al Qaeda, 2, 149, 153, 158, 165, 273

al-Dura video, 226–228
Amnesty International, 117, 118, 122
assassination, 9–10, 61, 81, 102–106, 124,
        274t, 275
    informers and collaborators and,
        108–110
    military and political targets in, 103–108
asymmetric war, 22, 63, 100, 159, 182, 264
    free riding and, 74
    human shielding and, 149

backfire, 14, 152, 241, 250–251
    Marmara incident and, 256–261
Basque guerrillas, 28, 192
boycott, divestiture and sanctions (BDS),
        255–256

Cassese, Antonio, 30, 38, 45
Cast Lead. See Gaza War (Cast Lead
        2008–9)
censorship, 232–234
Chaumont, Charles, 46–47, 130
Chechen guerrillas, 82, 83
    assassinations and, 104
    enemy uniforms and, 84
    executions by, 208
    right to fight of, 36
    Russian POWs and, 117, 123–124
    size of forces of, 52
    terrorism and, 154
Chechnya, 1–2, 15, 35, 158
chemical weapons, 174, 179
child soldiers, 58, 59
civil disobedience, 240–241, 256–259,
        274t. See also nonviolent resistance
civilian casualties, 8, 64–67, 99–100,
        138–141, 149
    economic warfare and, 192–194

civilian casualties (*cont.*)
  First Gulf War and, 187
  in Qana, Lebanon, 224
CNN, 229, 233
collaboration, 220
  assassination and, 108–110
  prosecution and, 110–114
collateral harm, 89–95, 98, 274,
      274t. *See also* direct harm,
      psychological harm
  economic sanctions and, 190
  indiscriminate weapons and, 86–88
  proportionate, 137, 138
combatants
  liability to harm and, 64–65
  military medical rescuers immunity
      and, 88–89
  participatory liability and, 67–69
  rights and duties of, 47, 49, 101
compatriot terrorism, 155–156.
      See also terrorism
  claim of necessity of, 160–161
  claim of not being absolutely necessary,
      161–162
conscription, 51–54
  consent and social ostracism
      and, 57–60
  human shields and, 136–138, 144
  morality of, 56–57
  of children, 58, 59
  overt coercion and, 60–61
  social, material and ideological
      incentives in, 54–56
consent, 6, 11, 246, 274t
  conscription and, 57–60
  human shielding and, 75, 130–131,
      136–138, 140, 275
  hunger strikes and, 261–264
  legitimate authority and, 23, 39, 40,
      41–42, 55
  nonviolent resistance and, 250,
      252–254, 259, 260
  taxation and, 198
counterfeiting, 204–207
Covenant on Civil and Political
      Rights, 24
crimes against humanity, 31–35, 76
Cuba, 194
cyber terrorism, 178–182
cyber weapons and warfare, 11–12, 65, 78,
      151, 166–168, 274t
  psychological harm and, 170–174,
      176–178
  Tallinn Manual on, 172, 176, 178

Democratic League of Kosovo (LDK),
      52, 265

deterrence, 93–96, 149, 274t
  defined, 140t
  human shielding and, 138–141
dignified life, 48, 205, 280–281
  just cause for war and, 7, 23
  just cause of war and, 275
  legitimate government authority and, 39
  prisoner's right to, 118
  right to fight and, 33, 34–35, 37, 46
  right to national self-determination and,
      6, 25, 27–28
dignity, 33, 75, 119, 180, 209, 279.
      See also dignified life; human rights
dilemma actions, 244, 250
direct harm. *See also* collateral harm;
      psychological harm
  journalists and, 234–235
  sublethal weapons and, 167–170
discrimination, 97, 100–101, 169, 272,
      274t. *See also* indiscriminate
      targeting; indiscriminate violence;
      indiscriminate weapons
disproportionate harm, 64–67, 98, 142–
      147, 274t. *See also* proportionality
distinction, principle of, 72, 95, 180
  defined, 51
  direct and indirect participation and, 133
  economic warfare and, 188
  morality and, 67
  selective and indiscriminate violence
      and, 160
  uniforms and, 69
  voluntary and involuntary human shields
      and, 131–132
drug trafficking, 13, 209–210, 274t
due process, 112–114, 156, 159, 160,
      274t, 276

East Timor guerrillas, 50, 165, 245. *See also*
      FALANTIL; Santa Cruz massacre
  Diplomatic and Clandestine fronts of, 52
  information campaign of, 221, 222, 224
  national and remedial claims to
      independence and, 36
  nonviolent resistance and, 248, 251
  representation and, 41
  repudiation of terrorism and, 153
economic sanctions, 78, 185–188,
      255–256, 274t
  in Iraq and Gaza, 186–188
economic warfare, 12–13, 151–152,
      185–190. *See also* war economies
  armed conflict and, 195
  civilian casualties and, 192–194
  effectiveness of, 186–187
  guerrilla economic warfare and,
      190–191

international law and, 185
kinetic force and, 191–192
legal and moral criteria for, 188–190
responsibility for effects of, 194–195
effectiveness, 23, 44, 101, 131
human shields and, 143
of economic sanctions, 188
terrorism and, 157–158
electromagnetic weapons, 174, 179
Eritrea, 2, 15–16, 23, 29, 161, 278
army of, 266
repudiation of terrorism and, 153
Eritrean People's Liberation Front (EPLF),
16, 17, 42, 76, 81, 83
assassination and, 104, 105
nonviolent resistance and, 264
POWs and, 116, 276
public works and, 52
repudiation of terrorism by, 273, 279
size of, 52
ethnic cleansing, 31, 157, 265

FALANTIL (guerrillas of East Timor), 50,
214, 248, 273
size of, 52
Free Aceh Movement (GAM), 40, 104,
112, 113, 193, 251. *See also* East
Timor guerrillas
free riding, 72–76
FRELIMO insurgents, 57

Gandhi, Mahatma, 252–254
Gaza War (Cast Lead, 2008–9), 66, 127,
169, 195
economic sanctions and, 186–188
Geneva Conventions, 1949, 75
prisoners of war and, 115–122, 276
Geneva Conventions, 1977, Protocols I
and II, 30, 46, 66, 230
direct participation and, 133
journalist immunity and, 236
military necessity and, 71
prisoners of war and, 115–122
uniforms and, 61, 62
genocide, 31–35, 265
Guantanamo Bay, hunger strikes at,
261–262
guerrilla warfare. *See also* conscription;
just guerrilla warfare; kinetic
weapons and warfare; non-kinetic
weapons and warfare; uniforms; war
economies
defined, 4
foreign support sources for, 221
guerilla defined, 3
hit and run tactics, 81–83
legitimate authority and, 37–44

material and legal asymmetry of, 4
nonviolent resistance in, 245–247
prosecution of informers and
collaborators by, 110–114
public works and propaganda and,
218–225, 228
repudiation of terrorism and, 17,
153–154, 279
rules of war adherence and, 74
size of guerrilla military
organizations, 51
social services for local populations
and, 52
tourism and, 191–192
Gulf War, First, 187, 267
Gulf War, Second, embedded journalists
and, 233

Hamas, 81, 139, 155, 169, 186, 194.
See also Palestinian guerrillas
Al Jazeera and, 221
assassination and, 105, 106
attack on police of, 90
charitable work of, 203
funding sources of, 199, 202
human shielding and, 127, 148
Israeli POWs and, 116, 117, 122–124
legitimate authority and, 41
Marmara blockade incident and,
256–261
missiles and, 92–93, 96, 97, 169
public works of, 220
size of, 51
smuggling and, 205
suicide bombing and, 158
terrorism and, 158
hard war, compared with soft war,
151–152
Hezbollah, 17, 83, 100, 117, 139.
See also Lebanon
drug trafficking and, 209
human shielding and, 141, 142, 149
Israeli POWs and, 122–124
media manipulation of, 213, 221, 224,
230, 231, 233
missiles and, 92–94, 95–97
prisoners and, 116, 118, 119
public diplomacy and, 214, 220
public works of, 52
size of, 51
smuggling and, 204
suicide bombing and, 158
uniforms and, 63
war economy of, 199, 202, 205, 209
hit and run guerrilla tactics, 81–83
hostage taking, 120–122, 274t.
See also kidnapping

human rights, 17, 42–43, 200, 222, 278.
    See also dignified life; dignity;
    genocide; humanitarian law
  for prisoners of war, 118, 119, 125
  human rights law, 46
  in post-WWII, 157
  information operations and, 224
  kidnapping and, 208
  lying and, 238
  media warfare and, 235
  prisoner exchanges, 122
  right to fight and, 45, 48, 50,
    68, 107
  right to self-determination and,
    23–27
  unilateral compliance and, 116
Human Rights Watch (HRW), 93, 96, 133,
    135, 166
human shielding, 10, 98, 126, 148–150,
    241, 272, 274t
  attacker responsibilities in, 142–143
  conscription and, 136–138, 144
  consent and, 130–131, 275
  defender responsibilities for and in,
    143–145
  defined, 128
  deterrence and, 138–141
  international law on, 126, 128–129
  Islamic law and, 149
  LTTE and, 147–148
  military advantage of, 130
  voluntary and involuntary and immunity
    from harm, 129, 131–136
  voluntary and involuntary and
    protection of war sustaining and
    fighting assets and, 145–148
humanitarian intervention, 18, 48
humanitarian law, 2, 161, 222.
    See also international law
  aid worker uniforms and, 84
  challenge of formulating for just
    guerrilla war and, 280
  economic warfare and, 185
  hostages and, 118
  human shielding and, 128, 149
  indiscriminate attacks and, 96
  just cause and, 22
  lying and, 231
  media warfare and public diplomacy
    and, 14
  minimizing harm to civilians and, 163
  permissible tactics and, 75
  prisoner's rights and, 117
  right to fight and, 46
hunger strikes, 261–264
Hussein, Saddam, 194, 206, 229

IEDs, 8–9, 81–82, 86–88, 100–101,
    274t, 275
  indiscriminate targets and, 93–95
  noncombatant harm and, 89–92
  successive attacks to disable rescuers
    and, 88–89
improvised explosive devices (IEDs).
    See IEDs
incitement, 226–228, 237, 238, 274t
indiscriminate harm, 274t
indiscriminate targets, 93–95, 101
indiscriminate violence, 159–162
indiscriminate weapons, 86–88, 93
  technological asymmetry as
    proportionality discount and,
    97–101
information operations.
    See also propaganda; public
    diplomacy
  censorship and, 232–234
  communication strategies in, 215
  guidelines for information owed
    compatriots, 234
  in East Timor, 222
  lying and deception in, 229–232
  Taliban and, 223–224
informers, prosecution of, 110–114
insurgents. *See* guerrilla warfare
International Committee of the Red Cross
    (ICRC), 125, 165
  force feeding of detainees and, 261
  indirect participation interpretive
    guidelines, 70
  noncombatant immunity and, 133–136
International Criminal Court, 33
international law, 101, 163, 272.
    See also Geneva Conventions;
    humanitarian law; International
    Committee of the Red Cross (ICRC)
  economic sanctions and, 188
  economic warfare and, 12, 185
  human shielding and, 128–129
  just guerrilla war and, 5–6
  mental suffering and, 173
  on excessive harm, 65
  on noncombatant immunity, 63–69
  perfidy and, 84
  propaganda and information operations
    and, 225
  prosecution of informers and
    collaborators and, 113
  right to fight and, 29–31, 46–47
  rules of war for guerrillas and, 7
  Tallinn Manual on Cyber Warfare and,
    172, 176, 178
  uniforms and, 62

International Security Assistance Force
(ISAF), 85, 89, 90, 91, 105,
213, 217
public works of, 220
IRA (Irish Republican Army), 168,
170, 240
British POWs and, 117
defensive propaganda of, 221
economic warfare and, 192
funding sources and, 199
hunger strikes and, 261
kidnappings of business executives
by, 208
lack of POW status for, 115
prosecution of informers by, 112
proxy bombing and, 87
Warrenpoint bombing and, 88
Iraq, 13, 102, 206, 238. *See also* Gulf War,
First; Gulf War, Second; Hussein,
Saddam
economic sanctions in, 186–188
Israel, 17, 127, 264. *See also* Hamas;
Hezbollah; Palestinian guerrillas
attack on Hamas police cadets, 90
attacks by against civilian
facilities, 166
deterrence and, 139
Hezbollah and, 83, 98, 106, 141, 149,
158, 213
Israeli POWs and, 116, 117, 122–124
Marmara blockade incident and, 256–261
missiles of, 95
noncombatant deaths and, 100
Palestinian terrorism and, 154, 158
Protocol I of Geneva Convention and, 115
Qana, Lebanon strikes, 92, 224
torture of detainees of, 116

journalists, 213, 214
destruction of media facilities and,
235–237
embedding, 233
ethics of third party vs. media workers
affiliated with, 239
liability of to direct harm and, 234–235
jus ad bellum (just cause of war), 21–23, 30
guerilla war economy and, 200
missiles use and, 100
rules of war and, 76–78
terrorism and, 162
jus in bello (just means of war), 5, 6, 7, 22,
51, 78, 225
just guerrilla warfare, 3, 5, 10, 45, 49, 51,
160, 271–272. *See also* guerrilla
warfare; *jus ad bellum* (just cause of
war); morality

enforcement and compliance with
international norms and, 278–279
guerilla war economy and, 200
human shielding and, 130, 149
national self-determination and, 15–18
nonviolent resistance and, 267
OAU Charter (1981) on, 5–6
practices and provisos of, 23, 44, 75, 273
prospects for, 279–281
right of self-determination and, 27
right to a fighting chance and, 49
terrorism and, 276
just war theory, 5, 98, 152, 271–272
civil disobedience and nonviolent
resistance and, 241
economic warfare and, 184
last resort principle and, 263–267
noncombatant and civil targets attack
prohibition in, 163
non-kinetic warfare and, 277
war propaganda and, 226

Kamajor militia, 55
kidnapping, 13, 207–209. *See also* hostage
taking
kinetic weapons and warfare, 8, 150,
273. *See also* assassination; human
shielding; IEDs; missiles; prisoners
of war
guerrilla economic warfare and,
191–192
KLA. *See* Kosovo Liberation Army (KLA)
Kosovo, 1–2, 15, 29, 36, 76, 125,
165, 250
assassinations and, 153
claims to independence and, 36
ethnic cleansing in, 157
humanitarian and military aid from US
and NATO, 199
LDK (Democratic League of Kosovo)
and, 52
nonviolent resistance and, 240, 243,
245, 246–248, 264–266
recruitment in, 55
taxes and, 201
Kosovo Liberation Army (KLA), 40, 76,
105, 107, 125, 154, 248, 265
public works and, 214
size of, 51
smuggling of, 204
Kurdistan Workers Party (PKK),
58, 114

Laqueur, Walter, 1, 72
Lasswell, Harold, 214
last resort principle, 23, 45, 263–267

law of armed conflict, 7, 64, 77, 98, 172
  guerrilla's obeying of, 74
  human shielding and, 128
  just cause and, 22
LDK. *See* Democratic League of
    Kosovo (LDK)
Lebanon, 2, 13, 86, 92, 165, 202, 203.
    See also Second Lebanon War
  First Lebanon War (1983), 122
  information operations in, 238
  smuggling and, 205
legitimate authority, 37–44, 49, 78,
    137, 234
  overt coercion and, 263
  representation and, 38–43
liability, 9, 10, 76, 124–126, 162,
    206, 207, 236, 274–276, 274t.
    See also noncombatant liability;
    participatory liability
  of combatants to harm, 64–65
  of journalists to direct harm and,
    234–235
  of military rescuers, 88
  terrorism and, 155–156
Liberation Tigers of Tamil Eelam (LTTE),
    16, 125, 147–148, 192, 252
Libya, 35
  destruction of TV stations in by NATO,
    235–237
Lieber, Francis, 103
looting, 204–207
LTTE. *See* Liberation Tigers of Tamil
    Eelam (LTTE)
lying and deception in propaganda,
    229–232, 238

Marmara blockade incident, 256–261
Maoist guerrillas, 56, 204
media warfare, 13, 152, 238.
    See also information operations;
    propaganda; public works
  al-Dura video and, 226–228
  destruction of TV stations and, 235–237
  ethics in, 213–214
  Hezbollah and, 233
  journalists and, 214
  public works and propaganda and,
    218–225
  staged media events and, 227
medical civic action programs
    (MEDCAP), 218
medical workers, 88–89, 101
mental suffering. *See* psychological harm
military necessity, 71, 129, 231
military targets, 8, 92, 98–100, 103–108,
    182, 274t
  defined, 81

missiles and rockets, 274t
  Hamas and, 92–93, 96, 97
  Hezbollah and, 92–94, 95–97
  indiscriminate targets and, 93–95
  indiscriminate weapons and, 86–88
  psychological harm following attacks of
    in Israel, 171
  rules for use of, 98
  technological asymmetry as
    proportionality discount and,
    97–101
morality, 81. *See also jus ad bellum* (just
    cause of war); just guerilla warfare;
    just war theory
  conscription and, 56–57
  economic warfare and, 188–190
  human shielding vs. deterrence,
    138–141
  journalist ethics vs. media workers
    affiliated with a conflict, 239
  lying and deception in propaganda and,
    229–232
  media warfare and, 14, 213–214
  of charitable giving, 202
  of IED use, 87
  principle of distinction and, 67
  prisoner's rights and, 116
  public diplomacy and, 14, 225, 237, 238
  right to fight and, 33–34, 37
  uniforms and, 84–86
Mozambique, 57

NATO, 76
  destruction of Libyan TV stations and,
    235–237
necessity, 23, 38, 76, 192, 210, 214
  collateral harm and, 99, 139
  cyber warfare and, 12
  economic sanctions and, 189
  human shielding and, 129, 143
  information operations and, 234
  lying and, 230, 232
  military necessity, 71, 129, 231
  participatory liability and, 71
  right to fight and, 44–45
  terrorism and, 156, 160–161
  uniforms and, 62, 131
Nepal, 56
noncombatant immunity, 47, 49, 61, 163
  civilian liability and harm and, 63–69
  defined, 51
  human shielding and, 131–136, 148
  indiscriminate harm and IEDs
    and, 89–92
  medical workers and, 88–89
  police and, 90–91
  uniforms and, 69

noncombatant liability, 162–164, 275
  civilians providing war sustaining aid
    and, 164–166
  of journalists to direct harm and,
    234–235
  terrorism and, 155–160
noncombatants. *See also* civilian casualties;
    participating civilians
  collateral harm and, 168t
  economic sanctions and, 190
non-kinetic weapons and warfare, 8,
    11, 150, 151–152, 273, 274,
    277. *See also* economic sanctions;
    nonviolent resistance; public
    diplomacy; terrorism
nonlethal weapons and warfare, 166–168,
    274t. *See also* cyber weapons and
    warfare; sublethal weapons and
    warfare
nonmilitary targets, 72
nonviolent resistance, 14, 152, 240–242,
    245–247, 264–267, 272, 274t
  boycott, divestiture and sanctions (BDS)
    and, 255–256
  defined, 241
  dilemma actions and, 250
  effectiveness of in East Timor, Palestine
    and Kosovo, 247–250
  hunger strikes and, 261–264
  Marmara blockade incident and,
    256–261
  outcomes in, 243–245
  provocation and, 251–255
  tactics and aims in, 242–244
nuclear deterrence, 138

OAU Charter (1981), 5–6
Obama, Barack, 271
Oslo peace accords, 1

Palestine Liberation Organization
    (PLO), 43
Palestinian guerrillas
  assassinations and, 104
  backfire and, 251
  boycott, divestiture and sanctions (BDS)
    and, 255–256
  nonviolent resistance of, 246–247,
    248–250
  prisoners of war and prisoner exchanges
    and, 122–124
  terrorism and, 154, 158
participating civilians, 68–71, 126, 162–
    167, 169, 174–176, 178, 180–182,
    206. *See also* participatory liability
  collateral harm and, 167
  psychological distress and, 172, 173

participatory liability, 68–72, 116,
    156, 165, 169, 170, 188.
    See also participating civilians
  journalists and, 214
perfidy, 84–86
Pictet, Jean, 22
PKK. *See* Kurdistan Workers Party (PKK)
police, 90–91
post-traumatic stress disorder (PTSD),
    170–173
poverty, 34–35
prisoners of war, 9–10, 114–120, 124,
    274t, 276
  hostage taking and, 120–122
  prisoner exchanges, 122–124
propaganda, 152, 215–217, 238,
    274t. *See also* information
    operations; media warfare; public
    diplomacy
  incitement and, 226–228
  lying and deception in, 229–232
proportionality, 38, 274t
  civilian deaths and, 65–67
  disabling of war sustaining facilities
    and, 164
  human shielding and, 98, 129, 135, 144,
    149, 150
  of IEDs and missiles, 91–93
  right to fight and, 44–45
  technological asymmetry in missiles and,
    97–101
proportionate harm, 137, 138,
    143–145, 148, 173.
    See also proportionality
Provisional Irish Republican Army (PIRA),
    16, 153. *See also* IRA
provocation, 251–255
proxy bombing, 87
psychological harm, 12, 92, 165, 170–174,
    181–182
  cyber warfare and, 176–178
  PTSD and, 170–173
public diplomacy, 13, 152, 155, 213–219,
    237–239. *See also* information
    operations; media warfare;
    propaganda; public works
  morality and, 225, 237, 238
  permissible components of, 238
  propaganda and, 215–217
public sphere, permissible attacks on,
    163–164
public works, 238
  medical civic action programs
    (MEDCAP), 217

Qaddafi, Muammar, 186, 199, 236
Qassam rockets, 169. *See also* missiles

recruitment. *See* conscription
remedial rights, 27, 29, 36, 37
representation, 25, 26, 32, 38–43, 51, 53,
     58, 60, 110, 137, 198
right to a fighting chance, 49, 101, 136
  human shielding and, 148, 150
right to fight, 31, 50–51
  aggression and, 35–37
  imminent threat and, 33–34
  international law on, 29–31
  legal non-interference and, 46–48
  material non-interference and, 47–49
  overt coercion and abduction and, 60
  proportionality and necessity and,
     44–45
  right violations and, 34–35
  uniforms and, 60–63
  vs. national self-determination, 37
rockets. *See* missiles
rogue regimes, 199
Rome Statute of the ICC, 120

sanctions. *See* economic sanctions
Santa Cruz massacre, 222, 227, 251, 253
Second Lebanon War (2006), 17, 51, 52,
     166, 213, 224, 232
selective violence, 160
self-determination, national
  as just guerrilla war case studies, 15–18
  nonviolent resistance and, 264–267
  repudiation of terrorism and, 153
  right to, 24–27, 35–37
  sovereignty vs. political independence
     and, 27–29
  vs. right to fight, 37
September 11, 2001 attacks, 170–171
Shalit, Corp. Gilad, 117
Sharp, Gene, 252
shields. *See* human shielding
Sierra Leone, 55
Sinn Fein, 240
smuggling, 204–207, 274t
soft power, 11
soft war, compared with hard war,
     151–152
Sri Lanka, 1–2. *See also* Liberation Tigers of
     Tamil Eelam (LTTE)
sublethal weapons and warfare, 166–168.
     See also nonlethal weapons and
     warfare
  direct harm and, 167–170
  psychological harm and, 170–174
Sudan, 2, 15, 23, 38, 43, 76, 193, 206,
     264, 278
  crimes against humanity and, 33
  kidnappings in, 207

Sudanese People's Liberation Army
     (SPLA), 42
  size of, 52
suicide bombing, 87–88, 89, 158

Taliban, 17, 204, 223–224
  human shielding and, 127
  night letters of to collaborators, 220
  prosecution of informers and, 113
  public works of, 220
  recruitment techniques of, 59
  size of, 51
Tallinn Manual on the International Law
     Applicable to Cyber Warfare, 172,
     176, 178
Tamil Tigers. *See* Liberation Tigers of Tamil
     Eelam (LTTE)
targeted killing. *See* assassination
targets. *See also* indiscriminate targets;
     military targets
  dual use, 90
terrorism, 2, 11, 91, 126, 155–158, 273,
     274t, 276
  claim of necessity of, 160–161
  claim of not being absolutely necessary,
     161–162
  compatriot terrorism, 155–156
  cyber terrorism, 11–12, 178–182
  free riding and, 72–76
  guerrilla groups repudiation of, 17,
     153–154, 273, 279
  ineffectiveness of, 157–158
  noncombatant liability
     and, 155–160
  Palestinians and, 154, 273
  selective violence and, 160
  tourism and, 191–192
therapeutic harm, 65
threats and intimidation, 220–221, 228,
     274t
torture, 116, 208
tourism, terrorism and, 191–192
treason and treachery, 110–114
truth, as legitimate casualty of just war,
     214, 229–232, 238, 239
Turkey, 1–2

UN Human Rights Council, 90
uniforms, 60–63, 69, 83–84, 130–131
  perfidy and, 84–86
United Nations, 38
  Article 51 of, 29–30
  Declaration on Principles of
     International Law (1970), 30
  UN Assistance Mission in
     Afghanistan, 74

United States, 17, 106, 115, 116, 237
attacks by against civilian facilities, 166

Vietnam War, 57
medical civic action programs
(MEDCAP) in, 218
violence
indiscriminate, 159–162
selective, 160

war crimes, 76, 120, 239, 274t
war economies, 12–13, 190–191,
199–200, 274t. *See also* economic
warfare
charitable organizations and business
operations, 202–204
drug trafficking and, 209–210
kidnapping and, 207–209
non-state revenue and, 197–199

smuggling, looting and counterfeiting
and, 204–207
state revenue and, 196–197
taxation as revenue and, 201–202
war fighting aid, 70, 133–135
war fighting facilities, voluntary shields
and, 147
war sustaining aid, 68–71, 133–135, 276
human shielding and immunity and,
131–136
noncombatant liability and, 164–166
war sustaining facilities, 71, 166–170, 274t
economic sanctions and, 189
human shielding and protection of war
sustaining and assets and, 145–148
Warrenpoint bombing, 88
wedge attacks, 85
Wilson, Woodrow, 24
World Drug Report (2012), 209